METAL FATIGUE IN ENGINEERING

UNIVERSITIES AT MEDWAY
DRILL HALL LIBRARY

This book must be returned or renewed by the date printed on your
self service receipt, or the last date stamped below if applicable.
Fines will be charged as soon as it becomes overdue.

31/7/2010

UNIVERSITIES
at MEDWAY

METAL FATIGUE IN ENGINEERING

Second Edition

RALPH I. STEPHENS
Professor, Mechanical Engineering Department
The University of Iowa

ALI FATEMI
Professor, Mechanical, Industrial, and Manufacturing Engineering
  Department
The University of Toledo

ROBERT R. STEPHENS
Associate Professor, Mechanical Engineering Department
The University of Idaho

HENRY O. FUCHS
Formerly Emeritus Professor
Mechanical Engineering Department
Stanford University

A Wiley-Interscience Publication
JOHN WILEY & SONS, INC.
New York / Chichester / Weinheim / Brisbane / Singapore / Toronto

We hope this book will be used in making decisions about the design and operation of machines and structures. We have tried to state the facts and opinions correctly, clearly, and with their limitations. But because of uncertainties inherent in the material and the possibility of errors we cannot assume any liability. We urge the readers to spend effort in tests in verification commensurate with the risks they will assume.

This book is printed on acid-free paper. ∞

Copyright © 2001 by John Wiley & Sons, Inc. All rights reserved.

Published simultaneously in Canada.

This publication is designed to provide accurate and authoritative information in regard to the subject matter covered. It is sold with the understanding that the publisher is not engaged in rendering professional services. If professional advice or other expert assistance is required, the services of a competent professional person should be sought.

Library of Congress Cataloging-in-Publication Data:

Metal fatigue in engineering / by Ralph I. Stephens . . . [et al.].—2nd ed.
   p. cm.
  Rev. ed. of: Metal fatigue in engineering / H.O. Fuchs, R.I. Stephens. c1980.
  "A Wiley-Interscience publication."
  Includes bibliographical references.
  ISBN 0-471-51059-9 (cloth : alk. paper)
   1. Metals—Fatigue. I. Stephens, R. I. (Ralph Ivan) II. Fuchs, H. O. (Henry Otten), 1907—Metal fatigue in engineering.

TA460.M4437 2000
620.1′66—dc21

Printed in the United States of America.

10 9 8 7 6 5

Dedication
We sincerely appreciate the exceptional understanding our wives,
Barbara, Shirin, and Marci, gave us while we wrote, along with our
children Lili, Sarah, and Joelle, who sometimes did not get the attention
they deserved. We thank our colleagues on the SAE Fatigue Design
and Evaluation (SAEFDE) Committee and the ASTM Committee E-08
on Fatigue and Fracture for many years of fruitful exchange of
fatigue information.

CONTENTS

PREFACE

The second edition of *Metal Fatigue in Engineering,* like the first is written for practicing engineers and engineering students concerned with the design, development, and failure analysis of components, structures, and vehicles subjected to repeated loading and for others who must make decisions concerning fatigue resistance/durability. It is intended as a textbook for a regular one-semester senior or graduate-level course in mechanical, civil, agricultural, aerospace, engineering mechanics, or materials engineering and for short courses on the subject. It is also intended for use and self-study by practicing engineers. The subjects covered are applicable to a wide variety of structures and machines including automobiles, aerospace vehicles, bridges, tractors, bio-implants, and nuclear pressure vessels. The key prerequisite for efficient usage of the book is knowledge of mechanics of materials and materials science.

The focus of this second edition, as with the first edition, is on applied engineering design with the intent to attain safe, reliable, and economical products. Adequate background information is provided on the different fatigue design methods to enable readers to judge the validity of the recommended models and procedures. Topics of interest primarily to researchers, yet not applied to actual fatigue design, have been omitted. The methods presented have been, and are, used successfully in fatigue design and life predictions/estimations.

Twenty years have passed since the publication of the first edition. This is due to the untimely death of Professor Fuchs in 1989, when an effort had begun to write the second edition. In the spring of 1998, we decided to begin again with two new authors. Wiley Interscience was in complete agreement with this decision, since the first edition was very successful and was still producing sales after nearly 20 years of usage. The first edition had even been translated into Chinese. The second edition is approximately 50 percent larger than the first edition due to extensive additional and updated text, figures, example problems, homework problems, chapter summaries, dos and don'ts in design, and references. A solutions manual for the second edition is also available to faculty using the book in a course.

New topics on fatigue in the second edition include computer-aided engineering and digital prototyping, enhanced discussion of micro/macro mechanisms and factors influencing fatigue behavior, small crack growth, elastic-plastic fracture mechanics, improved crack closure modeling, enhanced presentation of notch strain analysis that includes strain energy density or Glinka's rule, cracks emanating from notches, production and measurement of residual stresses, residual stress intensity factors, expanded presentation of cycle counting and crack nucleation and growth models for life prediction from variable amplitude loading, nonproportional multiaxial loading, critical plane approaches and mixed-mode fatigue crack growth for multiaxial fatigue, expanded coverage of environmental effects, BS 7608 fatigue design code for weldments, and a new chapter on statistical aspects of fatigue, along with expansion of the material property tables in the Appendix.

Four analytical methods of fatigue design and fatigue life predictions/estimations for metals are covered: nominal stress (S–N), local strain (ε–N), fatigue crack growth (da/dN–ΔK), and a two-stage method involving ε–N for nucleation of small cracks and da/dN–ΔK for continued crack growth until fracture. These four methods have been incorporated into many commercial, government, and in-house computer programs that have been used successfully in many complex fatigue design situations. These can involve notches, residual stresses, small and long cracks, variable amplitude loading, multiaxial states of stress, and environmental conditions. The four analytical methods complement the need for verification testing under simulated and real-life conditions, and form the basis of digital prototyping and computer-aided engineering for product durability.

Chapter 1 explains different mechanical failure modes and includes a historical overview of fatigue, including biographical sketches showing that the development of fatigue knowledge is a human endeavor. Chapter 2 introduces general fatigue design philosophies and procedures such as safe-life, fail-safe, and damage-tolerant design, synthesis, analysis, testing, digital prototyping, and inspection. Chapter 3 describes macroscopic and microscopic aspects of fatigue behavior and explains the mechanisms associated with fatigue crack nucleation and growth. The term "crack initiation" is not used, consistent with recent ASTM recommendations. Chapter 4 involves fatigue testing, ASTM and ISO fatigue test standards for metals, and nominal stress–life (S–N) behavior for unnotched constant amplitude loading. Chapter 5 describes monotonic and cyclic stress–strain deformation behavior and strain–life (ε–N) fatigue behavior and properties for unnotched constant amplitude loading. Chapter 6 introduces LEFM concepts and their application to fatigue crack growth (da/dN–ΔK) under constant amplitude loading. This includes crack closure, small and long cracks, and crack tip plasticity considerations. Chapter 7 discusses the importance of stress concentrations and incorporates their effects into fatigue design using all four fatigue design methods. Chapter 8 considers residual stresses with their production, relaxation, measurement, beneficial or detrimental effects, and residual and applied stress intensity factors for use in fatigue crack growth. Chapter 9 discusses cumulative damage and cycle counting and extends all four fatigue design methods developed in Chapters 4 through 7 to variable amplitude loading. Multiaxial fatigue including proportional and nonproportional loading, equivalent stresses and strains, critical plane approaches, and mixed-mode crack growth is covered in Chapter 10. Environmental effects including corrosion, fretting, low and high temperatures, and neutron irradiation are covered in Chapter 11. Chapter 12 covers the four fatigue design methods applied to fatigue of weldments, including the usage of British Standard BS 7608. Chapter 13 considers fatigue data scatter and probabilistic aspects of fatigue design.

In the Appendix, five tables on fatigue and fracture toughness properties for selected engineering materials have been updated. These tables are used in both the text and the problems. Example problems are included to better show the applicability of the concepts to engineering design situations. Unsolved problems are included at the end of each chapter, and in some cases more than one answer, or a range of answers, is reasonable because of different ways of modeling, methods of solution, material properties chosen, and the inherently empirical and nonexact nature of fatigue and fatigue design. The 166 problems often require the reader to indicate the significance of a solution and to comment on the assumptions made. Each chapter has significant references (424 total, an increase of about 40 percent) and figures (183 total), a Summary section, and a "Dos and Don'ts in Design" section. SI units are used with American units where appropriate. Most figures and tables have dual units.

We acknowledge our debt to many workers in the field of fatigue by reference to their publications. Many people who are not mentioned have also contributed to the study of fatigue. Discussions, debates, presentations, and standards development with colleagues, particularly with SAE and ASTM fatigue and fracture committees, have helped to refine our understanding of this complex subject. We thank Dr. Harold Reemsnyder of Bethlehem Steel, Professor Drew Nelson of Stanford University, Professor Darrell Socie of the University of Illinois at Urbana–Champaign, Dr. Gary Halford of NASA Glen, and Lee James, consultant, for their constructive comments during the writing of this second edition. We also thank our colleagues for providing source information and/or suggestions: Professor Gregory Glinka of the University of Waterloo, Professor David Hoeppner of the University of Utah, Dr. James Newman of NASA Langley, Professor Huseyin Sehitoglu of the University of Illinois at Urbana–Champaign, Keith Smith of the Association of American Railroads, Professor Steve Tipton of the University of Tulsa, and Professor Tim Topper of the University of Waterloo. We owe many thanks to our wives and children, who endured the diversion of our energies while we wrote this second edition.

BIOGRAPHICAL SKETCHES

RALPH I. STEPHENS is Professor of Mechanical Engineering at the University of Iowa, Iowa City, Iowa. He received his B.S. in General Engineering (1957) and M.S. in T&AM (1960) at the University of Illinois and his Ph.D. in Engineering Mechanics (1965) at the University of Wisconsin. In 1965 he joined the University of Iowa. Professor Stephens has taught many different engineering courses at the Universities of Illinois, Wisconsin, and Iowa. These courses have involved primarily solid mechanics, mechanical systems design, capstone design, fatigue, and fracture mechanics. He has received several departmental or student designated outstanding teaching awards. His principal field of research and publications is fatigue and fracture mechanics, and his consulting practice involves products liability. He has directed more than 60 M.S. and Ph.D. graduate student theses. Dr. Stephens has published more than 130 papers on fatigue and fracture mechanics, has edited 4 special technical publication books with ASTM and SAE, and has presented more than 200 papers, lectures, or seminars in more than 25 countries. He is a member of Sigma ZI, the ASTM Committee E08 on Fatigue and Fracture, and the SAE Fatigue Design and Evaluation (SAEFDE) Committee. Professor Stephens has received awards of appreciation from ASME, SAE, and ASTM. He is coordinator of the annual SAEFDE–University of Iowa short course on Fatigue Concepts in Design.

ALI FATEMI is Professor of Mechanical, Industrial, and Manufacturing Engineering at the University of Toledo. He received his B.S. (1979) and M.S. (1980) in Civil Engineering, specializing in structural analysis and design, and his Ph.D. in Mechanical Engineering (1985) from the University of Iowa. He joined the faculty of Mechanical Engineering at Purdue University in Fort Wayne in 1985 and then moved to the University of Toledo, Toledo, Ohio, in 1987. Professor Fatemi has taught many engineering subjects in the areas of solid mechanics, engineering materials, and mechanical design. At the graduate level, he has been teaching courses on mechanics of composites, fatigue of materials and structures, fracture mechanics, experimental mechanics, and durability analysis. Professor Fatemi's primary field of research involves materials mechanical behavior, including fatigue and fracture mechanics, and he has published over 60 papers dealing with these topics. He has directed the theses and dissertations of more than 25 students at the masters and doctoral levels. Professor Fatemi has been the Principal Investigator for many research projects with more than $1 million dollars in funding from industrial sponsors. He was the recipient of the College of Engineering's Outstanding Faculty Research Award in 1998. Professor Fatemi is a member of ASME, the ASTM Committee E08 on Fatigue and Fracture, and the SAE Fatigue Design and Evaluation Committee.

ROBERT R. STEPHENS is Associate Professor of Mechanical Engineering at the University of Idaho, Moscow, Idaho. He received his B.S. in Mechanical Engineering (1985) at the University of Iowa and his M.S. (1987) and Ph.D. (1991) at the University of Utah. In 1992 he joined the University of Idaho. Dr. Stephens has taught many different courses at the Universities of Utah and Idaho, mainly in the field of solid mechanics. These include statics, strength of materials, machine component design, advanced strength of materials, fatigue, fracture mechanics, materials science, and capstone design. He received the College of Engineering's Outstanding Young Faculty Award in 1996 and the Faculty Excellence Award from the Naval ROTC in 1996. Dr. Stephens' research interest, along with involving his graduate students, is in fatigue and fracture mechanics, with an emphasis on fatigue crack growth under constant and variable amplitude loading. Much of his research has been associated with the aerospace industry. He has authored and coauthored many technical papers on fatigue crack growth. Several of these papers were coauthored with his father, Ralph Stephens. He is a member of ASM, ASME, and the ASTM Committee E08 on Fatigue and Fracture.

HENRY O. FUCHS (1907–1989) was Professor and then Professor Emeritus of Mechanical Engineering at Stanford University, Palo Alto, California. He received his Diploma of Engineering (1929) and Dr. of Engineering degree (1932) from Karlshrue Technical University, Germany. In 1933 he joined General Motors in Detroit, where he became a colleague of J. O. Almen. In 1945 he left GM to design railway car suspensions and assemblies in Los Angeles. He founded a shot-peening business, Metal Improvement Co., in his garage in 1946 with his son-in-law. Professor Fuchs taught at the University of Detroit and UCLA and joined Stanford University in 1964 as a Professor, becoming Professor Emeritus in 1972. He taught courses in solid mechanics, design, and fatigue. He emphasized the case study method in teaching design and chaired the ASEE case study program. Dr. Fuchs published about 65 technical papers, with substantial emphasis on case studies, residual stresses, notches, and multiaxial fatigue and held 24 U.S. patents. He was a member of ASEE, ASME, ASTM, and SAE and received awards for his many contributions to engineering from all four societies. He was the primary catalyst in forming the SAEFDE–University of Iowa short course on Fatigue Concepts in Design and coauthored the first edition of *Metal Fatigue in Engineering* with R. I. Stephens. SAE founded the H. O. Fuchs memorial award program for a university student to present her or his research at each of the biannual SAEFDE meetings. Professor Fuchs preferred to call residual stresses "self-stresses" because he felt that "residual" suggested something unimportant and left over. Henry Fuchs was a wonderful engineer, teacher, researcher, colleague, and friend to many.

CHAPTER 1

INTRODUCTION AND HISTORICAL OVERVIEW

1.1 MECHANICAL FAILURE MODES

Mechanical failures have caused many injuries and much financial loss. However, relative to the large number of successfully designed mechanical components and structures, mechanical failures are minimal. Mechanical failures involve an extremely complex interaction of load, time, and environment, where environment includes both temperature and corrosion. Loads may be monotonic, steady, variable, uniaxial, or multiaxial. The loading duration may range from centuries to years, as in steel bridges, or to seconds or milliseconds, as in firing a handgun. Temperatures can vary from cryogenic with rocket motor fuels, to ambient with household kitchen chairs, to over 1000°C with gas turbine engines. Temperatures may be isothermal or variable. Corrosive environments can range from severe attack with automobile engine exhaust and salt water exposure to essentially no attack in vacuum or inert gas. The interaction of load, time, and environment along with material selection, geometry, processing, and residual stresses creates a wide range of synergistic complexity and possible failure modes in all fields of engineering. Table 1.1 provides a list of possible mechanical failure modes in metals that can occur with smooth, notched, and cracked components or structures. A brief description of these failure modes along with examples follows.

Excess deformation or yielding is probably the most commonly studied failure mode and is based upon the maximum shear stress criterion or the octahedral shear stress (energy of distortion) criterion. An example of an area where excess deformation with appreciable plasticity has been used successfully in design is in the ground vehicle industry, with off-road tractors using

1

TABLE 1.1   Mechanical Failure Modes of Metals

1. Excess deformation—elastic, yielding, or onset of plasticity
2. Ductile fracture—substantial plasticity and high-energy absorption
3. Brittle fracture—little plasticity and low-energy absorption
4. Impact or dynamic loading—excess deformation or fracture
5. Creep—excess deformation or fracture
6. Relaxation—loss of residual stress or external loading
7. Thermal shock—cracking and/or fracture
8. Wear—many possible failure mechanisms
9. Buckling—elastic or plastic
10. Corrosion, hydrogen embrittlement, neutron irradiation*
11. Stress corrosion cracking (environmental assisted cracking)
12. Esthetic aspects*
13. Fatigue—repeated loading
    a. Fatigue crack nucleation
    b. Fatigue crack growth
    c. Constant or variable amplitude loading
    d. Uniaxial/multiaxial loading
    e. Corrosion fatigue
    f. Fretting fatigue
    g. Creep-fatigue
       1. Isothermal
       2. Thermomechanical
    h. Combinations of a to g

* These are not mechanical failure modes, but they usually interact with mechanical aspects.

rollover protection systems (ROPS) to protect operators during rollover accidents. Failure by excess deformation may also be elastic, such as in rotating machinery, where seizure can occur. Ductile fracture involves significant plasticity and is associated with high-energy absorption, with fracture occurring primarily by microvoid coalescence (ductile dimpling) formed at inclusions and/or secondary phase particles. Brittle fracture contains little macro or micro plasticity, and involves low-energy absorption and very high crack growth velocity during fracture. Transcrystalline cleavage and intercrystalline fracture are the most common micro mechanisms of brittle fracture in metals. The many brittle fractures of World War II welded Liberty ships were attributed to the interaction in weld regions of sharp stress concentrations, residual tensile stresses, multiaxial stress state, and low fracture toughness. Impact or dynamic loading conditions that create high strain rates in metals tend to cause lower tensile toughness, fracture toughness, and ductility. An example of a controlled impact fracture involves the catapult pin used on aircraft carriers in launching aircraft. The pin is designed to restrain the throttled aircraft prior to launch and to fracture during the impact launch. Creep and relaxation of metals are most predominant at elevated temperatures. Creep can cause significant permanent deformation and/or fracture and is usually intercrystalline. Creep failures have occurred in gas turbine engine blades due to centrifugal forces. These have been significantly overcome by using single-crystal turbine blades. Relaxation is primarily responsible for loss of residual stress and loss of external load that can occur in bolted fasteners at elevated or ambient temperature. Thermal shock tends to promote cracking and/or brittle fracture. This failure mode has resulted from the quenching operation during heat treatment of metals. Wear occurs at all temperatures and includes many different crack nucleation and crack growth behaviors. Wear failures are dominant in roller or taper bearings along with gear teeth surfaces. Buckling failure can be induced by external loading or thermal conditions and can involve elastic or plastic instabilities. These failures are most dominant in columns and thin sheets subjected to compressive loads. Corrosion by itself involves pitting and crack nucleation. Crack growth can occur due to interaction with applied and/or residual stresses and the corrosive environment. This interaction is called stress corrosion cracking (SCC) or environmental assisted cracking (EAC). Examples of stress corrosion cracking occur in many brass components such as plumbing fixtures and ammunition shell cases. These failures occur even without use due to residual stresses and corrosive environment interaction. Hydrogen embrittlement is most susceptible in high-strength steels and causes intercrystalline cracking that can lead to brittle fracture. This type of fracture has occurred in SAE grade 8 bolts used in truck frame connections. An example of an esthetic failure is the surface appearance changes through environmental degradation of highway corrugated steel retaining fences.

The last failure mode listed in Table 1.1 is failure due to fatigue, that is, from repeated loading. At least half of all mechanical failures are due to fatigue. No exact percentage is available, but many books and articles have suggested that 50 to 90 percent of all mechanical failures are fatigue failures; most of these failures are unexpected. They include everyday simple items such as door springs, tooth brushes, tennis racquets, electric light bulbs, and repeated bending of paper clips. They also include more complex components and structures involving ground vehicles, ships, aircraft, and human body implants. Examples are automobile steering linkage, engine connecting rods, ship propeller shafts, pressurized airplane fuselage, landing gears, and hip replacement prostheses. The different aspects of fatigue listed in Table 1.1, and their interaction and/or synergistic effects, are the subjects of this book.

1.2 IMPORTANCE OF FATIGUE CONSIDERATIONS IN DESIGN

Even though the number of mechanical failures relative to the number of successful uses of components and structures is minimal, the cost of such failures is enormous. A comprehensive study of the cost of fracture in the United States indicated a cost of $119 billion (in 1982 dollars) in 1978, or about 4 percent of the gross national product [1]. The investigation emphasized that this cost could be significantly reduced by using proper and current technology in design, including fatigue design. Many approaches to fatigue design exist. They can be simple and inexpensive or extremely complex and expensive. A more complete fatigue design procedure may initially be more expensive, but in the long run it may be the least expensive. Thus, an important question in fatigue design is how complete the synthesis/analysis/testing procedure should be. Current product liability laws have placed special emphasis on explicitly documented design decisions, suggesting that a more complete design procedure is needed by law.

Currently, proper fatigue design involves synthesis, analysis, and testing. Fatigue testing alone is not a proper fatigue design procedure, since it should be used for product durability determination, not for product development. Analysis alone is also insufficient fatigue design, since current fatigue life models, including commercial and government computer-aided engineering (CAE) software programs, are not adequate for safety critical parts. They are only models and usually cannot take into consideration all the synergistic aspects involved in fatigue, such as temperature, corrosion, residual stress, and variable amplitude loading. Thus, both analysis and testing are required components of good fatigue design. The more closely analysis and testing simulate the real situation, the more confidence one can have in the results.

Safety factors are often used in conjunction with or without proper fatigue design. Values that are too high may lead to noncompetitive products in the global market, while values that are too low can contribute to unwanted failures. Safety factors are not replacements for proper fatigue design procedures, nor should they be an excuse to offset poor fatigue design procedures.

In Section 1.1 it was indicated that fatigue failures can occur in both simple, inexpensive products and complex, expensive products. The consequences of each type of product fatigue failure may be minimal or catastrophic. For example, the fatigue failure of an inexpensive front wheel bearing can cause complete loss of control of an automobile. It can be a nuisance failure as the driver comes to an unexpected stop, or it can be catastrophic if the fracture occurs while rounding a highway curve. Likewise, an aircraft engine failure may delay takeoff, but it could also result in complete loss of passengers and aircraft. Fatigue failures occur in every field of engineering and also in interdisciplinary engineering fields. They include thermal/mechanical fatigue failure in electrical circuit boards involving electrical engineers, bridges involving civil engineers, automobiles involving mechanical engineers, farm tractors involving agricultural engineers, aircraft involving aeronautical engineers, heart valve implants involving biomedical engineers, pressure vessels involving chemical engineers, and nuclear piping involving nuclear engineers. Thus, all fields of engineering are involved with fatigue design of metals.

1.3 HISTORICAL OVERVIEW OF FATIGUE

Fatigue of materials is still only partly understood. What we do know has been learned and developed step by step and has become quite complex. To gain a general understanding, it is best to start with a brief historical review of fatigue developments. This shows a few basic ideas and indicates very briefly how they were developed by the efforts of many people.

The first major impact of failures due to repeated stresses involved the railway industry in the 1840s. It was recognized that railroad axles failed regularly at shoulders [2]. Even then, the elimination of sharp corners was recommended. Since these failures appeared to be quite different from normal ruptures associated with monotonic testing, the erroneous concept of "crystallization" due to vibration was suggested but was later refuted. The word "fatigue" was introduced in the 1840s and 1850s to describe failures occurring from repeated stresses. This word has continued to be used for the normal description of fracture due to repeated stresses. In Germany during the 1850s and 1860s, August Wöhler[a*] performed many laboratory fatigue tests under repeated stresses. These experiments were concerned with railway axle failures and are considered to be the first systematic investigation of fatigue. Thus, Wöhler has been called the "father" of systematic fatigue testing. Using stress versus life $(S-N)$ diagrams, he showed how fatigue life decreased with higher stress amplitudes and that below a certain stress amplitude, the test specimens did not fracture. Thus, Wöhler introduced the concept of the $S-N$ diagram and the fatigue limit. He pointed out that for fatigue, the range of stress is more important than the maximum stress [3]. During the 1870s and 1890s, additional researchers substantiated and expanded Wöhler's classical work. Gerber and others investigated the influence of mean stress, and Goodman[b] proposed a simplified theory concerning mean stresses. Their names are still associated with diagrams involving alternating and mean stresses. Bauschinger[c] [4] in 1886 showed that the yield strength in tension or compression was reduced after applying a load of the opposite sign that caused inelastic deformation. This was the first indication that a single reversal of inelastic strain could change the stress–strain behavior of metals. It was the forerunner of understanding cyclic softening and hardening of metals.

In the early 1900s, Ewing and Humfrey [5] used the optical microscope to pursue the study of fatigue mechanisms. Localized slip lines and slip bands leading to the formation of microcracks were observed. Basquin [6] in 1910 showed that alternating stress versus number of cycles to failure $(S-N)$ in the finite life region could be represented as a log-log linear relationship. His equation, plus modifications by others, are currently used to represent finite life fatigue behavior. In the 1920s, Gough[d] and associates contributed greatly to the understanding of fatigue mechanisms. They also showed the combined

---

* Biographical sketches are found in Section 1.6 for the individuals designated by superscripts a through u.

effects of bending and torsion (multiaxial fatigue). Gough published a compre-hensive book on fatigue of metals in 1924 [7]. Moore[e] and Kommers[f] [8] published the first comprehensive American book on fatigue of metals in 1927. In 1920 Griffith[g] [9] published the results of his theoretical calculations and experiments on brittle fracture using glass. He found that the strength of glass depended on the size of microscopic cracks. If S is the nominal stress at fracture and a is the crack size at fracture, the relation is $S\sqrt{a}$ = constant. With this classical pioneering work on the importance of cracks, Griffith developed the basis for fracture mechanics. He thus became the "early father" of fracture mechanics. In 1924 Palmgren [10] developed a linear cumulative damage model for variable amplitude loading and established the use of the B_{10} fatigue life based upon statistical scatter for ball bearing design. McAdam [11] in the 1920s performed extensive corrosion fatigue studies in which he showed significant degradation of fatigue resistance in various water solutions. This degradation was more pronounced in higher-strength steels. In 1929/1930 Haigh[h] [12] presented a rational explanation of the difference in the response of high tensile strength steel and of mild steel to fatigue when notches are present. He used the concepts of notch strain analysis and residual stresses, which were more fully developed later by others. During the 1930s, an impor-tant practical advance was achieved by the introduction of shot-peening in the automobile industry. Fatigue failures of springs and axles, which had been common, thereafter became rare. Almen[i] [13] correctly explained the spectacular improvements by compressive residual stresses produced in the surface layers of peened parts, and promoted the use of peening and other processes that produce beneficial residual stresses. Horger [14] showed that surface rolling could prevent the growth of cracks. In 1937 Neuber[j] [15] intro-duced stress gradient effects at notches and the elementary block concept, which states that the average stress over a small volume at the root of the notch is more important than the peak stress at the notch. In 1939 Gassner [16] emphasized the importance of variable amplitude testing and promoted the use of an eight-step block loading spectrum for simulated testing. Block testing was prominent until closed-loop electrohydraulic test systems became available in the late 1950s and early 1960s.

During World War II the deliberate use of compressive residual stresses became common in the design of aircraft engines and armored vehicles. Many brittle fractures in welded tankers and Liberty ships motivated substantial efforts and thinking concerning preexisting discontinuities or defects in the form of cracks and the influence of stress concentrations. Many of these brittle fractures started at square hatch corners or square cutouts and welds. Solutions included rounding and strengthening corners, adding riveted crack arresters, and placing greater emphasis on material properties. In 1945 Miner [17] formu-lated a linear cumulative fatigue damage criterion suggested by Palmgren [10] in 1924. This criterion is now recognized as the Palmgren-Miner linear damage rule. It has been used extensively in fatigue design and, despite its many shortcomings, remains an important tool in fatigue life predictions. The formation of the American Society for Testing and Materials (ASTM) Committee E-09 on Fatigue in 1946, with Peterson[k] as its first chairman, provided a forum for fatigue testing standards and research. Peterson emphasized that the fatigue notch factor, K_f, was a function of the theoretical stress concentration factor, K_t, the notch and component geometry, and the ultimate tensile strength [18]. In 1953 he published a comprehensive book on stress concentration factors [19] and an expanded version [20] in 1974.

The Comet, the first jet-propelled passenger airplane, started service in May 1952 after more than 300 hours of flight tests. Four days after an inspection in January 1954, it crashed into the Mediterranean Sea. After much of the wreckage had been recovered and exhaustive investigation and tests on components of the Comet had been made, it was concluded that the accident was caused by fatigue failure of the pressurized cabin. The small fatigue cracks originated from a corner of an opening in the fuselage. Two Comet aircraft failed catastrophically. The Comet had been tested thoroughly. The cabin pressure at high altitudes was 57 kPa (8.25 psi) above outside pressure. By September 1953, a test section of the cabin had been pressurized 18 000 times to 57 kPa in addition to 30 prior cycles between 70 and 110 kPa. The design stress for 57 kPa was 40 percent of the tensile strength of the aluminum alloy. Probably the first 30 high load levels induced sufficient residual stresses in the test section so as to falsely enhance the fatigue life of the test component and provide overconfidence. All Comet aircraft of this type were taken out of service, and additional attention was focused on airframe fatigue design. Shortly after this, the first emphasis on fail-safe rather than safe-life design for aircraft gathered momentum in the United States. This placed much more attention on maintenance and inspection.

Major contributions to the subject of fatigue in the 1950s included the introduction of closed-loop servohydraulic test systems, which allowed better simulation of load histories on specimens, components, and total mechanical systems. Electron microscopy opened new horizons to a better understanding of basic fatigue mechanisms. Irwin[l] [21] introduced the stress intensity factor K_I, which has been accepted as the basis of linear elastic fracture mechanics (LEFM) and of fatigue crack growth life predictions. Irwin coined the term "fracture mechanics," and because of his many important contributions to the subject at this time, he is considered the modern "father of fracture mechanics." The Weibull[m] distribution [22] provided both a two- and a three-parameter statistical distribution for probabilistic fatigue life testing and analysis.

In the early 1960s, low-cycle strain-controlled fatigue behavior became prominent with the Manson[n]-Coffin[o] [23,24] relationship between plastic strain amplitude and fatigue life. These ideas were promoted by Topper[p] and Morrow[q] [25,26] and, along with the development of Neuber's rule [27] and rainflow counting by Matsuishi and Endo [28] in 1968, are the basis for current notch strain fatigue analysis. The formation of the ASTM's Special Committee on Fracture Testing of High-Strength Steels in the early 1960s was the starting point for the formation of ASTM Committee E-24 on Fracture Testing in 1964. This committee has contributed significantly to the field of fracture mechanics and fatigue crack growth and was combined with ASTM Committee E-09 in 1993 to form Committee E-08 on Fatigue and Fracture. Paris[r] [29] in the early 1960s showed that the fatigue crack growth rate, da/dN, could best be described using the stress intensity factor range ΔK_I. In the late 1960s, the catastrophic crashes of F-111 aircraft were attributed to brittle fracture of members containing preexisting flaws. These failures, along with fatigue problems in other U.S. Air Force planes, laid the groundwork for the requirement to use fracture mechanics concepts in the B-1 bomber development program of the 1970s. This program included fatigue crack growth life considerations based on a preestablished detectable initial crack size. Schijve[s] [30] in the early 1960s emphasized variable amplitude fatigue crack growth testing in aircraft, along with the importance of tensile overloads in the presence of cracks that can cause significant fatigue crack growth retardation. In 1967 the Point Pleasant Bridge at Point Pleasant, West Virginia, collapsed without warning. An extensive investigation [31] of the collapse showed that a cleavage fracture in an eyebar caused by the growth of a flaw to a critical size was responsible. The initial flaw was due to fatigue, stress corrosion cracking, and/ or corrosion fatigue. This failure has had a profound influence on subsequent design requirements established by the American Association of State and Highway and Transportation Officials (AASHTO).

In 1970 Elber[t] [32] demonstrated the importance of crack closure on fatigue crack growth. He developed a quantitative model showing that fatigue crack growth was controlled by an effective stress intensity factor range rather than an applied stress intensity factor range. The crack closure model is commonly used in current fatigue crack growth calculations. In 1970 Paris [33,34] demonstrated that a threshold stress intensity factor could be obtained for which fatigue crack growth would not occur. During the 1970s, an international independent and cooperative effort formulated several standard load spectra for aircraft, offshore structures, and ground vehicle usage [35,36]. These standard spectra have been used by many engineers in a variety of applications. In July 1974 the U.S. Air Force issued Mil A-83444, which defines damage tolerance requirements for the design of new military aircraft. The use of fracture mechanics as a tool for fatigue was thus thoroughly established through practice and regulations. This practice also emphasized the increased need for an improved quantitative, nondestructive inspection capability as an integral part of the damage tolerance requirements.

During the 1980s and 1990s, many researchers were investigating the complex problem of in-phase and out-of-phase multiaxial fatigue. The critical plane method suggested by Brown and Miller[u] [37] motivated a new philosophy concerning this problem, and many additional critical plane models were developed. The small crack problem was noted during this time, and many workers attempted to understand the behavior. The small crack problem was complex and important, since these cracks grew faster than longer cracks based upon the same driving force. Definitions became very confusing. Interest in the fatigue of electronic materials increased, along with significant research on thermomechanical fatigue. Composite materials based on polymer, metal, and ceramic matrices were being developed for many different industries. The largest accomplishments and usage involved polymer and metal matrix composites. These developments were strongly motivated by the aerospace industry but also involved other industries. During this time, many complex, expensive aircraft components designed using safe-life design concepts were routinely being retired with potential additional safe usage. This created the need to determine a retirement for cause policy. From a fatigue standpoint, this meant significant investigation and application of nondestructive inspection and fracture mechanics. In 1988 the nearly fatal accident of the Aloha Boeing 737, after more than 90 000 flights, created tremendous concern over multisite damage (MSD) and improved maintenance and nondestructive inspection. Corrosion, corrosion fatigue, and inadequate inspection were heavy contributors to the MSD problem that existed in many different airplane types. Comprehensive investigations were undertaken to understand the problem better and to determine how best to cope with it and resolve it. Also during the 1980s and 1990s, significant changes in many aspects of fatigue design were attributed to advances in computer technology. These included software for different fatigue life (durability) models and advances in the ability to simulate real loadings under variable amplitude conditions with specimens, components, or full-scale structures. This brought significantly more field testing into the laboratory. Integrated CAE involving dynamic simulation, finite element analysis, and life prediction/estimation models created the idea of restricting testing to component durability rather than using it for development. Increased digital prototyping with less testing has become a goal of twenty-first-century fatigue design.

Additional readings on the history of fatigue can be found in [38,39].

1.4 SUMMARY

Many different mechanical failure modes exist in all fields of engineering. These failures can occur in simple, complex, inexpensive, or expensive components or structures. Failure due to fatigue, i.e., repeated loading, is multidisciplinary and is the most common cause of mechanical failure. Even though the number of mechanical failures compared to successes is minimal, the cost in lives, injuries, and dollars is too large. Proper fatigue design can reduce these undesirable losses. Proper fatigue design includes synthesis, analysis, and testing. The closer the simulated analysis and testing are to the real product and its usage, the greater confidence in the engineering results.

Applicable fatigue behavior and fatigue design principles have been formulated for nearly 150 years since the time of Wöhler's early work. These principles have been developed, used, and tested by engineers and scientists in all disciplines and in many countries. The current capability of computers and simulated testing has a pronounced influence on the efficiency and quality of today's fatigue design procedures. However, in proper fatigue design, both computer synthesis and analysis must be integrated with proper simulated and field testing, along with continued evaluation of product usage and maintenance, including nondestructive inspection.

1.5 DOS AND DON'TS IN DESIGN

1. Do recognize that fatigue failures are the most common cause of mechanical failure in components, vehicles, and structures and that these failures occur in all fields of engineering.
2. Do recognize that proper fatigue design methods exist and must be incorporated into the overall design process when cyclic loadings are involved.
3. Don't rely on safety factors in attempting to overcome poor design procedures.
4. Do consider that good fatigue design, with or without computer-aided design, incorporates synthesis, analysis, and testing.
5. Do consider that fatigue durability testing should be used as a design verification tool rather than as a design development tool.
6. Don't overlook the additive or synergistic effects of load, environment, geometry, residual stress, time, and material microstructure.

1.6 BIOGRAPHICAL SKETCHES

These sketches are provided to ensure that readers are aware that the concepts, principles, models, and material behavior in fatigue were formulated/obtained by dedicated engineers and scientists.

a. August Wöhler (1819–1914). After graduating from the Technical University of Hanover, Germany, and working on railways, he became chief of rolling stock of the Berlin to Breslau railroad in 1847. From 1847 to 1889 he was director of Imperial Railroads in Strasbourg. In 1870 he stated that the stress range is decisive for fatigue failures. His exhibit of fatigue test results at the Paris exhibition of 1867 was perceptively reviewed in *Engineering,* Vol. 2, 1867, p. 160.
b. John Goodman (1862–1935). He was Professor of Civil and Mechanical Engineering at the University of Leeds, England, and published the widely used textbook *Mechanics Applied to Engineering* (1st ed., 1904; 8th ed., 1914), in which he said, "it is assumed that the varying loads applied to test bars by Wöhler and others produce the same effects as

suddenly applied loads." This statement has been modified for application to actual behavior and gives what is called the "modified Goodman diagram" for mean stress.

c. Johann Bauschinger (1833–1893). He was Director of the Materials Testing Laboratory and Professor of Mechanics at Munich Polytechnic Institute. In 1884 he organized the first International Congress on Materials Testing.

d. Herbert J. Gough (1890–1965). He received his engineering degrees from University College School and London University in England, including the D.Sc. and Ph.D. He joined the National Physical Laboratory (NPL) in England in 1914 but then spent the next five years involved in World War I. Gough returned to the scientific staff at NPL and became Superintendent of the Engineering Department from 1930 to 1938. During World War II he was Director of Scientific Research in the War Department, followed by an appointment in the Ministry of Supply. He was President of the Institute of Mechanical Engineers in 1949 and published more than 80 papers on fatigue of metals, as well as giving many international lectures and receiving several awards.

e. Herbert F. Moore (1875–1960). Moore was an Instructor and Assistant Professor at the University of Wisconsin from 1903 to 1907. He was then a member of the faculty at the University of Illinois from 1907 to 1944. He had also been a mechanical engineer at Riehle Brothers Test Machine Company. His research on fatigue and engineering materials was the forerunner of continuous accomplishments in fatigue research and teaching of fatigue throughout most of the twentieth century at the University of Illinois. Moore was a member of ASTM, American Society for Metals (ASM), American Society for Engineering Education (ASEE), and the British Institute of Metals.

f. Jesse B. Kommers (1884–1966). Kommers was involved in teaching and research in the Mechanics Department at the University of Wisconsin from 1907 to 1953, except for 1 1/2 years spent at the University of Illinois (1919–1921). He rose from Instructor of Mechanics (1907–1913) to Professor of Mechanics in 1927 and chaired the Mechanics Department from 1946 to 1953, when he retired. Kommers published many papers on the subject of fatigue. His stay at the University of Illinois brought him in contact with Professor Herbert F. Moore, leading to their coauthorship of the book *Fatigue of Metals*. He was a member of ASTM, American Society of Professional Engineers (ASPE), Society for Experimental Stress Analysis (SESA), and ASEE.

g. Alan A. Griffith (1893–1963). He graduated from the University of Liverpool, England, in 1921 with the B. Eng., M. Eng., and D. Eng. degrees. He entered the Royal Aircraft factory in 1915 and advanced through a workshop traineeship followed by other positions to become senior scientific officer in 1920. In 1917, together with G. I. Taylor, he published a pioneering paper on the use of soap films in solving torsion problems, and in 1920 he published his famous paper on the theory of brittle fracture. He then worked on the design theory of gas turbines. Griffith was Head of the Engine Department of the Royal Aircraft Establishment in 1938 and joined Rolls Royce as research engineer in 1939. He worked first on the conceptual design of turbojet engines and later on vertical takeoff aircraft design. He retired in 1960 but continued working as a consultant for Rolls Royce.

h. Bernard P. Haigh (1884–1941). Haigh was born in Edinburgh, Scotland, and received engineering training at the University of Glasgow. He received the D.Sc. degree. During World War I he served with the Admiralty Service, and beginning in 1921 he was Professor of Applied Mechanics at the Royal Naval College. He produced many inventions and published many scientific papers on engineering subjects, including significant contributions on fatigue of metals.

i. John Otto Almen (1886–1973). Born in a sod cabin in North Dakota, he graduated from Washington State College in 1911 and became a prolific inventor. He joined General Motors Research Laboratories in 1926. Almen said that he turned to the study of residual stresses to escape from being second-guessed by administrators who thought they could improve his mechanical inventions. According to Almen, "fatigue failures are tensile failures."

j. Heinz Neuber (1906–1989). He was a graduate of Technical University in Munich, Germany, Professor of Applied Mechanics there, and director of its Mechanical Technology Laboratory.

k. Rudolph E. Peterson (1901–1982). Peterson received his B.S. and M.S. degrees from the University of Illinois in 1925 and 1926, respectively. In 1926 he joined Westinghouse Electric Corporation in Pittsburgh, where he became manager of the Mechanics Department. He was very active in leadership and research roles in ASME, ASTM, and SESA and received many awards for his contributions. He was the first chairman of ASTM's Committee E-09 on Fatigue, chairman of the ASME Applied Mechanics Division and president of SESA.

l. George R. Irwin (1907–1998). He received his Ph.D. in physics from the University of Illinois in 1937 and joined the Naval Research Laboratory (NRL), where he was a research scientist and became supervisor of the Mechanics Division. Upon retiring from the NRL in 1967, he became a professor at Lehigh University until 1972 and then at the University of Maryland. He continually provided both knowledge and leadership to ASTM's Committee E-24 on Fracture Testing and to Committee E-08 on Fatigue and Fracture. Many ASTM awards, as well as awards from other professional societies, were bestowed upon him for his contributions to fracture mechanics.

m. Waloddi Weibull (1887–1979). He received his B.A. degree in 1912, his M.S. in 1924, and his Ph.D. in 1932 from the University of Uppsala, Sweden. He was Director of Research at the NKA Ball Bearing Company in Goteborg, Sweden, from 1916 to 1922. Weibull was then Professor of Mechanical Engineering at the Royal Institute of Technology, Stockholm, from 1923 to 1941 and Professor of Applied Physics from 1941 to 1953. He was a member of the Royal Swedish Academies of Science, Engineering Science, Military Science, and Naval Science. He remained active in engineering and reliability after his retirement in 1953.

n. S. S. Manson (1919). Manson was born in Jerusalem, and his family immigrated to New York in his early years. He received a BSME degree from Cooper Union Institute of Technology in 1941 and an MS degree in mechanical and aeronautical engineering from the University of Michigan in 1942, and did additional graduate studies at Case Institute of Technology. He joined NACA (which became the National Aeronautics and Space Administration, NASA) in 1942 and advanced from mechanical engineer to Chief of the Materials and Structures Division, where he was responsible for a staff of 150 to 200 people from 1956 to 1974. In 1974, after 32 years at NASA, he took early retirement and became Professor of Mechanical and Aerospace Engineering at Case Western Reserve University until his second retirement in 1995. Manson is a member of ASTM and ASME, as well as a fellow and past president of SESA and a fellow of the Royal Aeronautical Society. He has published over 100 papers and has received many awards and given many invited key lectures. His book *Thermal Stress and Low-Cycle Fatigue,* published in 1966, was the first book on the subject.

o. Louis F. Coffin Jr. (1917). Coffin received his BSME from Swathmore College and his Sc.D. from the Massachusetts Institute of Technology. His many contributions and awards in fatigue occurred from his research as an engineer/scientist at the General Electric Company in Schenectady, New York. Upon retirement he became Distinguished Research Professor in Mechanical Engineering at Rensselear Polytechnic Institute. Coffin is a fellow of ASME, ASTM, and ASM and a member of American Institute of Mining, Metallurgical, and Petroleum Engineers (AIME). He was elected member of the National Academy of Engineering in 1975. From 1974 to 1978, he was chairman of ASTM's Committee E-09 on Fatigue. He has obtained many patents and has published more than 150 papers on plasticity, flow and fracture, fatigue, friction, and wear. Both he and S. S. Manson are associated independently with formulating low-cycle fatigue concepts.

p. Timothy H. Topper (1936). He was born in Kleinburg, Ontario, Canada, and received his BSCE from the University of Toronto in 1959 and his Ph.D. from Cambridge University in 1962. He joined the University of

Waterloo in 1963, became Professor in 1969, was Associate Chairman of the department from 1966 to 1972, Chairman from 1972 to 1978, and became Distinguished Professor Emeritus and Research Professor in 1996. Topper also held the position of Professor of Mechanical Engineering at the Federal University of Paraiba in Brazil from 1975 to 1988, and during this time he was Director of the Waterloo-Brazil program to develop the Science and Engineering Center at University of Paraiba. He has directed the research of many graduate students in diverse areas of fatigue and fracture mechanics. Topper is a member of the SAE Fatigue Design and Evaluation Committee and ASTM Committee E-08.

q. JoDean Morrow (1929). Morrow received his BSCE from the Rose-Hulman Institute of Technology in 1953 and his M.S. in 1954 and Ph.D. in 1957 in Theoretical and Applied Mechanics from the University of Illinois. He remained on the University of Illinois's professorial faculty, engaged in research and teaching of engineering mechanics, mechanical behavior, and properties of materials until his retirement in 1984. His major research contributions involved the promotion of the local strain approach to fatigue design. He formulated the terms and definitions of σ_f', ε_f', b, and c that have been used internationally. He was an excellent motivator of students, and many of his Ph.D. students also contributed to, and promoted, knowledge of the local strain approach. Morrow has been a member of SAE, ASTM (fellow), and Japan Society of Material Science (JSMS) and has received awards from SAE and ASTM for his contributions on cyclic deformation and fatigue properties of materials.

r. Paul C. Paris (1930). He received his B.S. degree in engineering mechanics from the University of Michigan in 1953 and his M.S. and Ph.D. degrees in applied mechanics from Lehigh University in 1955 and 1962, respectively. He was Assistant Professor at the University of Washington (1957–1962), Assistant Director of the Institute of Research at Lehigh University (1962–1964), and Program Director at the National Science Foundation (1964–1965). Paris returned to Lehigh University in 1965 and became Professor of Mechanics. From 1974 to 1976 he was Visiting Professor of Mechanical Engineering at Brown University and then accepted his current position as Professor of Mechanical Engineering at Washington University in St. Louis, Missouri. His idea of relating the fatigue crack growth rate to the stress intensity factor range was formulated in 1957 when he was a Faculty Associate at the Boeing Company. His concept was compared to data in 1959 but was not published until 1961 because of rejections by reviewers for ASME, American Institute of Aeronautics and Astronautics (AIAA), *and Philosophical Magazine.* The reviewers did not believe that the elastic stress intensity factor range could be the driving force for fatigue crack growth that involved crack tip plasticity. They turned out to be in error. Paris founded the annual ASTM National Fracture Mechanics Symposium in 1967. He has received many awards for his contributions to fatigue and fracture mechanics.

s. Jacobus (Jaap) Schijve (1927). He was born in Oostburg, the Netherlands. He received the MS degree in Aerospace Engineering and the Sc.D. at Delft University in 1953 and 1964, respectively. He joined the National Aerospace Laboratory, NLR, at Amsterdam/Noordoostpolder in 1953 and rose to head the Division of Structures and Materials. There he was involved in fatigue and fracture of aircraft materials and structures, flight simulation and full-scale fatigue testing, fatigue crack growth, accident investigation, fractography, and nondestructive testing. In 1973 Schijve became Professor of Aircraft Materials at Delft University, and in 1992 he became Emeritus Professor. He continues to perform fatigue related research. Over 130 M.S. students did their thesis research with him, along with 17 Ph.D. students. He has published over 150 papers and has been a member of the editorial boards of three international journals. He is a member of ASTM and the International Committee on Aeronautical Fatigue (ICAF), where he was the Netherlands' National Delegate (1967–1979) and General Secretary (1979–1992). Schijve has received many awards for his accomplishments.

t. Wolf Elber (1941). Elber was born in Quellendorf, Germany, and received his B.S. and Ph.D. degrees in civil engineering from the University of New South Wales, Australia. While working on his Ph.D., he studied the phenomenon of crack closure. In 1969 he joined the German equivalent of NASA, and in 1970 he came to NASA's Langely Research Center as a postdoctoral fellow to continue his pioneering research on crack closure. He became a permanent NASA employee in 1972 and later headed the Fatigue and Fracture Branch at Langely. He then became the Director of the Army Aviation Directorate and has served as Director of the U.S. Army Research Laboratory's Technology Center since 1992. Elber has received awards for his outstanding achievements and performance.

u. Keith J. Miller (1932). Miller received his Sc.D. from Cambridge University and was a fellow of Trinity College. In 1982 he left Cambridge to become Professor and Head of Mechanical Engineering at the University of Sheffield, and from 1987 to 1989 he was Dean of the Faculty of Engineering. He is currently the President of the Structural Integrity Research Institute at the University of Sheffield. Miller has directed the research of many postgraduate students, has published more than 200 papers, and has edited several books on fatigue and fracture research. He is Editor-in-Chief of the *International Journal of Fatigue and Fracture of Engineering Materials and Structures*. He has received many prizes and awards for his accomplishments in fatigue and fracture.

REFERENCES

1. R. P. Reed, J. H. Smith, and B. W. Christ, "The Economic Effects of Fracture in the United States," U.S. Department of Commerce, National Bureau of Standards, Special Publication 647, March 1983.

2. R. E. Peterson, "Discussion of a Century Ago Concerning the Nature of Fatigue, and Review of Some of the Subsequent Researches Concerning the Mechanism of Fatigue," *ASTM Bull.*, No. 164, 1950, p. 50.

3. "Wöhler's Experiments on the Strength of Metals," *Engineering,* August 23, 1967, p. 160.

4. J. Bauschinger, "On the Change of the Position of the Elastic Limit of Iron and Steel Under Cyclic Variations of Stress," *Mitt. Mech.-Tech. Lab.*, Munich, Vol. 13, No. 1, 1886.

5. J. A. Ewing and J. C. W. Humfrey, "The Fracture of Metals Under Repeated Alterations of Stress," *Phil. Trans. Roy. Soc.,* London, Vol. CC, 1903, p. 241.

6. O. H. Basquin, "The Experimental Law of Endurance Tests," *Proc. ASTM,* Vol. 10, Part II, 1910, p. 625.

7. H. J. Gough, *The Fatigue of Metals,* Scott, Greenwood and Son, London, 1924.

8. H. F. Moore and J. B. Kommers, *The Fatigue of Metals,* McGraw-Hill Book Co., New York, 1927.

9. A. A. Griffith, "The Phenomena of Rupture and Flow in Solids," *Trans. Roy. Soc.,* London, Vol. A221, 1920, p. 163.

10. A. Palmgren, "Die Lebensdauer von Kugellagern," *ZDVDI,* Vol. 68, No. 14, 1924, p. 339 (in German).

11. D. J. McAdam, "Corrosion Fatigue of Metals," *Trans. Am. Soc. Steel Treating,* Vol. 11, 1927, p. 355.

12. B. P. Haigh, "The Relative Safety of Mild and High-Tensile Alloy Steels Under Alternating and Pulsating Stresses," *Proc. Inst. Automob. Eng.,* Vol. 24, 1929/1930, p. 320.

13. J. O. Almen and P. H. Black, *Residual Stresses and Fatigue in Metals,* McGraw-Hill Book Co., New York, 1963.

14. O. J. Horger and T. L. Maulbetsch, "Increasing the Fatigue Strength of Press Fitted Axle Assemblies by Cold Rolling," *Trans. ASME,* Vol. 58, 1936, p. A91.

15. H. Neuber, *Kerbspannungslehre,* Springer-Verlag, Berlin, 1937 (in German); also *Theory of Notches,* J. W. Edwards, Ann Arbor, MI, 1946.

16. E. Gassner, "Festigkeitsversuche mit Wiederholter Beanspruchung im Flugzeugbau," *Deutsche Luftwacht, Ausg. Luftwissen,* Vol. 6, 1939, p. 43 (in German).

17. M. A. Miner, "Cumulative Damage in Fatigue," *Trans. ASME, J. Appl. Mech.,* Vol. 67, 1945, p. A159.

18. R. E. Peterson, "Notch Sensitivity," *Metal Fatigue,* G. Sines and J. L. Waisman, eds., McGraw-Hill Book Co., New York, 1959, p. 293.

19. R. E. Peterson, *Stress Concentration Design Factors,* John Wiley and Sons, New York, 1953.

20. R. E. Peterson, *Stress Concentration Factors,* John Wiley and Sons, New York, 1974.

21. G. R. Irwin, "Analysis of Stresses and Strains Near the End of a Crack Traversing a Plate," *Trans. ASME, J. Appl. Mech.* Vol. 24, 1957, p. 361.

22. W. Weibull, "A Statistical Distribution Function of Wide Applicability," *J. Appl. Mech.*, Sept. 1951, p. 293.

23. S. S. Manson, "Discussion of Ref. 24," *Trans. ASME, J. Basic Eng.*, Vol. 84, No. 4, 1962, p. 537.

24. J. F. Tavernelli and L. F. Coffin, Jr., "Experimental Support for Generalized Equation Predicting Low Cycle Fatigue," *Trans. ASME, J. Basic Eng.*, Vol. 84, No., 4, 1962, p. 533.

25. T. H. Topper, B. I. Sandor, and J. Morrow, "Cumulative Fatigue Damage Under Cyclic Strain Control," *J. Materials*, Vol. 4, No. 1, 1969, p. 189.

26. T. H. Topper, R. M. Wetzel, and J. Morrow, "Neuber's Rule Applied to Fatigue of Notched Specimens," *J. Materials*, Vol. 4, No. 1, 1969, p. 200.

27. H. Neuber, "Theory of Stress Concentration for Shear-Strained Prismatic Bodies with Arbitrary Nonlinear Stress-Strain Laws," *Trans. ASME, J. Appl. Mech.*, Vol. 28, 1961, p. 544.

28. M. Matsuishi and T. Endo, "Fatigue of Metals Subjected to Varying Stress," Presented to Japan Society of Mechanical Engineers, Fukuoka, Japan, March 1968.

29. P. C. Paris, M. P. Gomez, and W. E. Anderson, "A Rational Analytical Theory of Fatigue," *Trend Eng.*, Vol. 13, No. 9, 1961, p. 9.

30. J. Schijve and D. Broek, "Crack Propagation Tests Based on a Gust Spectrum with Variable Amplitude Loading," *Aircraft Eng.*, Vol. 34, 1962, p. 314.

31. J. A. Bennett and H. Mindlin, "Metallurgical Aspects of the Failure of the Point Pleasant Bridge," *J. Test. Eval.*, Vol. 1, No. 2, 1973, p. 152.

32. W. Elber, "Fatigue Crack Closure under Cyclic Tension," *Eng. Fracture Mech.*, Vol. 2, 1970, p. 37.

33. P. C. Paris, "Testing for Very Slow Growth of Fatigue Cracks," *Closed Loop*, Vol. 2, No. 5, 1970.

34. R. A. Schmidt and P. C. Paris, "Threshold for Fatigue Crack Propagation and the Effects of Load Ratio and Frequency," *Progr. Flaw Growth Fract. Toughness Testing*, ASTM STP 536, ASTM, West Conshohocken, PA, 1973, p. 79.

35. W. Schütz, "Standardized Stress-Time Histories—An Overview," *Development of Fatigue Loading Spectra*, J. M. Potter and R. T. Watanabe, eds., ASTM STP 1006, ASTM, West Conshohocken, PA, 1989, p. 3.

36. L. E. Tucker and S. L. Bussa, "The SAE Cumulative Fatigue Damage Test Program," *Fatigue Under Complex Loading*, R. M. Wetzel, ed., SAE, Warrendale, PA, 1977, p. 1.

37. M. W. Brown and K. J. Miller, "A Theory for Fatigue Failure Under Multiaxial Stress-Strain Conditions," *Proc. Inst. Mech. Eng.*, Vol. 187, No. 65, 1973, p. 745.

38. P. C. Paris, "Fracture Mechanics and Fatigue: A Historical Perspective," *Fat. Fract. Eng. Mat. Struct.*, Vol. 21, No. 5, 1998, p. 535.

39. W. Schütz, "A History of Fatigue," *Eng. Fract. Mech.*, Vol. 54, No. 2, 1996, p. 263.

PROBLEMS

1. What different failure modes could exist for the following components? (*a*) helical gear/shaft set, (*b*) ship propeller, (*c*) handlebar on a mountain bike, (d) airplane landing gear strut, (e) automotive engine connecting rod, and (f) key to your house door.

2. Name three components that could be considered (a) simple and inexpensive and (b) complex and expensive. What failure modes exist for these choices?

3. Give an example of the following failure modes: (a) brittle fracture, (b) ductile fracture, (c) excess deformation, (d) creep rupture, (e) wear, (f) stress corrosion cracking, (g) esthetic failure, and (h) fatigue.

4. List the significant failure modes for a bicycle and indicate their likely locations.

5. Give three examples of synergistic failure modes.

6. Give four actual fatigue failures that you have been involved with or are aware of.

7. Write a one-page paper indicating why you believe that the cost of fracture in the United States as a percentage of the gross national product has increased or decreased since the 1978 survey.

8. Obtain a copy of one of the 39 references given in this chapter and write a one-page paper on its significance.

CHAPTER 2

FATIGUE DESIGN METHODS

2.1 STRATEGIES IN FATIGUE DESIGN

Fatigue design methods have many similarities but also differences. The differences exist because a component, structure, or vehicle may be safety critical or nonsafety critical, simple or complex, expensive or inexpensive, and failures may be a nuisance or catastrophic. Only a single end product may be desired or perhaps thousands or millions of the end product are to be produced. The product may be a modification of a current model or a new product. Significant computer-aided engineering (CAE) and computer-aided manufacturing (CAM) capabilities may or may not be available to the design engineer. In all of the above situations, the commonality of fatigue design can be represented by the fatigue design flow chart shown in Fig. 2.1. The flow chart clearly brings out the many aspects of fatigue design applicable to any of the above different product situations. Figure 2.1 indicates the iterative nature of fatigue design and the need for significant input items (top row) such as geometry, load history, environment, design criteria, material properties, and processing effects. With these inputs, fatigue design is performed through synthesis, analysis, and testing. This requires selecting the configuration, material, and processes, performing stress analysis, choosing a fatigue life and a cumulative damage model, and making a computational life prediction/estimation. (We will use the words "prediction" or "estimation" interchangeably throughout the book; both terms are common in fatigue literature.) This is followed by durability testing, which can suggest modification or the decision to accept and manufacture the product and put it into service. Evaluation of service usage and success is part of the fatigue design method.

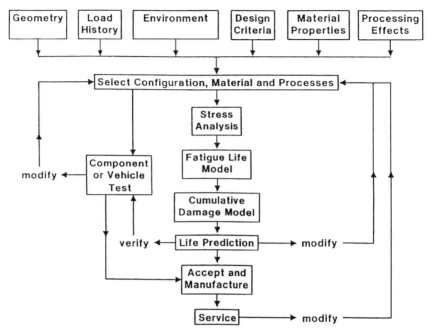

Figure 2.1   Fatigue design flow chart originated by H. S. Reemsnyder from Bethlehem Steel Corp. and slightly modified by H. O. Fuchs. It was created for use by the Society of Automotive Engineers Fatigue Design and Evaluation (SAEFDE) Committee– University of Iowa's annual short course on Fatigue Concepts in Design.

Choosing the fatigue life model is a significant decision. Currently four such models exist for design engineers. These are:

1. The nominal stress–life (S–N) model, first formulated between the 1850s and 1870s.
2. The local strain–life (ε–N) model, first formulated in the 1960s.
3. The fatigue crack growth (da/dN–ΔK) model, first formulated in the 1960s.
4. The two-stage model, which consists of combining models 2 and 3 to incorporate both macroscopic fatigue crack formation (nucleation) and fatigue crack growth.

As noted, the S–N model has been available for about 150 years, while the other models have been available only since the 1960s. The nominal S–N model uses nominal stresses and relates these to local fatigue strengths for notched and unnotched members. The local ε–N model deals directly with local strain at a notch, and this is related to smooth specimen strain-controlled fatigue behavior. Several analytical models can be used to determine local strains from global or nominal stresses or strains. The fatigue crack growth $da/dN-\Delta K$ model requires the use of fracture mechanics and integration of the fatigue crack growth rate equation to obtain the number of cycles required to grow a crack from a given length to another length and/or to fracture. This model can be considered a total fatigue life model when it is used in conjunction with information on the existing initial crack size following manufacture. The two-stage method incorporates the local $\varepsilon-N$ model to obtain the life to the formation of a small macrocrack, followed by integration of the fatigue crack growth rate equation for the remaining life. The two lives are added together to obtain the total fatigue life. All four of these fatigue life models are covered in this book, and each has areas of best applicability.

Depending on the purpose of the design and the different conditions discussed above, the design engineer will proceed through Fig. 2.1 in different ways. For the purpose of illustration, we look at four of the many different possible situations:

1. Designing a device, perhaps a special bending tool or a test rig, to be used in the plant where it was designed. We call it an "in-house tool."
2. Changing an existing product by making it larger or smaller than previously, using a different material or different shapes, perhaps a linkage and coil spring in place of a leaf spring. We call it a "new model."
3. Setting up a major project that is quite different from past practice. A spacecraft or an ocean drilling rig or a new type of tree harvester are examples. We call it a "new product."
4. Designing a highway bridge or a steam boiler. The expected loads, acceptable methods of analysis, and permissible stresses are specified by the customer or by a code authority. We call it "design to code."

2.1.1 The In-House Tool

If part of a tool is subjected to repeated loads, as, for instance, a ratchet mechanism or a rotating shaft carrying a pulley, it must be designed to avoid fatigue failure. For the in-house tool, provided that the designer knows the expected load-time history to which the part will be subjected in service, he or she will start with a shape that avoids stress concentrations as much as possible, will determine the stresses, and will select a material and treatments, depending on the requirements for weight, space, and cost. The design may have a suitable margin between the stress that corresponds to a 50 percent probability of failure at the desired life and the permitted stresses. A second and a third iteration may be required to balance the conflicting factors of weight or space, expected life, and cost. If the expected loadings are not uniformly repeated, the designer will consider cumulative damage.

The differences between design for fatigue resistance and design for a few loadings are greater attention to the details of shape and treatments and the need to decide on a required lifetime of the part. The designer may prevent serious consequences of failure by making the part accessible for inspection and replacement, by providing a fail-safe design or by using larger safety factors, and by performing appropriate fatigue tests.

2.1.2 The New Model

For a new model, more certainty may be required and more data should be available from service records or previous models. In addition to the steps outlined in Section 2.1.1, tests are needed to confirm the assumptions and calculations. Broken parts from previous models provide the most useful data. They can be used to adjust the test procedures so that testing produces failures similar in location and appearance to service failures. Tests that produce other types of failure probably have a wrong type of loading or wrong load amplitude. From experience with previous models, one sometimes also knows what type of accelerated uniform cycle test is an index of satisfactory performance. In testing passenger car suspension springs, for instance, it has been found that 200 000 cycles of strokes from full rebound to maximum possible deflection is an acceptable test that is much quicker and less expensive than a spectrum test with different amplitudes. Data on loads encountered by the parts may be available directly from previous models or by analogy with previous models. Instead of doing a complete stress analysis, it may be possible to determine the relation of significant stresses to loads from measurements on previous satisfactory models and to reproduce the same relation in the new model.

2.1.3 The New Product

This requires the greatest effort in fatigue design. Predicting future loads is the most important factor. No amount of stress analysis can overcome an erroneous load prediction. After the loads or load spectra have been obtained, one can analyze the fatigue worthiness of all parts. Many computer software programs are available to do this. The results are verified by component fatigue tests, which may lead to design modifications. Whenever possible, prototypes or pilot models are used to confirm functional performance and the predicted loads.

2.1.4 Design to Code

Many industries provide data on permissible stresses. The American Welding Society and the British Standards Institution, for instance, have published curves that show recommended stresses as a function of the desired life for various types of weldments. The ASME Boiler and Pressure Vessel Code has recommended fatigue design criteria based upon current fatigue models and many fatigue test data. Such codes permit the designer to use data based on the experience of many others. As a rule, a design according to code is a conservative, safe design. However, in case of a product liability lawsuit, U.S. courts do not accept compliance with a code as sufficient to exonerate the manufacturer or seller of a product that eventually failed.

2.2 FATIGUE DESIGN CRITERIA

Criteria for fatigue design have evolved from so-called infinite life to damage tolerance. Each of the successively developed criteria still has its place, depending on the application. The criteria for fatigue design include usage of the four fatigue life models ($S–N$, $\varepsilon–N$, $da/dN–\Delta K$, and the two-stage method) discussed in Section 2.1.

2.2.1 Infinite-Life Design

Unlimited safety is the oldest criterion. It requires local stresses or strains to be essentially elastic and safely below the pertinent fatigue limit. For parts subjected to millions of cycles, like engine valve springs, this is still a good design criterion. However, most parts experience significant variable amplitude loading, and the pertinent fatigue limit is difficult to define or obtain. In addition, this criterion may not be economical or practical in many design situations. Examples include excessive weight of aircraft for impracticality and global competitiveness for cost effectiveness.

2.2.2 Safe-Life Design

Infinite-life design was appropriate for the railroad axles that Wöhler investigated, but automobile designers learned to use parts that, if tested at the maximum expected stress or load, would last only hundreds of thousands of cycles instead of many millions. The maximum load or stress in a suspension spring or a reverse gear may occur only occasionally during the life of a car; designing for a finite life under such loads is quite satisfactory. The practice of designing for a finite life is known as "safe-life" design. It is used in many other industries too—for instance, in pressure vessel design and in jet engine design.

The safe life must, of course, include a margin for the scatter of fatigue results and for other unknown factors. The calculations may be based on stress–life, strain–life, or crack growth relations. Safe-life design may be based solely or partially on field and/or simulated testing. Examples of products in which field and simulated testing play a key role in safe-life determination are jet engines, gun tubes, and bearings. Here appropriate, regular inspections may not be practical or possible; hence, the allowable service life must be less than the test life or calculated life. For example, the U.S. Air Force historically has required that the full-scale fatigue test life of production aircraft/parts be four times longer than the expected or allowable service life. With gun tubes, the U.S. Army has required both actual firing tests and simulated laboratory pressure fatigue tests of six or more tubes to establish the allowable service life as a fraction of the mean test life. Ball bearings and roller bearings are noteworthy examples of safe-life design. The ratings for such bearings are often given in terms of a reference load that 90 percent of all bearings are expected to withstand for a given lifetime—for instance, 3000 hours at 500 RPM or 90 million revolutions. For different loads or lives or for different probabilities of failure, the bearing manufacturers list conversion formulas. They do not list any load for infinite life or for zero probability of failure at any life.

The margin for safety in safe-life design may be taken in terms of life (e.g., calculated life = 20 × desired life), in terms of load (e.g., assumed load = 2 × expected load), or by specifying that both margins must be satisfied, as in the ASME Boiler and Pressure Vessel Code.

2.2.3   Fail-Safe Design

When a component, structure, or vehicle reaches its allowable safe life, it must be retired from service. This can be inadequate since all the fleet must be retired before the average calculated life or test life is attained. This practice is very costly and wasteful. Also, testing and analysis cannot predict all service failures. Thus fail-safe fatigue design criteria were developed by aircraft engineers. They could not tolerate the added weight required by large safety factors, or the danger to life created by small safety factors, or the high cost of safe-life design. Fail-safe design requires that if one part fails, the system does not fail. Fail-safe design recognizes that fatigue cracks may occur, and structures are arranged so that cracks will not lead to failure of the structure before they are detected and repaired. Multiple load paths, load transfer between members, crack stoppers built at intervals into the structure, and inspection are some of the means used to achieve fail-safe design. This philosophy originally applied mainly to airframes (wings, fuselages, control surfaces). It is now used in many other applications as well. Engines are fail-safe only in multiengine planes. A landing gear is not fail-safe, but it is designed for a safe life.

2.2.4   Damage-Tolerant Design

This philosophy is a refinement of the fail-safe philosophy. It assumes that cracks will exist, caused either by processing or by fatigue, and uses fracture mechanics analyses and tests to determine whether such cracks will grow large enough to produce failures before they are detected by periodic inspection. Three key items are needed for successful damage-tolerant design: residual strength, fatigue crack growth behavior, and crack detection involving nondestructive inspection. Of course, environmental conditions, load history, statistical aspects, and safety factors must be incorporated in this methodology.

Residual strength is the strength at any instant in the presence of a crack. With no cracks, this could be the ultimate tensile strength or yield strength, depending upon the failure criteria chosen. As a crack forms and grows under cyclic loading, the residual strength decreases. This decrease as a function of crack size is dependent upon material, environment, component and crack configuration, location, and mode of crack growth. Residual strength is usually obtained using fracture mechanics concepts. Fatigue crack growth behavior is also a function of the previous parameters and involves fracture mechanics concepts. Crack detection methods, using several different nondestructive inspection techniques and standard procedures, have been developed. Inspection periods must be laid out such that as the crack grows, the applied stresses remain below the residual strength. Cracks need to be repaired or components replaced before fracture occurs under the service loads. This philosophy looks for materials with slow crack growth and high fracture toughness. Damage-tolerant design has been required by the U.S. Air Force. In pressure vessel design, "leak before burst" is an expression of this philosophy.

Retirement for cause is a special situation requiring damage-tolerant usage. Imagine the number of jet engine turbine blades that have been retired from service because they have reached their designed safe-life service life based upon analytical and test results. This cost is enormous since most of these blades could have significant additional service life. To allow for possible extended service life, damage-tolerant methodology based upon both analytical considerations and additional blade testing is required. In the case of jet engine turbine blades, this is not an easy task because of the safety-critical situation and the many complex parameters involved. Retirement for cause, using damage-tolerant procedures, can be applicable to many engineering situations involving products already designed by safe-life methods or with new designs.

2.3   ANALYSIS AND TESTING

Analysis and testing are both key aspects of fatigue design, as indicated in the flow chart of Fig. 2.1. How much time and money should be put into each is an important engineering decision. A more complete and correct analysis involving iteration and optimization can provide prototypes that are closer to the final product and thus require less testing. Insufficient or incorrect analysis may result in too much dependence upon testing and retesting, creating both time and cost inefficiencies. Analysis capabilities are largely dependent upon the computer capabilities available to the engineer. Complete computer programs are available for taking a product from an input such as a road profile, or a strain or load spectrum, to a final calculated fatigue life. However, the engineer must realize that these calculations are for the models; the key to confidence in these results is how closely the models represent the real product and its usage. For example, environmental influence and nonproportional multiaxial loading conditions are not usually properly integrated into the calculations, along with the fact that the results have varied from excellent to fair to poor. Thus, even the best analysis should not necessarily be the final product design, particularly with safety critical products. However, analysis is a must in proper fatigue design and should lead to a very reasonable prototype design. A design based on analysis alone, without fatigue testing, requires either a large margin for uncertainty or an allowance for some probability of failure. A probability of failure of a few percent can be permitted if failures do not endanger lives and if replacement is considered a routine matter, as in automobile fan belts. In most other situations, analysis needs to be confirmed by tests.

Fatigue testing has involved enormous differences in complexity and expense and has ranged from the simple constant amplitude rotating beam test of a small specimen to the simulated full-scale, complex, variable amplitude thermomechanical cycling of the Concord supersonic aircraft structure in the 1970s or the Boeing 777 aircraft structure in the 1990s. The objective of fatigue testing may be to obtain the fatigue properties of materials, aid in product development, determine alterations or repairs, evaluate failed parts, establish inspection periods, or determine the fatigue durability of components, subassemblies, or the full-scale product. Durability testing requires a representative product to test and therefore occurs late in the design/development process. Parts manufactured for fatigue testing should be processed just like production parts because differences in processing (for instance, cut threads instead of rolled threads or forged parts instead of cast parts) may have a major effect on fatigue resistance. Test specimens may be considered one-dimensional, as with small cylindrical specimens used for baseline material characterization under well-controlled environmental conditions. They may be considered two-dimensional in simple component testing that may include geometrical discontinuities and surface finish such as an engine connecting rod. Three-dimensional specimens would include subassembly structures such as a truck suspension system including joints and multiple parts to full-scale structures such as the Concorde and the Boeing 777 aircraft.

Since the introduction of closed-loop servohydraulic test systems in the late 1950s, significant emphasis and success have occurred in bringing the test track or proving ground into the laboratory. Current simulation test systems are capable of variable amplitude load, strain, or deflection with one channel or multiple channels of input. Road simulators provide principally one-dimensional input through the tires or three-dimensional input through each axle shaft/spindle. Test systems are, or can be, available for almost every engineering situation, discipline, or complexity. One such full-scale simulated fatigue test is shown in Fig. 2.2, where an automobile is subjected to three-dimensional variable amplitude load inputs at each wheel spindle. Simulated laboratory testing and test track or proving ground testing can be performed at the same time. They both provide significant input to final product decisions.

Figure 2.2 Full-scale simulated road fatigue test of an automobile (courtesy of MTS Corporation).

Fatigue or durability testing for design verification, or development testing, is an art. It is far more demanding than the art of fatigue testing for research because it requires the engineer to make the test represent conditions of use, but it is far less demanding than fatigue testing for research in its requirements for precision. Loading and environment conditions similar to those encountered in service are prime requirements for simulated fatigue testing. Determining the service loads may be a major task. Multichannel data acquisition systems are available to obtain the load, torque, moment, strain, deflection, or acceleration versus time for many diverse components, structures, and vehicles subjected to service usage.

Acceleration of simulation or field testing poses problems. It is often required in order to bring products to market before the competition or to find a method for improving marginal products and controlling test costs. Three common methods to accelerate testing involve increasing test frequency, using higher test loads, and/or eliminating many small load cycles from the load spectrum. All three methods have the significant advantages of less test time and lower costs, but each has disadvantages. Increased frequency may have an effect on life and may not provide enough time for environmental aspects to operate fully. Increasing loads beyond service loads accelerate tests but may produce misleading results; residual stresses that might have remained in service may be changed by excessive test loads. Fretting and corrosion may not have enough time to produce their full effects. Eliminating many small load cycles from the test load spectrum is common, and several analytical methods exist to aid in eliminating so-called nondamaging cycles. We prefer to call these "low-damaging" rather than "nondamaging" cycles because we cannot be sure if these small cycles will or will not influence fatigue life. Elimination of low-damaging cycles may hide the influence of both fretting and corrosion. In fatigue testing, the more closely the simulation or field test represents the service conditions, the more confidence one can have in the results.

Proof testing involves a single loading of a component or structure to a level usually slightly higher than the maximum service load. It can provide information on maximum crack size that could exist at the time of proof testing, which can be helpful in damage-tolerant design situations and in determining inspection periods. Proof testing may alter fatigue resistance by creating desirable and/or undesirable residual stresses. Periodic proof testing at low temperature has been used on F-111 airplane wings as part of the routine in-service inspection procedure to ensure continuous damage-tolerant fleet readiness.

2.4 PROBABILISTIC DESIGN AND RELIABILITY

Fatigue behavior of simple or complex specimens, components, structures, or vehicles involves variability. The variability in life has ranged from almost a factor of 1 to several orders of magnitude for a given test or service condition. These two extremes, however, are not the norm. Fatigue data available at present permit probabilistic design for a few situations down to a probability of failure of about 10 percent. For lower probabilities we hardly ever have the necessary data. Extrapolation of known probability data to lower probabilities of failure requires large margins for uncertainty or safety factors. Fatigue reliability can be determined from service experience. However, design without data from service experience cannot, with our present knowledge, provide quantitative reliability figures in the ranges that are required or desired for service. Chapter 13 on statistical aspects of fatigue covers these topics.

2.5 CAE AND DIGITAL PROTOTYPING

CAE involves computer usage to perform much of the iteration of synthesis and analysis in the design procedure. "Digital prototyping" refers to a computer-generated realistic prototype model near or at the final state of the product. Thus, this means that a computer prototype is formed by analysis and synthesis only. This procedure could also be called "digital testing." Its goals are to reduce product development time and cost and to provide a nearly optimal product. The computational scheme could begin at several entry levels. These include known or assumed road or terrain profiles for ground vehicles, load–time histories, and stress– or strain–time histories. The computational procedures and information needed depend upon the entry level. The most extensive computational requirements occur for the road input entry level, and the least extensive ones occur for the stress– or strain–time history entry level. Computational schemes can provide fatigue life prediction analysis, reliability analysis, design sensitivity analysis, and design optimization. This can require three-dimensional graphics for shape determination, component/structure/vehicle modeling, rigid or flexible multibody kinematics and dynamics for velocity, acceleration, or load-time history determination, material properties, processing effects, and fatigue life prediction methodology. Design sensitivity analysis and optimization can be accomplished based upon stress, strain, stiffness, and so on, or fatigue life. The use of CAE and digital prototyping is a very important and rapidly growing key segment of fatigue design.

2.6 IN-SERVICE INSPECTION AND ACQUISITION OF RELEVANT EXPERIENCE

Imperfections of design will eventually become known. Either the part is too weak and fails too often or it is too strong and a competitor can produce it more economically. Part of engineering responsibility involves efforts to find and correct weaknesses before customers and competitors find them. Obtaining records of loads through continuous in-service monitoring of customer usage, field testing, and from proving grounds, and deciding which loads are frequent, which are occasional, which are exceptional, and how much greater loads than those measured can occur are important. Past experience aids in this determination.

In-service inspection is also a way to avoid surprises. Many companies put an early production model into severe service with a friendly user and inspect it very carefully at frequent intervals to find any weaknesses before others find them. Determining suitable inspection intervals and procedures of in-service inspection is often a key part of fatigue design. In damage-tolerant design, inspection for cracks is mandatory. This inspection must be nondestructive in order to be meaningful. The ASTM, ASM International, and the American Society for Nondestructive Testing (ASNT) have published significant information on nondestructive inspection (NDI) and nondestructive testing (NDT) [1–3]. Additional books on these subjects and on nondestructive evaluation (NDE) are also available [4–6]. ASTM Committee E-07 on Nondestructive Testing is responsible for Vol. 03.03 [1], which includes over 100 standards on the following nondestructive testing/inspection techniques applicable to crack detection: acoustic emission, electromagnetic (eddy current), gamma and x-radiology, leakage, liquid penetrant and magnetic particles, neutron radiology, ultrasonic, and emerging NDT methods. Both ASM and

ASNT handbooks [2,3] also have significant information on these same methods that range from simple to very sophisticated in terms of required user capability. Some methods provide only qualitative information on crack existence, while others provide quantitative size measurements. Excessive inspection is wasteful and expensive, and inspection delayed too long may be fatal, yet tough inspection decisions need to be made. A simple, nondestructive procedure involves railway inspectors hitting each axle of express trains with long-handled hammers to detect fatigue cracks by sound before the cracks become large enough to produce fractures.

2.7 SUMMARY

The fatigue design process is determined by design objectives and by the extent of the available knowledge. It is an iterative process involving synthesis, analysis, and testing. The extent to which these three processes are used depends upon the situation and the computational and testing capabilities available. Four different analytical or computational fatigue life models—S–N, ε–N, da/dN–ΔK, and a two-stage model—are available to the engineer, and each has been used successfully (and unsuccessfully). Four different fatigue design criteria exist involving infinite-life, safe-life, fail-safe, and damage-tolerant design. Each of these criteria has specific goals and significant differences. In-service inspection and continuous monitoring of customer usage are part of fatigue design.

Optimum analysis and testing is a major decision in fatigue design from the standpoint of time, cost, and product reliability. Current digital prototyping can produce computer prototypes formed at different levels of input, such as road profile, load, stress, or strain histories. These computer prototypes can be optimized using fatigue life predictions. However, analytical or computational fatigue life predictions should not be considered sufficient, particularly for safety critical situations. They can, however, provide excellent prototype designs. Testing for final product durability but not for development then becomes an important aspect of fatigue design. The total cost of design, testing, and manufacturing must be balanced against the cost (in money, goodwill, reputation, or even lives) of fatigue failures.

2.8 DOS AND DON'TS IN DESIGN

1. Do recognize that fatigue design is an iterative process involving synthesis, analysis, and testing.
2. Do recognize that different fatigue design criteria and different analytical fatigue life models exist and that no one criterion or analytical model is best for all situations.

3. Don't forget that damage-tolerant design may be necessary due to the existence or development and growth of cracks in safety critical structures.

4. Don't consider computational/analytical fatigue life predictions/estimations as the end of the fatigue design process.

5. Do emphasize digital prototyping and rely on fatigue testing primarily for product durability determination rather than product development.

6. Do place more emphasis on bringing the test track or proving ground into the laboratory, but keep in mind that the more closely testing simulates the real in-service conditions, the greater the confidence in the results.

7. Don't neglect the advantages and limitations of accelerated fatigue testing.

8. Do pursue inspection of in-service components, structures, and vehicles and continue to monitor customer usage.

9. Don't neglect the importance of environmental conditions in both analytical and testing aspects of fatigue design.

REFERENCES

1. *Annual Book of ASTM Standards, Nondestructive Testing,* Vol. 03.03, ASTM, West Conshohocken, PA, 2000.

2. *ASM Handbook, Nondestructive Testing and Quality Control,* Vol. 17, ASM International, Materials Park, OH, 1989.

3. *Nondestructive Testing Handbook,* 2nd ed., Vol. 1-10, ASNT, Columbus, OH, 1982–1993.

4. D. E. Bray and R. K. Stanley, *Nondestructive Evaluation: A Tool in Design, Manufacturing, and Service,* rev. ed., CRC Press, Boca Raton, FL, 1997.

5. D. E. Bray and D. McBride, eds., *Nondestructive Testing Techniques,* Wiley Interscience, New York, 1992.

6. H. Berger and L. Mordfin, eds., *Nondestructive Testing Standards–Present and Future,* ASTM STP 1151, ASTM, West Conshohocken, PA, 1992.

PROBLEMS

1. What safety critical parts on your automobile are (*a*) fail-safe and (*b*) safe-life? How could the critical safe-life parts be made fail-safe? Is this needed?

2. Why is damage-tolerant design used less in the automotive industry than in the aerospace industry?

3. What fatigue design considerations must be kept in mind when converting (*a*) a regular commercial jet aircraft to the "stretch" version and (*b*) a regular automobile to a "stretch" limousine?

4. What types of loading modes (e.g., axial, torsion, bending, combined torsion/bending, pressure) exist for the following components:

   (*a*) Hip replacement prosthesis
   (*b*) Jet engine turbine blade
   (*c*) A rear leg of a chair you frequently use
   (*d*) Motorcycle front axle
   (*e*) Alaska pipeline

5. Sketch a reasonable load spectrum for the components of Problems 4*a* through 4*e*. How would you determine the actual service load spectrum for each component?

6. For the components of Problems 4*a* through 4*e*, how would you integrate analysis and testing for each component? What testing would you recommend?

7. For the components of Problems 4*a* through 4*e*, what design criteria from Section 2.2 would be best suited for each component and why?

8. For the following four components, what fatigue life model (i.e., S–N, ε–N, da/dN–ΔK, or the two-stage method) would you recommend for (*a*) an automobile axle without stress concentrations, (*b*) a gear subjected to periodic cyclic overloads, (*c*) a plate component with an edge crack, and (*d*) a riveted plate such as an airplane wing? Explain why you chose a particular fatigue life model for each of the four cases.

9. Write a one- to two-page review paper on one of the different NDI techniques used to measure crack sizes or to determine if cracks exist.

CHAPTER 3

MACRO/MICRO ASPECTS OF FATIGUE OF METALS

By common usage, the word *fatigue* refers to the behavior of materials under the action of repeated stresses or strains, as distinguished from their behavior under monotonic or static stresses or strains. The definition of fatigue, as currently stated by ASTM, follows [1]:

> The process of *progressive localized permanent* structural change occurring in a material subjected to conditions that produce *fluctuating* stresses and strains at some point or points and that may culminate in *cracks* or complete *fracture* after a sufficient number of fluctuations.

Six key words have been italicized in the definition for emphasis. The word *progressive* implies that the fatigue process occurs over a period of time or usage. A fatigue failure is often very sudden, with no obvious warning; however, the mechanisms involved may have been operating since the beginning of the component's or structure's usage. The word *localized* implies that the fatigue process operates at local areas rather than throughout the entire component or structure. These local areas can have high stresses and strains due to external load transfer, abrupt changes in geometry, temperature differentials, residual stresses, and material imperfections. The engineer must be very concerned with these local areas. The word *permanent* implies that once there is a structural change due to fatigue, the process is irreversible. The word *fluctuating* implies that the process of fatigue involves stresses and strains that are cyclic in nature and requires more than just a sustained load. However, the magnitude and amplitude of the fluctuating stresses and strains must exceed certain material limits for the fatigue process to become critical. The word *cracks* is often the one most misunderstood and misused on the subject of fatigue. It somehow seems repulsive or promotes an "I don't believe it" attitude that too many engineers and managers have concerning the role of a crack or cracks in the fatigue process. The ultimate cause of all fatigue failures is that a crack has grown to a point at which the remaining material can no longer tolerate the stresses or strains, and sudden fracture occurs. The crack had to grow to this size because of the cyclic loading. In fact, as indicated in Section 2.2.4, because of costly fatigue failures in the aerospace fields, current aircraft design criteria for certain safety critical parts require the assumption that small crack-like discontinuities exist prior to initial use of the aircraft [2]. The fatigue life of these safety critical parts is thus based solely on crack growth. The word *fracture* indicates that the last stage of the fatigue process is separation of a component or structure into two or more parts.

3.1 FATIGUE FRACTURE SURFACES AND MACROSCOPIC FEATURES

Before looking at the microscopic aspects of the fatigue process, we will examine a few representative macroscopic fatigue fracture surfaces. Many of these fracture surfaces have common characteristics, and the words "typical fatigue failure" are often found in the literature and in practice. However, there are also many "atypical" fatigue failures. Both are discussed in this section.

Figure 3.1 shows two typical fatigue fracture surfaces of threaded members. Figure 3.1*a* shows a typical fatigue fracture surface of a 97.5 mm (3.84 in.) square thread column from a friction screw press [3]. The thread was not rounded off in the roots, and the flanks exhibited numerous chatter marks, particularly at the lower region A. The poor machining increased the stress concentration on the thread region, which contributed to the fatigue failure. The fracture surface appears to have two distinct regions. The smaller, light, somewhat coarse area at the top of the fracture surface is the remaining cross section that existed at the time of fracture. The other cross-sectional area consists of the fatigue crack (or cracks) region. Many initial cracks can be seen near the lower left outer perimeter; these are shown by the somewhat radial lines, often called "ratchet marks," extending around the lower left perimeter. It is this region where the initial cracking process began. At first, the numerous small fatigue cracks (microcracks) propagated at an angle of about 45° for a few millimeters before turning at right angles to the column longitudinal axis, which is the plane of maximum tensile stress. Growth of the small cracks along the 45° angles cannot be identified in this photograph, as a higher magnification inspection is needed to make this observation. As the component was subjected to additional cyclic stresses, these small initial cracks grew and joined together such that primarily one fatigue crack grew across about 80 percent of the surface. Somewhat wavy darker and lighter bands are

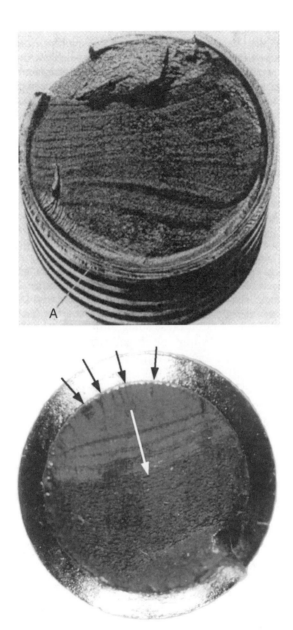

Figure 3.1 Typical fatigue failures of threaded members. (*a*) Square thread column [3] (reprinted with permission of *Der Maschinenschaden*). (*b*) Mountain bike seat post bolt (courtesy of R. R. Stephens).

evident in the main fatigue crack region. These markings are often called "beach marks," "clam shell markings," "arrest lines," or "conchoidal marks." "Beach marks" is the most widely used term and has arisen because of the similarity of the fracture pattern to sand markings left after a wave of water leaves a sandy beach. These markings are due to the two adjacent crack surfaces that open, close, and rub together during cyclic loading, and to the crack's starting and stopping and growing at different rates during the variable (or random) loading spectrum. The interaction of a crack and a corrosive environment under variable loading will also tend to produce distinct beach marks.

Figure 3.1*b* shows a fracture surface of an 8 mm (0.3 in.) diameter threaded bolt. This bolt was used to fasten a seat to a mountain bike seat post. You can imagine the surprise of the bike rider when the bolt fractured and the seat fell off, leaving very little to sit on but the seat post itself! The loading condition imposed on the threaded bolt was predominantly axial but also included a small bending component. Similar to Fig. 3.1*a,* many initial cracks can be seen near the upper outer perimeter, as indicated by the black arrows. However, the number of these radial lines is not as great as in Fig. 3.1*a.* The dark radial lines emphasize that these small initial cracks nucleated and grew at slightly different heights within the thread root. Again, these small cracks grew and eventually joined together to form one large crack, which then grew across approximately 40 percent of the surface in the direction indicated by the white arrow. The point of the white arrow identifies the end of fatigue crack growth and the beginning of final fracture. Beachmarks are again evident, and upon close inspection of the fracture surface, the dark beachmarks showed discoloration and corrosion debris indicating an environmental contribution to crack growth. The final fracture region is shown in the lower portion of the figure. The lowest right region of the final fracture is where the pieces of the bolt rubbed or slid against each other when the bolt fractured. This behavior caused the somewhat flat, featureless appearance.

In comparing Figs. 3.1*a* and 3.1*b,* typical fatigue features show different characteristics. For example, the number of beachmarks observed on each fracture surface is significantly different. During the life of the two threaded members, conditions that tend to produce beachmarks occurred more frequently for the square thread column than for the seat post bolt. The seat post bolt was in service for less than 2 years and experienced several distinct seasonal (environmental) changes. This contributed to the number of beachmarks observed. The number of crack nucleation sites, identified by the radial lines along the outer perimeter of each threaded member, reveals something about the manufacturing process. The seat post bolt exhibited a relatively smooth surface at the thread root, while the square thread column had a poorly machined surface. The poorly machined surface included more surface discontinuities, which led to a greater number of crack nucleation sites along the root of the thread. Therefore, surface conditions are an important aspect of fatigue and will be discussed in greater detail in later chapters. Perhaps the most obvious difference between the two fracture surfaces is the percentage of the cross section where the fatigue crack grew prior to final fracture. The difference can be attributed to either the cyclic load levels imposed on each member or the fracture toughness of the material used for each component. These concepts will be addressed in Chapter 6.

If the adjacent fractured parts were placed together (something that should not be done to actual failures, since this can obscure metallurgical markings and thus hamper proper fractographic analysis), they would fit very neatly together and indicate very little gross permanent deformation. Because of this small gross permanent visual deformation, fatigue failures are often called "brittle failures." However, this term should be modified since substantial plastic deformation occurs in small local regions near the fatigue crack tip and at the crack nucleation sites. Many fatigue failures do not have appreciable visual permanent deformation, but the mechanisms of crack nucleation and growth involve small local regions of plastic deformation. Failures similar to those in Fig. 3.1 are often called "typical" fatigue failures because they exhibit the following common features:

1. Distinct crack nucleation site or sites.
2. Beach marks indicative of crack growth.
3. Distinct final fracture region.

Figure 3.2 shows a fatigue failure of an end bearing from an air compressor in a high-capacity locomotive that originated at a metallurgical defect containing slag inclusions and an oxide film near the lower edge [3]. Under the influence of this defect, the high cyclic service stresses led to a fatigue failure. Metallurgical discontinuities, such as inclusions and voids, are inherent in metals and often act as sites for cracks to nucleate. The top final fracture

Figure 3.2 Fatigue failure of an end bearing due to slag inclusions [3] (reprinted with permission of *Der Maschinenschaden*).

surface of Fig. 3.2 shows a very thin lip around the perimeter. This lip is often called a "shear lip" because appreciable permanent deformation can exist here. This shear lip can be quite extensive, very small, or essentially nonexistent. Its occurrence depends on the type of loading, on the environment, and on the ductility of the material. Shear lips are more prominent in materials that behave in a ductile manner. The total fracture surface shows the crack growth region indicated by the elliptically shaped beach marks, along with the coarse final fracture and shear lip regions.

Figure 3.3a shows a torsion fatigue failure of a 25 mm (1 in.) circular shaft [3]. The fatigue crack nucleated at the top surface and propagated along a 45° helix. The crack path is shown schematically in Fig. 3.3c on a 45° plane. This is the plane of maximum tensile stress for torsional loading, which indicates that the fatigue crack grew primarily in the plane of maximum tensile stress. However, many materials, when loaded in torsion or in multiaxial directions, favor crack nucleation and/or crack growth along the maximum shear planes. This behavior is addressed in Chapter 10 on multiaxial fatigue. The close-up in Fig. 3.3b shows that the shape of the smooth fatigue crack was semielliptical, a very common surface fatigue crack shape. Note that beach marks are not evident in this smooth, semielliptical fatigue crack surface. The final fracture region has a fibrous appearance, with radial lines essentially perpendicular to the perimeter of the elliptical fatigue crack. These radial or "river" patterns are often seen on the final fracture surfaces and point to the origin of the crack.

A comparison of fatigue fracture surfaces of cast SAE 0030 steel and hot-rolled SAE 1020 steel, resulting from the same programmed variable amplitude load spectrum using 8.2 mm (0.325 in.) thick keyhole compact-type specimens, is shown in Fig. 3.4 [4]. The direction of crack growth is from the lower edge to the upper edge. Again, several fatigue cracks have nucleated at the keyhole edge, and upon subsequent cycling, these initial cracks coalesced into one major crack that eventually extended across the specimen thickness. This major crack then grew along the length of the specimen and occupied most of the crack growth region. Beach marks are not evident on the much coarser cast steel fatigue fracture surface; a few can be seen at longer crack lengths in the hot-rolled steel. Thus, fatigue fracture surfaces can be void of beach marks in both cast and wrought materials.

The major crack growth path for both steels in Fig. 3.4 was flat and perpendicular to the maximum tensile stress. Fig. 3.5 shows a schematic drawing of fatigue crack growth paths often found in sheet or plate specimens and components [5]. At short crack lengths where crack tip plastic zones are small, the crack path is usually flat. As the crack grows, the crack tip plastic zone increases in size and the crack plane can turn to a 45° shear or slant mode. This can be either single shear, as shown in Fig. 3.5a, or double shear, as shown in Fig. 3.5b. The mode of fatigue crack growth in sheets or plates is governed primarily by the size of the crack tip plastic zone relative to the

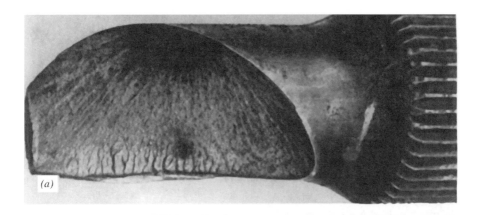

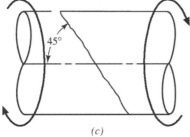

Figure 3.3 Fatigue failure of a torsion shaft [3] (reprinted with permission of *Der Maschinenschaden*). (*a*) Total fracture surface. (*b*) Close-up. (*c*) Idealized form.

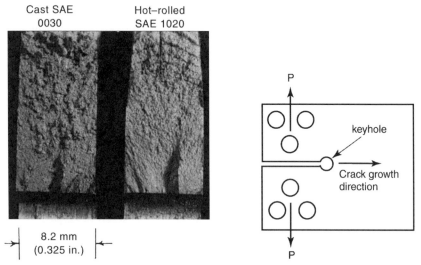

Figure 3.4 Fatigue failure of cast and hot-rolled low carbon steel keyhole specimens subjected to programmed variable amplitude load spectrum [4] (reprinted by permission of the American Society for Testing and Materials).

thickness of the sheet. The yield strength of the material, the magnitude of the applied loads, and the length of the crack principally determine this behavior.

A general schematic summary of the many types of macroscopic fatigue fracture surfaces as a function of load magnitude and geometry is shown in Figs. 3.6 [6] and 3.7 for axial and bending loads. The hashed regions in Fig. 3.6 identify final fracture regions, while lighter regions with lines and arrows

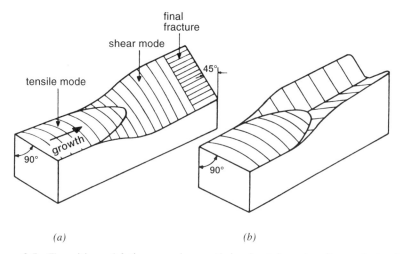

Figure 3.5 Transition of fatigue crack growth in sheet from tensile mode to shear mode [5] (reprinted with permission of the Noordhoff International Publishing Co.). (*a*) Single shear. (*b*) Double shear.

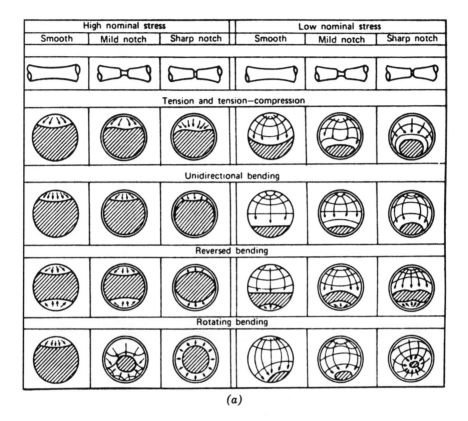

(a)

(b)

Figure 3.6 Schematic fatigue fractures for axial and bending loads [6] (reprinted with permission of the Society for Experimental Stress Analysis). (a) Round specimens. (b) Sheet or plate specimens.

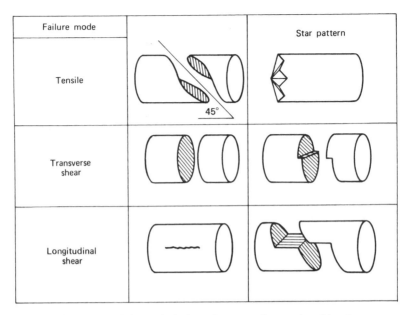

Figure 3.7 Schematic fatigue fractures for torsional loads.

identify fatigue crack regions. The beginnings of the arrows typically show where the crack or cracks nucleated. Figure 3.6 shows fracture surfaces due to axial loadings and three types of bending; unidirectional, reversed, and rotating. In each case, the fatigue cracks nucleate at the surface or corner and eventually grow in the plane of maximum tensile stress. For reversed bending, cracks usually nucleate at opposite sides, since both sides experience repeated tensile stresses. Figure 3.7 shows the several crack patterns commonly found in torsion fatigue. You should be able to identify the distinct similarity between some of the schematic drawings in Figs. 3.6 and 3.7 and the fracture surfaces shown in Figs. 3.1–3.4. For example, Fig. 3.1a shows a mild threaded notch with a large percentage of the surface showing crack growth. Based on the percentage of the cross section showing fatigue crack growth, it is likely that a low nominal stress was applied to this member. However, for the seat post bolt shown in Fig. 3.1b, the threaded notch appears much deeper, and a smaller percentage of the cross section shows crack growth. Thus, it is likely that a high nominal stress was applied to this bolt. However, the material fracture toughness can also play a significant role in final crack size. Similar analyses can be performed for Figs. 3.2–3.4.

We can summarize the above aspects of fatigue fracture surfaces as follows:

1. The entire fatigue process involves the nucleation and growth of a crack or cracks to final fracture.

2. The fatigue crack at fracture can be very small or very large, occupying less than 1 percent of the fracture surface up to almost 100 percent, depending on the magnitude of the applied stresses and the fracture toughness of the material.

3. Often the fatigue crack region can be distinguished from the final fracture region by beach marks, smoothness, and corrosion. However, there are many exceptions.

4. Fatigue cracks usually nucleate at the surface where stresses are highest and where a corrosive environment and changes in geometry exist.

5. Microscopic fatigue cracks usually nucleate and grow on planes of maximum shear.

6. Macroscopic fatigue cracks often grow in the plane of maximum tensile stress. However, for torsional and multiaxial loading, macroscopic fatigue cracks have also been observed to grow on planes of maximum shear.

3.2 FATIGUE MECHANISMS AND MICROSCOPIC FEATURES

During the late nineteenth century, Bauschinger [7] first observed that the stress-strain behavior in a monotonic tension or compression test could be quite different from the stress-strain behavior obtained after cyclic loads were applied. He showed that by applying a load in one direction that caused inelastic (plastic or irreversible) deformation, the magnitude of the yield strength in the opposite direction could be changed. A detailed discussion of the material response under cyclic stress and strain is presented in Section 5.3. However, the basic micro mechanisms of cyclic deformation are introduced here so that the reader will gain a better understanding of the onset of fatigue damage.

Cyclic applications of inelastic strain to a metal can cause continuous changes until cyclic stability is reached. This means that the metal becomes either more or less resistant to the applied strain, i.e., the material either cyclic hardens or cyclic softens. Some materials may never stabilize under cyclic applications of inelastic strain, while others are cyclically stable from the onset. The latter condition implies there is no softening or hardening. Based on the monotonic stress-strain curve, we are sometimes able to determine which alloy will harden or soften during cyclic loading [8]. Perhaps of greatest importance here is why metals harden or soften during cyclic deformation. The answer is related to the density and arrangement of the dislocation structure and substructure of the metal. While the concept of dislocations is often viewed as unimportant to most practicing design engineers, it is probably the most important conceptual development in metallurgy and material science, as dislocations are basic to the understanding of the mechanical response of metals. Therefore, it is desirable to be familiar with the fundamentals of dislocation movement, as this will help explain the mechanisms involved in cyclic deformation and the development of fatigue damage. The relationship between dislocations and inelastic deformation is typically defined by "slip," which is a shear deformation of the material. Slip occurs in metals within individual grains by dislocations moving along crystallographic planes. For materials that are initially soft, the dislocation density is low. As inelastic deformation occurs as a result of stress or strain cycling, the density of dislocations increases rapidly. This increase in dislocation density leads to a decrease in dislocation mobility. Therefore, if dislocations are constrained to move or glide and slip is minimized, the material is said to cyclic harden and the cyclic yield strength becomes greater than the monotonic yield strength. For materials that are initially hard or have been hardened, inelastic strain cycling causes the existing dislocation structure to rearrange into a configuration such that there is less resistance to deformation. Reconfiguration of the dislocation structure tends to promote greater dislocation mobility. Therefore, dislocations are able to circumnavigate around microstructural barriers that generally tend to restrict deformation, such as precipitates and grain boundaries. Thus, the material cyclic softens, and the cyclic yield strength is less than the monotonic yield strength. These two cyclic material characteristics, hardening and softening, can occur either at a global scale, such as a net section area, or at a very localized region, say in the vicinity of metallurgical discontinuities, notches, or cracks. Notches, metallurgical discontinuities, and cracks can increase local stress or strain and cause inelastic deformation. The scale of inelastic deformation is very localized, with the bulk of the material experiencing elastic or reversible deformation. Additional readings on the mechanisms associated with cyclic hardening and softening can be found elsewhere [9–11].

Metals are crystalline in nature, which means that atoms are arranged in an ordered manner. Most structural metals are polycrystalline and thus consist of a large number of individual ordered crystals or grains. Each grain has its own particular mechanical properties, ordering direction, and directional properties. Some grains are oriented such that planes of easy slip or glide (dislocation movement) are in the direction of the maximum applied shear stress. The onset of slip creates an appearance of one or more planes within a grain sliding relative to each other. Slip occurs under both monotonic and cyclic loading and is the localization of plastic strain.

The degree of slip, or cyclic deformation, is primarily related to the crystallographic structure of the metal, i.e., the ductility of the metal. In metals that show brittle behavior, dislocations are practically immobile and the extent of slip is very limited. In metals that behave in a ductile fashion, there is little restriction on dislocation movement and slip is abundant. For metals that show intermediate ductile behavior, dislocations are mobile but are restricted to a limited number of slip planes. Thus, crack nucleation mechanisms vary, depending on the type of metal under consideration. Words such as crack "initiation," "formation," and "nucleation" have been used interchangeably throughout the engineering community to describe the early stages of the fatigue damage process. ASTM Committee E-08 recently adopted the use of either "nucleation" or "formation" to describe the early stages of the fatigue process, and dropped the use of the word "initiation." The purpose of this change was to try to minimize some of the misunderstanding associated with these words when describing the early stages of the fatigue damage process. In this book, the words "nucleate" or "nucleation" will primarily be used.

Let us look at the general situation in which crack nucleation occurs due to slip under fatigue loading conditions. Figure 3.8a shows a schematic of an edge view of coarse slip normally associated with monotonic loading. Under cyclic loading fine slip occurs, as shown in Fig. 3.8b. Coarse slip can be considered an avalanche of fine movements in which only a few adjacent parallel slip bands move relative to each other. The slip lines shown in Fig. 3.8 appear as parallel lines or bands within a grain when viewed perpendicular to the free surface. Both fine and coarse slip are studied with prepolished, etched specimens using optical and electron microscopy techniques. Figure 3.8c shows the progressive development of an extrusion/intrusion pair under cyclic loading. The vertical arrows shown in Fig. 3.8c indicate the loading direction (tension or compression), and the horizontal arrows identify the progression of slip deformation. Forsyth [12] showed that both slip band intrusions and extrusions (the valleys and peaks, respectively, shown in Figs. 3.8b and 3.8c) occurred on the surface of metals when they were subjected to cyclic loading. Slip band intrusions form stress concentrations, which can be the location for cracks to develop.

The introduction of the scanning electron microscope in the 1950s allowed significant advances to be made in understanding fatigue mechanisms. Optical microscopes have magnification limitations of approximately 1000× and a very shallow depth of field. Therefore, using optical microscopes at high magnification with surfaces that are not extremely flat usually results in poor-quality micrographs. Scanning electron and transmission electron microscopes provide excellent depth of field at magnifications in excess of 10 000×, thus providing excellent images of fatigue artifacts. These are the primary fractographic tools of choice for most metallurgists, material scientists, and engineers studying the micro and macro mechanisms of fatigue.

A single slip band crack on the surface of a nickel-base superalloy specimen is shown in Fig. 3.9. The slip band crack terminated at the grain boundary within which it grew, as indicated by the two vertical white arrows. This was because of the unique crystalline orientation of that specific grain with respect to adjacent grains. The continuous light grain boundary precipitates identify grain boundaries. The white material located along the slip band crack has been extruded from the slip band microcrack during the fatigue process. Shear stresses or shear strains rather than normal stresses or normal strains, as indicated by the schematic shown in the micrograph, primarily control slip. The higher the shear stress or shear strain amplitude or the larger the number of repetitions, the greater the slip.

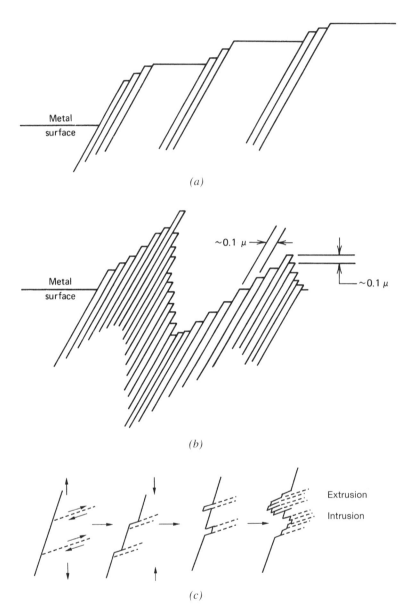

Figure 3.8 Schematic of slip due to external loads. (*a*) Static (steady) stress. (*b*) Cyclic stress. (*c*) Fatigue progression in the formation of an extrusion/intrusion pair.

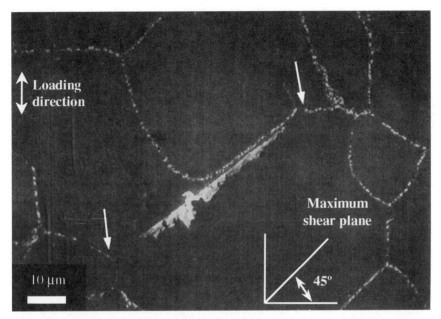

Figure 3.9 Surface fatigue microcrack along a slip band in nickel-base superalloy Waspaloy (courtesy of R. R. Stephens).

The progressive nature of slip line intensification as a function of applied cycles for pure polycrystalline nickel is shown in the three-stage microphotograph in Fig. 3.10 [13]. Each photograph is from the same microscopic region, but they were taken at 10^4, 5×10^4, and 27×10^4 cycles, respectively. Grain boundaries are identified by the closed contours. In Fig. 3.10a, after 10^4 cycles, very little evidence of slip is seen. However, it has occurred in some grains. Figure 3.10b shows a large increase in the number of slip lines after 5×10^4 cycles and also an increased thickness of some slip lines. As cycling continues, more slip lines occur and continue to thicken, as shown in Fig. 3.10c. These intensified slip lines or bands are the sources for actual fatigue cracks. Note that the slip lines are primarily contained within each grain. A slightly different orientation exists in adjacent grains because of the different directional ordering (crystallographic orientation) of each grain. This slip occurs at local regions of high stress and strain, while much of the component or structure may contain very little slip, similar to that in Fig. 3.10a, even at fracture.

The two-dimensional slip lines in Fig. 3.10 are really three-dimensional, since they go into the surface to varying depths, as shown in the schematic of Fig. 3.8b. Most of the slip bands can be eliminated by removing several microns (>0.003 mm) from the surface by electropolishing. However, a few slip bands may become more distinct and have therefore been labeled "persistent slip bands." It has been found that fatigue cracks grow from these persistent slip bands [12]. Fatigue life has been substantially improved by removing

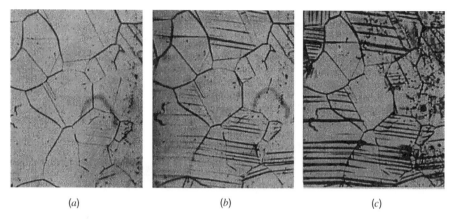

(a)   (b)   (c)

Figure 3.10   The progressive nature of slip in nickel subjected to cyclic loading [13] (reprinted with permission of John Wiley and Sons, Inc.). (a) After 10^4 cycles, general fine slip has developed. (b) After 5×10^4 cycles, some of the original slip lines have intensified and broadened. (c) After failure at 27×10^4 cycles, continued development of the intense slip lines has occurred, with some grains and some regions still showing very little evidence of slip.

the persistent slip bands. In fact, intermittent cycling and electropolishing have been employed to cause indefinite life extension such that fatigue failures do not occur [13]. This establishes the fact that early stages of fatigue are primarily a surface phenomenon; hence, the surface and the environment play important roles in fatigue life. The preferential nucleation of fatigue cracks at or near the surface is explained by the fact that inelastic deformation is easier at the surface and that slip steps (intrusions and extrusions) are able to develop on the surface. Also, in most loading situations, the stresses or strains developed in a component or structure due to external loads are greatest on the surface. Surface effects are emphasized in later chapters.

Fatigue cracks that nucleate in local slip bands initially tend to grow in a plane of maximum shear stress range. This growth is quite small, usually on the order of several grains. Growth of microcracks is strongly influenced by the slip characteristics of the material, the material grain size, and the extent of plasticity near the crack tip. When the size of the crack or the plastic deformation near the crack tip is confined within a few (typically within 10) grain diameters, fatigue crack growth occurs predominantly by a shear process. This cracking behavior is usually referred to as "microcrack growth." Therefore, the physical length that a crack grows due to shear may vary from one material to the next, depending on the grain size of the material. Figure 3.11a shows a microcrack that formed along a slip plane and grew into a few surrounding grains. Notice that the microcrack did not grow on one specific plane but changed direction as the crack grew into adjacent grains. However, microcrack growth still occurred on planes oriented near the maximum shear

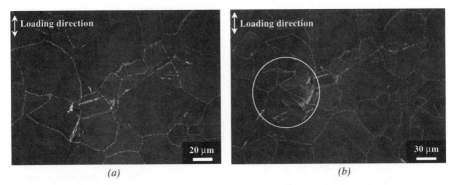

Figure 3.11 Surface fatigue microcracks. (*a*) Slip band microcracks linked up between adjacent grains. (*b*) After further cycling, continued development and additional formation of slip lines as the microcrack extended into adjacent grains (courtesy of R. R. Stephens).

plane. Figure 3.11*b* shows the same crack after additional fatigue cycling. The development of additional slip lines is seen in the vicinity of the tip of the primary microcrack and is identified within the circled region. Notice that the slip lines do not exist in Fig. 3.11*a*.

Not all fatigue cracks nucleate along slip bands, although in many cases slip bands are at least indirectly responsible for microcracks forming in metals. Under fatigue loading conditions, fatigue cracks may nucleate at or near material discontinuities on or sometimes just below the metal surface. Discontinuities include inclusions, second-phase particles, corrosion pits, grain boundaries, twin boundaries, pores, voids, and slip bands. Although cracks may start at a discontinuity other than a slip band, slip bands are often impinged at the boundary of the discontinuity, leading to higher localized inelastic strain that results in microcracking. Microcracks in high-strength or brittle-behaving metals are often formed directly at inclusions or voids and then grow along planes of maximum tensile stress. A number of crack nucleation mechanisms are shown in Fig. 3.12. Cracks often nucleate at grain boundaries at temperatures exceeding approximately 50 percent of the absolute melting point or also at large strain amplitudes. Figure 3.12*a* shows a crack that nucleated along a grain boundary, shown between the two vertical white arrows, of the nickel-base superalloy Waspaloy at 700°C. This crack nucleated in a specimen that was subjected to a dwell (hold) period at the maximum load of the loading cycle. At moderate temperatures the grain boundary precipitates strengthen the grain boundaries. However, at higher temperatures, these precipitates can reduce the grain boundary strength, and with the introduction of a dwell period, when creep-like mechanisms can dominate, grain boundary cracking (cavitation and sliding) and intergranular crack growth can occur. The fatigue behavior associated with elevated temperatures is discussed in detail in Section 11.4. Impurities that segregate to grain boundaries can also cause embrittlement in many metals, leading to crack

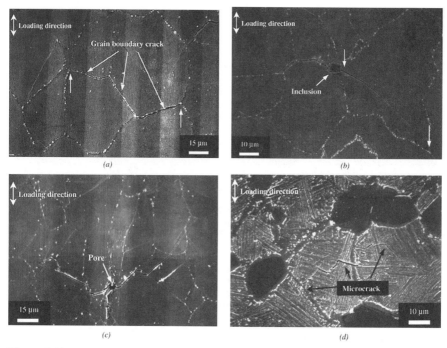

Figure 3.12   Fatigue microcracks that nucleated due to various mechanisms. (*a*) Elevated temperature (700°C) grain boundary crack in Waspaloy. (*b*) Surface inclusion/slip band crack in Waspaloy. (*c*) Elevated temperature (500°C) surface pore/slip band in Waspaloy. (*d*) Cracking within α and β phases in lamellar structure in titanium alloy IMI834 (courtesy of R. R. Stephens).

nucleation and growth at grain boundaries. Cracks can also nucleate at inclusions, pores, and other microstructural features (Fig. 3.12*b–d*). Figure 3.12*b* shows a crack that also nucleated in Waspaloy. However, this crack nucleated at an inclusion located on the surface. The slip band crack to the right of the inclusion, indicated between the two vertical white arrows, was impinged at the inclusion, which increased the localized plastic strain within the slip line. This led to cracking of both the slip band and the inclusion. The inclusion cracked predominantly along the maximum tensile stress plane, while the rest of the microcrack grew predominantly along the maximum shear planes. Cracking of the inclusion was probably a low-energy process, which will be addressed later in this section. Figure 3.12*c* shows a crack that nucleated at/near a pore on the surface of a Waspaloy specimen tested at 500°C. The symmetry of the microcrack with respect to the pore indicates that the crack probably started at the pore and then extended to the left and the right. Again, notice that on both sides of the pore the microcrack grew along planes near the maximum shear planes. Figure 3.12*d* shows a microcrack that nucleated in a two-phase (α + β) titanium alloy IMI834. The lamellar phase (lighter region) consists of "needles" of α within a β matrix. Delamination, or interfacial cracking between the α and β phases, occurred. The microcrack meandered through the $\alpha + \beta$ phase, often growing along the needles (between the two phases) and also across the α needles. Microcracks also can be present in metals prior to any cyclic load applications due to material processes and treatments. Thus, the nucleation phase of fatigue can be nonexistent in some cases, as a crack may exist prior to the first loading cycle.

Once a microcrack or microcracks are present and cycling continues, fatigue cracks tend to coalesce and grow along the plane of maximum tensile stress range, as described in Section 3.1. The two stages of fatigue crack growth are called "stage I" (shear mode) and "stage II" (tensile mode) [12]. Fatigue crack growth is shown schematically as a microscopic edge view in Fig. 3.13. A fatigue crack is shown to nucleate at the surface and grow across several grains, controlled primarily by shear stresses and shear strains, and then to grow in a zigzag manner essentially perpendicular to, and controlled primarily by, the maximum tensile stress range. Slip line progression precedes the fatigue crack tip vicinity as the crack grows across the material. Most fatigue cracks grow across grain boundaries (transcrystalline), as shown in Figs. 3.13 and 3.14. However, they may also grow along grain boundaries (intercrystalline), depending on the material, load, and environmental conditions (Fig. 3.12a).

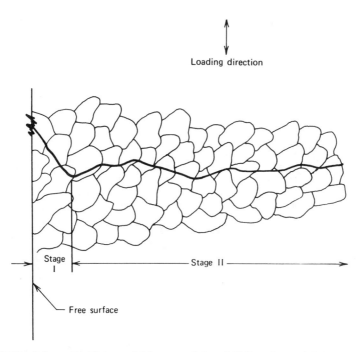

Figure 3.13 Schematic of stages I (shear mode) and II (tensile mode) transcrystalline microscopic fatigue crack growth.

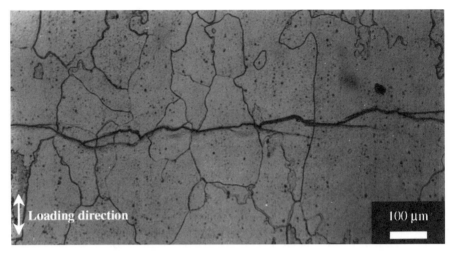

Figure 3.14 Surface crack profile of transcrystalline fatigue crack growth in aluminum–lithium alloy ML377. The direction of crack growth is from left to right (courtesy of R. R. Stephens).

An electron microscopic analysis of fracture surfaces, such as those in Figs. 3.1 to 3.4, can reveal a wide range of fatigue crack growth mechanisms. Figure 3.15, obtained by Crooker et al. [14], shows three of the more common modes: (a) striation formation, (b) microvoid coalescence, and (c) microcleavage. Materials that exhibit ductile behavior often display appreciable striations and microvoid coalescence (MVC). The ripples in Fig. 3.15a are called "fatigue striations." These striations are not the beach marks described in Section 3.1. Actually, one beach mark can contain thousands of striations [15]. Electron

Figure 3.15 Examples of microscopic fatigue crack growth in 17-4 PH stainless steel [14] (reprinted with permission of the American Society for Testing and Materials). (a) Striation formation. (b) Microvoid coalescence. (c) Microcleavage.

microscopic magnification between 1000× and 50 000× must be used to view striations. Frequently, they are rather dispersed throughout the fracture surface and may not be seen clearly because of substantial surface rubbing and pounding during repeated loading. They are also difficult to find in high-strength materials. In many studies, each striation has been shown to represent one load cycle, the striation forming by a plastic crack tip blunting mechanism during the loading and unloading portion of the fatigue cycle. However, several other studies have shown that there may not be a one-to-one correspondence between a single striation and each cycle. Thus, a combination of other fracture mechanisms along with striation formation may be responsible for advancing the crack front. MVC takes place by the nucleation, growth, and coalescence of microvoids during plastic deformation. MVC is shown in Fig. 3.15*b*. The formation of these voids, which evolve into "dimples," has been attributed largely to the interfacial cracking between precipitate particles or inclusions and the surrounding material matrix. Thus, the size and population density of dimples are generally related to the size and distribution spacing of the inclusions or precipitates inherent in the metal. The process of MVC is generally considered a high-energy process, and in fatigue it usually occurs at high crack growth rates. The fracture surface due to MVC usually has a dull and rough or fibrous appearance. Microcleavage crack growth is considered a lower-energy process and therefore an undesirable fatigue crack growth mechanism. It is likely that this was the mechanism of cracking for the "brittle" inclusion shown in Fig. 3.12*b*. Cleavage or microcleavage involves fracture along specific crystallographic planes and is transcrystalline in nature. Cleavage facets are usually flat and often contain several parallel ridges or cleavage planes, as shown in Fig. 3.15*c*. Cleavage fracture is more susceptible in metals having a body-centered cubic (BCC) or hexagonal close-packed (HCP) crystalline structure as opposed to a face-centered cubic (FCC) crystalline structure. The appearance of the fracture surface from cleavage is usually bright and appears shiny due to the high reflectivity of the flat cleavage facets. Cleavage is regarded as the most brittle form of fracture in crystalline materials, and the likelihood of cleavage is increased whenever plastic flow is restricted, such as at low temperatures, at high strain rates, or in notched components. MVC and cleavage are fracture mechanisms that can occur under either monotonic or cyclic loading conditions. Striations, however, do not occur under monotonic loading conditions, as their formation relies on the cyclic nature of fatigue. A comprehensive and extensive review of fractography and fractographic analysis of fatigue failures can be found in the ASM Handbook on fractography [16].

The typical stages of the fatigue damage process described in this section are summarized schematically in Fig. 3.16 [17]. In general, slip occurs first, followed by fine cracks that can be seen only at high magnification. These cracks continue to grow under cyclic loading and eventually become visible to the unaided eye. The cracks tend to combine such that just a few major cracks grow. These cracks (or crack) reach a critical size, and sudden fracture

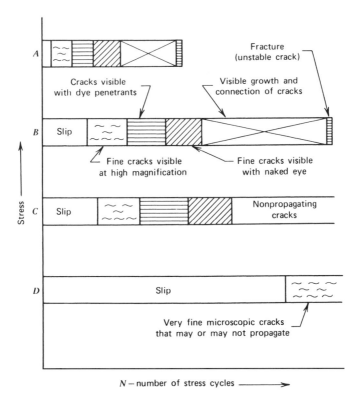

Figure 3.16 Schematic representation of the fatigue process [17] (reprinted with permission of McGraw-Hill Book Co.).

occurs. The higher the stress magnitude, the sooner all processes occur. Cracks may also stop without further growth as a result of compressive residual stress fields or as a crack grows out of a high-stress region such as a notch. Thus, fatigue generally consists of crack nucleation, growth, and final fracture. At high stress levels, a large portion of the total fatigue life is associated with microcrack and macrocrack growth. At low stress levels, a great deal of the total fatigue life is spent in the nucleation of the crack and microcrack growth. The crack nucleation stage depends on the level of thinking. A physicist interested in atomic levels could consider all fatigue as crack growth. A metallurgist may think that a crack exists due to the formation of a slip band. A design engineer may think of a crack as existing when it can be seen by the unaided eye. Thus, the termination of the crack nucleation stage is arbitrary, as it is a process that is defined differently by the various sciences. To avoid this difficulty, the engineer might best think in terms of nucleation of microcracks, followed by growth of these microcracks to the macrocrack level, and then macrocrack growth to unstable fracture. Thus a key concept, no matter which definitions are used, is that fatigue substantially involves the nucleation and growth of cracks. Only the nucleation stage can be absent in a fatigue fracture. Thus, we can overcome fatigue failure by preventing the nucleation and growth of cracks.

From the perspective of alloy design, conditions that favor good crack nucleation resistance and microcrack growth under fatigue loading may not favor good macrocrack growth resistance and vice versa. For example, fine grain sizes tend to offer the best resistance to crack nucleation and microcrack growth. Grain boundaries tend to act as crack stoppers or deflectors, thus reducing fatigue crack growth rates. This was shown in Figs. 3.9, 3.11, and 3.12. However, as the crack grows and macrocrack growth occurs, fine grain materials promote a flatter crack path that tends to promote higher crack growth rates. Coarse grain materials tend to promote a rougher crack path. A rougher crack path usually offers greater resistance to macrocrack growth through crack closure and crack tip deflection mechanisms. However, with the many inherent variables encountered in fatigue and the various fatigue life models, there are always exceptions.

3.3 SUMMARY

The aspects and mechanisms of the fatigue process are quite complex and are still only partially understood. In this chapter, we have attempted to describe the mechanisms associated with crack nucleation, microcrack growth, macrocrack growth, and final fracture. In general, the entire fatigue process involves the nucleation and growth of a crack or cracks to final fracture. Cracks tend to nucleate along slip lines oriented in the planes of maximum shear. Cracks can also nucleate at grain boundaries, inclusions, pores, and other microstructural features or discontinuities. Crack growth usually consists of microcrack growth along maximum shear planes followed by macrocrack growth along the maximum tensile stress plane. However, conditions exist in which macrocrack growth may occur on planes of maximum shear. Depending on the material and the stage of the fatigue process, crack growth can proceed by a number of mechanisms such as striation formation, microvoid coalescence, and cleavage. Surface features such as ratchet marks, beach marks, and river patterns help to identify a failure as a fatigue failure.

The characteristics of each stage of the fatigue process are somewhat different and are affected differently by the metal in question. Certain material characteristics may favor good crack nucleation resistance, good microcrack growth resistance, or good macrocrack growth resistance, but not necessarily all three. Thus, the selection of a material for a given application may be dictated by the importance of the various stages of the fatigue process. Thus, in fatigue design, a thorough understanding of the fatigue damage process that includes crack nucleation, microcrack growth, and macrocrack growth is very important.

3.4 DOS AND DON'TS IN DESIGN

1. Do recognize that fatigue is a localized, progressive, and permanent behavior involving the nucleation and growth of cracks to final, usually sudden fracture.

2. Do recognize that fatigue cracks nucleate primarily on planes of maximum shear and usually grow on the plane of maximum tensile stress.

3. Do examine fracture surfaces as part of a postfailure analysis, since substantial information concerning the cause of the fracture can be gained. The examination can involve a small magnifying glass or greater magnification up to that of the electron microscope.

4. Don't put fracture surfaces back together again to see if they fit or allow corrosive environments (including rain and moisture from fingers) to reach the fracture surface. These can obliterate key fractographic details.

5. Do consider that stress–strain behavior at notches or cracks under repeated loading may not be the same as that observed under monotonic tensile or compressive loading.

6. Do take into consideration that your product will very likely contain cracks during its design lifetime.

7. Do recognize that most fatigue cracks nucleate at the surface, and therefore that surface and manufacturing effects are extremely important.

8. Don't assume that a metal that has good resistance to crack nucleation also has good resistance to crack growth and vice versa.

REFERENCES

1. "Standard Terminology Relating to Fatigue and Fracture," Testing ASTM Designation E1823, Vol. 03.01, ASTM, West Conshohocken, PA, 2000, p. 1034.

2. M. D. Coffin and C. F. Tiffany, "New Air Force Requirements for Structural Safety, Durability, and Life Management," *J. Airc.,* Vol. 13, No. 2, 1976, p. 93.

3. E. J. Pohl, ed., *The Face of Metallic Fractures,* Munich Reinsurance Co., Munich, 1964.

4. R. I. Stephens, P. H. Benner, G. Mauritzson, and G. W. Tindall, "Constant and Variable Amplitude Fatigue Behavior of Eight Steels," *J. Test. Eval.,* Vol. 7, No. 2, 1979, p. 68.

5. D. Broek, *Elementary Engineering Fracture Mechanics,* 4th ed., Kluwer Academic Publications, Dordrecht, the Netherlands, 1986.

6. G. Jacoby, "Fractographic Methods in Fatigue Research," *Exp. Mech.,* Vol. 5, 1965, p. 65.

7. J. Bauschinger, "On the Change of the Position of the Elastic Limit of Iron and Steel Under Cyclic Variations of Stress," *Mitt. Mech.-Tech. Lab.,* Munich, Vol. 13, No. 1, 1886.

8. S. S. Manson and M. H. Hirschberg, *Fatigue: An Interdisciplinary Approach,* Syracuse University Press, Syracuse, NY, 1964, p. 133.

9. "ASM Handbook," *Fatigue and Fracture,* Vol. 19, ASM International, Materials Park, OH, 1996.

10. S. Suresh, *Fatigue of Materials,* 2nd ed., Cambridge Solid Science Series, Cambridge University Press, Cambridge, 1998.

11. M. Klensil and P. Lucas, *Fatigue of Metallic Materials,* 2nd rev. ed., Elsevier Science Co., Amsterdam, 1992.

12. P. J. E. Forsyth, *The Physical Basis of Metal Fatigue,* American Elsevier Publishing Co., New York, 1969.

13. A. J. Kennedy, *Processes of Creep and Fatigue in Metals,* John Wiley and Sons, New York, 1963.

14. T. W. Crooker, D. F. Hasson, and G. R. Yoder, "Micromechanistic Interpretation of Cyclic Crack Growth Behavior in 17-4 PH Stainless Steel," *Fractogr.-Microsc. Cracking Proc.,* ASTM STP 600, ASTM, West Conshohocken, PA, 1976, p. 205.

15. R. W. Hertzberg, *Deformation and Fracture Mechanics of Engineering Materials,* 4th ed., John Wiley and Sons, New York, 1996.

16. "ASM Handbook," *Fractography,* Vol. 12, ASM International, Materials Park, OH, 1987.

17. R. H. Christensen, "Fatigue Cracking, Fatigue Damage, and Their Detection," *Metal Fatigue,* G. Sines and J. L. Waisman, eds., McGraw-Hill Book Co., New York, 1959, p. 376.

PROBLEMS

1. What features distinguish fatigue fracture surfaces from monotonic fracture surfaces? When might you expect fatigue fractures to look similar to monotonic fractures with the unaided eye?

2. Two engineers were discussing fatigue failures, and one stated, "Fatigue failures are brittle," while the other stated, "Fatigue failures are ductile." Discuss the pros and cons of the two arguments.

3. If a material cyclic softens, would one expect a crack to nucleate more quickly or more slowly than if it cyclic hardens and why?

4. List three material or microstructural characteristics that favor good fatigue crack nucleation resistance.

5. List three material or microstructural characteristics that favor good fatigue crack growth resistance.

6. Identify a component where the control of fatigue crack nucleation is most important and explain why.

7. Identify a component where the control of fatigue crack growth is most important and explain why.

8. How might you differentiate between cyclic slip and an actual micro fatigue crack?

9. In what materials and conditions might fatigue cracks grow by microcleavage, which is undesirable? How can you avoid this in real structures and materials selection?

10. In line with current product liability litigation, how should we inform our management and legal officers that our products may contain small crack-like discontinuities before the products leave the factory? Also, provide a list of such possible discontinuities and their presumed size.

CHAPTER 4

FATIGUE TESTS AND THE STRESS–LIFE (*S–N*) APPROACH

4.1 FATIGUE LOADING, TEST MACHINES, AND SPECIMENS

4.1.1 Fatigue Loading

Components, structures, and vehicles are subjected to quite diverse load histories. At one extreme, their histories may be rather simple and repetitive; at the other extreme, they may be completely random. The randomness, however, may contain substantial portions of more deterministic loading. For example, the ground–air–ground cycle of an aircraft has substantial similarity from flight to flight. A schematic load history of one such flight is shown in Fig. 4.1a. The history consists primarily of flight with gusts, landing, and taxiing loads. Standardized test histories have been developed to aid in comparative test programs for aircraft [1]. Three typical load histories obtained by the SAE Fatigue Design and Evaluation Committee from actual ground vehicle components are shown in Fig. 4.1b. These load histories were also used in comparative test programs to better establish fatigue life prediction techniques [2]. A typical load history of short-span bridges that has been used in bridge fatigue life studies is shown in Fig. 4.1c [3]. These five load histories shown in Fig. 4.1 are typical of those found in real-life engineering situations. Fatigue from variable amplitude loading involving histories such as these is discussed in Chapter 9, and constant amplitude loading is introduced in this chapter. Constant amplitude loading is used to obtain material fatigue behavior/properties for use in fatigue design, and some real-life load histories can occasionally be modeled as essentially constant amplitude.

Nomenclature used in fatigue design has been superimposed on the constant amplitude stress versus time curve in Fig. 4.2. Definitions of alternating, S_a,

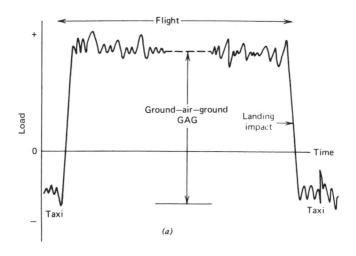

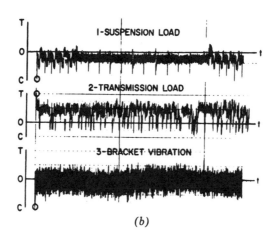

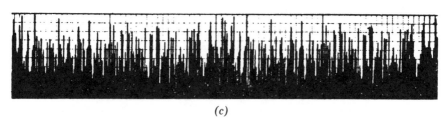

Figure 4.1 Typical in-service load spectra (reprinted by permission of the Society of Automotive Engineers and the American Society of Testing and Materials). (*a*) Schematic ground–air–ground flight spectrum. (*b*) SAE representative test spectra [2]. (*c*) Load spectrum for short-span bridges from the National Cooperative Highway Research Program [3].

60

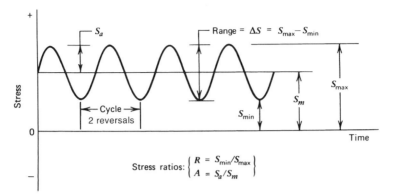

Figure 4.2 Nomenclature for constant amplitude cyclic loading.

mean, S_m, maximum, S_{max}, minimum, S_{min}, and range, ΔS, of stress are indicated. The algebraic relationships among these terms are:

$$S_a = \frac{\Delta S}{2} = \frac{S_{max} - S_{min}}{2}$$

$$S_m = \frac{S_{max} + S_{min}}{2}$$

$$S_{max} = S_m + S_a$$

$$S_{min} = S_m - S_a$$

(4.1)

The stress range, ΔS, is twice the alternating stress. Tensile or compressive stresses are taken algebraically as positive and negative, respectively. Alternating stress is an absolute value. The stress ratio, R, and the alternating stress ratio, A, are used frequently in fatigue literature, where

$$R = \frac{S_{min}}{S_{max}} \quad \text{and} \quad A = \frac{S_a}{S_m}$$

(4.2)

$R = -1$ and $R = 0$ are two common reference test conditions used for obtaining fatigue properties. $R = -1$ is called the "fully reversed" condition since S_{min} is equal to $-S_{max}$; $R = 0$ where $S_{min} = 0$ is called "pulsating tension." One cycle is the smallest segment of the stress versus time history that is repeated periodically, as shown in Fig. 4.2. Under variable amplitude loading, the definition of one cycle is not clear and hence reversals of stress are often considered. In constant amplitude loading, one cycle equals two reversals, while in variable amplitude loading a defined cycle may contain many reversals.

Stresses in Fig. 4.2 and Eqs. 4.1 and 4.2 can be replaced with load, moment, torque, strain, deflection, or stress intensity factors.

Figure 4.1 shows examples of different mean loadings. The aircraft flight history, Fig. 4.1*a,* the transmission history, Fig. 4.1*b,* and the bridge history, Fig. 4.1*c,* indicate significant tensile mean stresses. The suspension history, Fig. 4.1*b,* shows significant compressive mean stress loading, while the bracket history, Fig. 4.1*b,* is dominated by essentially fully reversed, $R = -1$, loading. A thin- or thick-walled pressure vessel subjected to cyclic internal pressure represents a component subjected to mean tensile stresses. Helical compression springs are actually under torsion, but the applied cyclic forces involve compressive mean forces. A cantilever beam deflected at the free end and then released to vibrate represents a damped vibration with essentially zero mean stress. Thus, both tensile or compressive mean loads and fully reversed loads are prevalent in all fields of engineering.

4.1.2   Fatigue Test Machines

Systematic, constant amplitude fatigue testing was first initiated by Wöhler on railway axles in the 1850s. Figure 4.3 schematically shows several common constant amplitude fatigue test machines. Rotating bending machines are shown in Figs. 4.3*a* and 4.3*b.* The test machine in Fig. 4.3*b* produces a uniform, pure bending moment over the entire test length of the specimen, while the cantilever test machine in Fig. 4.3*a* has a nonuniform bending moment along the specimen's length. These test machines are known as "constant load amplitude machines" because, despite changes in material properties or crack growth, the load amplitudes do not change. A constant deflection amplitude cantilever bending test machine that produces a nonuniform bending moment along the specimen's length is shown schematically in Fig. 4.3*c.* Since the rotating eccentric crank produces constant deflection amplitude, the load amplitude changes with specimen cyclic hardening or softening and decreases as cracks in the specimen nucleate and grow. For a given initial stress amplitude, constant deflection test machines may give longer fatigue life than load-control machines because of the decrease in load amplitude. The eccentric crank test machines, however, do have an advantage over the rotating bending test machines in that the mean deflection, and hence the initial mean stress, can be varied. Figure 4.3*d* shows a schematic of an axial loaded fatigue test machine capable of applying both mean and alternating axial loads in tension and/or compression. A common test setup for combined in-phase torsion and bending with or without mean stress loading is shown in Fig. 4.3*e.* This test machine provides a uniform torque and a nonuniform bending moment along the specimen's length. Many additional test machines have been designed over the years, but the most important contribution to fatigue testing has been the closed-loop servohydraulic test system. A modern servohydraulic test system utilizing its own personal computer is shown in Fig. 4.4. The principle of operation includes generating an input signal of load, strain, or displacement

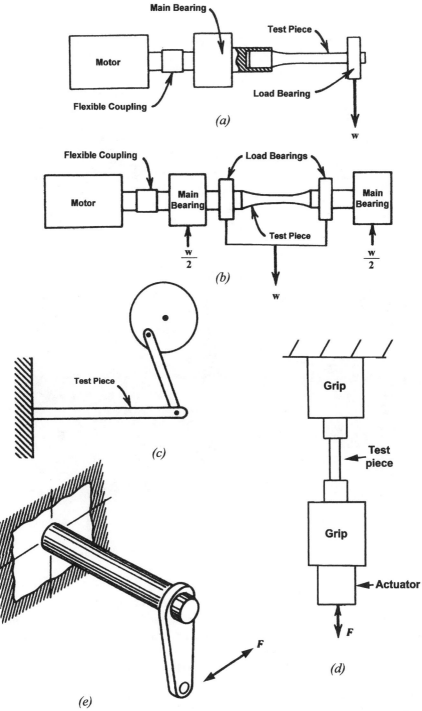

Figure 4.3 Fatigue testing machines. (*a*) Cantilever rotating bending. (*b*) Rotating pure bending. (*c*) Bending cantilever eccentric crank. (*d*) Axial loading. (*e*) Combined torsion and bending.

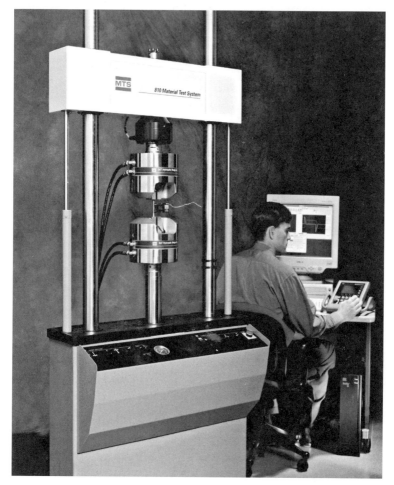

Figure 4.4   Closed-loop servohydraulic test system using personal computer control (courtesy of MTS Corporation).

using a function generator and applying this input through a hydraulic actuator; measuring the specimen response via a load cell, a clip gage, or a linear variable differential transducer (LVDT); and comparing this with the specific input. The difference drives the system. Control and test data outputs are usually through a personal computer and commercial or in-house software. The test frequency can range from mHz to kHz. These test systems can perform constant or variable amplitude load, strain, displacement, or stress intensity factor controlled tests on small specimens or can be utilized with hydraulic jacks for components, subassemblies, or whole structures. Two or more control systems are used for multiaxial testing.

TABLE 4.1   ASTM Standard Practices Related to Fatigue Testing of Metals [4]

E466	Conducting Force Controlled Constant Amplitude Axial Fatigue Tests of Metallic Materials
E467	Verification of Constant Amplitude Dynamic Forces in an Axial Fatigue Testing System
E468	Presentation of Constant Amplitude Fatigue Test Results for Metallic Materials
E606	Strain-Controlled Fatigue Testing
E647	Measurement of Fatigue Crack Growth Rates
E739	Statistical Analysis of Linear or Linearized Stress–Life (S–N) and Strain–Life (ε–N) Fatigue Data
E1012	Verification of Specimen Alignment Under Tensile Loading (under the jurisdiction of Committee E-28 on Mechanical Test Methods)
E1049	Cycle Counting in Fatigue Analysis
E1823	Standard Terminology Relating to Fatigue and Fracture Testing

Standard fatigue test methods and procedures for metals are available from ASTM [4]. These pertinent standards are under the direction of ASTM Committee E-08 and are given in Table 4.1. Both ASTM standard designations and titles are given in Table 4.1. International Organization for Standardization (ISO) draft standards on fatigue testing of metals are available through the ISO, Geneva, Switzerland. Three separate draft standards currently available are given in Table 4.2 with both designations and titles. Additional ISO fatigue testing draft standards are being developed.

4.1.3   Fatigue Test Specimens

Common test specimens for obtaining fatigue data are shown in Fig. 4.5. The specimens shown in Fig. 4.5*a–f* have been used to obtain total fatigue life, that includes crack nucleation life and crack growth life. These specimens usually have finely polished surfaces to minimize surface roughness effects.

TABLE 4.2   ISO Draft Standards Related to Fatigue Testing of Metals

ISO/DIS 12106	Metallic Materials-Fatigue Testing-Axial Strain-Controlled Method
ISO/DIS 12107	Metallic Materials-Fatigue Testing-Statistical Planning and Analysis of Data
ISO/DIS 12108	Metallic Materials-Fatigue Testing-Fatigue Crack Growth Method

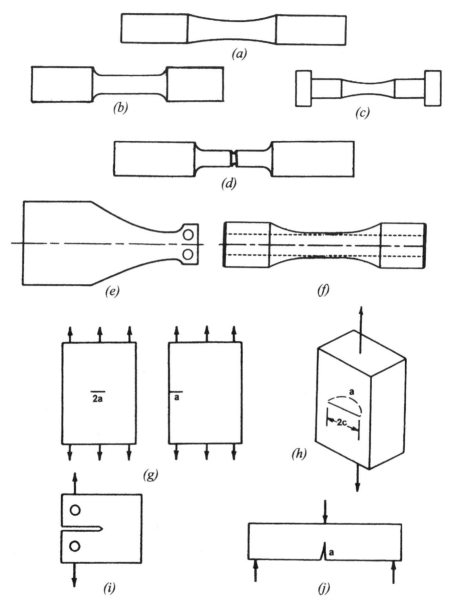

Figure 4.5 Fatigue test specimens. (*a*) Rotating bending. (*b*) Axial uniform. (*c*) Axial hourglass. (*d*) Axial or bending with circumferential groove. (*e*) Cantilever flat sheet/plate. (*f*) Tubular combined axial/torsion with or without internal/external pressure. (*g*) Axial cracked sheet/plate. (*h*) part-through crack. (*i*) Compact tension. (*j*) Three-point bend.

No distinction between crack nucleation and growth is normally made with these specimens, but it can be done with special care and observation or measurement. The keyhole specimen in Fig. 3.4 was designed specifically to monitor both fatigue crack nucleation and fatigue crack growth lives. Bending, axial, torsion, and combined axial/torsion specimens are included in Fig. 4.5*a–f*. The specimen in Fig. 4.5*f* is a thin-walled tube designed for torsion and combined axial/torsion, with the possibility of adding internal and/or external pressure. This multiaxial loading can be performed in-phase or out-of-phase. The thin-walled tube allows for essentially uniform elastic or inelastic normal and shear stresses in the cross-sectional area, making it very advantageous for multiaxial loading. Bending and axial specimens with solid circular cross sections usually have diameters between about 3 and 10 mm (1/8 to 3/8 in.). Stress concentration influence can be studied with most of these specimens by machining in notches, holes, or grooves, as shown with the specimen in Fig. 4.5*d*. Surface preparation guidelines for fatigue specimens can be found in ASTM Standards E466, E606, and E647 listed in Table 4.1. Careful alignment is needed for axial loaded specimens to minimize bending, and methods for verification of specimen alignment can be found in ASTM Standard E1012 listed in Table 4.1.

The specimens shown in Fig. 4.5*g–j* have been used to obtain fatigue crack growth data. In all cases a thin slit, notch, or groove with a very small root radius is machined into the specimen. A small pretest fatigue crack is then formed at this root radius by cycling at a low stress intensity factor range. After this sharp pretest fatigue crack has been formed, the real fatigue test can begin. Fatigue crack growth testing is discussed in Section 6.4, and recommended specimens, preparation, test procedure, and data reduction methods can be found in ASTM Standard E647 listed in Table 4.1. Control and recording of both humidity and temperature are recommended by ASTM for all fatigue tests.

4.2   STRESS–LIFE (*S–N*) CURVES

4.2.1   General *S–N* Behavior

Two typical schematic *S–N* curves obtained under axial load or stress control test conditions with smooth specimens are shown in Fig. 4.6. Here *S* is the applied nominal stress, usually taken as the alternating stress, S_a, and N_f is the number of cycles or life to failure, where failure is defined as fracture. N_f will be used rather than N in the text except when referring to the general *S–N* curve. The constant amplitude *S–N* curves of these types are plotted on semilog or log-log coordinates and often contain fewer data points than shown in Fig. 4.6*b*. *S–N* curves obtained under torsion or bending load-control test conditions often do not have data at the shorter fatigue lives (say, 10^3 or 10^4 cycles or less) due to significant plastic deformation; the torsion and bending

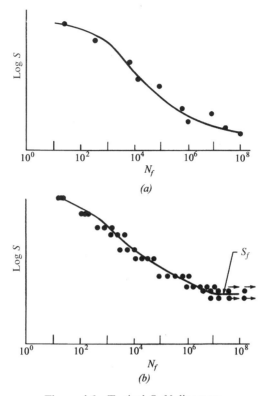

Figure 4.6   Typical *S–N* diagrams.

stress equations $\tau = Tr/J$ and $\sigma = My/I$ can be used only for nominal elastic behavior. Figure 4.6*b* shows typical variability, with less variability in life at the shorter lives and greater variability in life at the longer lives. From Fig. 4.6*b*, the variability in life for a given stress level can range from less than a factor of 2 to more than one or two orders of magnitude. Variability and statistical aspects of fatigue data are discussed in detail in Chapter 13. Figure 4.6*a* shows a continuously sloping curve, while Fig. 4.6*b* shows a discontinuity or "knee" in the *S–N* curve. This knee has been found in only a few materials (notably the low- and medium-strength steels) between 10^6 and 10^7 cycles under noncorrosive conditions. Most materials do not contain a knee even under carefully controlled environmental conditions. In corrosive environments all *S–N* data invariably have a continuously sloping curve. When sufficient data are taken at several stress levels, *S–N* curves are usually drawn through median lives and thus represent 50 percent expected failures. Common terms used with the *S–N* diagram are "fatigue life," "fatigue strength," and "fatigue limit." ASTM definitions of these terms, taken from Standard E1823 (Table 4.1), are given below [4].

The fatigue life, N_f, is the number of cycles of stress or strain of a specified character that a given specimen sustains before failure of a specified nature occurs. Fatigue strength, S_{Nf}, is a hypothetical value of stress at failure for exactly N_f cycles, as determined from an *S–N* diagram. The fatigue limit, S_f, is the limiting value of the median fatigue strength as N_f becomes very large. The above ASTM definitions are all based on median lives or 50 percent survival. The endurance limit is not defined by ASTM but is often implied as being analogous to the fatigue limit. We shall use ASTM terminology.

In Chapter 3 it was emphasized that fatigue typically consists of crack nucleation, growth, and final fracture. Figure 4.6 does not separate crack nucleation from growth, and only the total life to fracture is given. Let us assume a reasonable crack nucleation life defined by a crack length of 0.25 mm (0.01 in.). This dimension, which is easily seen at 10× to 20× magnification, can relate to engineering dimensions and can represent a small macrocrack. The number of cycles required to form this small crack in smooth, unnotched or notched fatigue specimens and components can range from a small percentage to almost the entire life. This is illustrated schematically in Fig. 4.7, where applied stress amplitude is plotted versus number of cycles to failure (fracture) and number of cycles to crack nucleation. Linear scales are implied so as not to disguise axis compression due to log-log scales. Here it is seen that a larger fraction of life for crack growth, the hatched area, occurs at higher stress levels, while a larger fraction of life for crack nucleation occurs at lower stress levels. Based on many factors covered in this book, fatigue crack nucleation, growth, and final fracture lives can be significantly altered. When fatigue crack growth life is significant, then fracture mechanics, as discussed in Chapter 6, should be used. The size of the final crack at fracture depends on the fracture toughness of the material and the stress level. The higher stress levels have shorter critical crack sizes and the lower stress levels have larger critical crack sizes. Assuming that test specimens are between 3 and 10 mm (1/8 and 3/8 in.) in diameter, and that critical crack lengths can

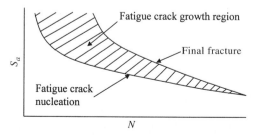

Figure 4.7 *S–N* schematic of fatigue crack nucleation, growth, and final fracture.

vary from 10 to 50 percent of the diameter, critical crack lengths for these small test specimens vary from about 0.25 to 5 mm (0.01 to 0.2 in.). These dimensions have significant importance in later chapters on fatigue life prediction of components.

4.2.2   Fatigue Limit Under Fully Reversed Uniaxial Stressing

The fatigue limit has historically been a prime consideration in long-life fatigue design. For a given material the fatigue limit has an enormous range, depending on surface finish, specimen or component size, type of loading, temperature, corrosive and other aggressive environments, mean stresses, residual stresses, and stress concentrations. Considering the fatigue limit based on a nominal alternating stress, S_a, this value has ranged from essentially 1 to 70 percent of the ultimate tensile strength. An example of a case in which the fatigue limit may be approximately 1 percent of S_u is a high-strength steel with a sharp notch subjected to a high mean tensile stress in a very corrosive atmosphere. An example of a case in which the fatigue limit might approach 70 percent of S_u is a medium-strength steel, in an inert atmosphere, containing appreciable compressive residual stresses.

Most long-life $S–N$ fatigue data available in the literature consist of fully reversed ($S_m = 0$) uniaxial fatigue strengths or fatigue limits of small, highly polished, unnotched specimens based on 10^6 to 5×10^8 cycles to failure in laboratory air environment. Representative monotonic tensile properties and bending fatigue limits of selected engineering alloys obtained under the above conditions are given in Table A.1. Most of these values are fairly independent of test frequency between about 1 and 200 Hz in noncorrosive environments. The data for the aluminum alloys were originally stamped "NOT FOR DE-SIGN" by the Aluminum Association [5]. This apparently was due to product liability concern and also to the fact that these data were obtained from small rotating bending, unnotched specimens, not from components subjected to actual service or field conditions and environments. The fatigue limits given in Table A.1 must be substantially reduced in most cases before they can be used in design situations [6–8]. For example, a 10 to 25 percent reduction in these values for the size effect alone is not unreasonable for bend specimens more than 10 mm (0.4 in.) in diameter [8].

A common procedure to partially compare materials for their fatigue resistance is to plot the unnotched, fully reversed fatigue limit, S_f, obtained under similar ideal laboratory conditions described above versus the ultimate tensile strength, S_u. The S_f/S_u ratio is called the "fatigue ratio." Figure 4.8 shows plots for many tests of steels, irons, aluminum alloys, and copper alloys subjected to rotating bending tests with fatigue limits or fatigue strengths based on 10^7 to 10^8 cycles [8]. Lines of constant fatigue ratio, S_f/S_u, equal to 0.35, 0.5, and 0.6 are superimposed on data in Fig. 4.8*a*, and 0.35 and 0.5 fatigue ratio lines are superimposed on data in Figs. 4.8*b–d*. It is quite clear that S_f/S_u varies from about 0.25 to 0.65 for these data. There is a tendency to generalize that S_f

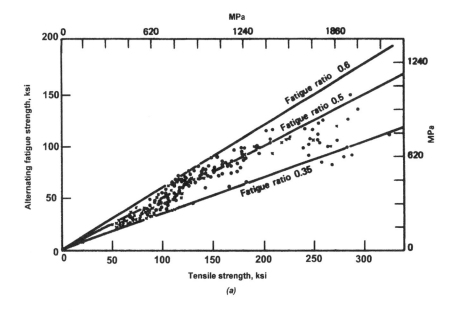

(a)

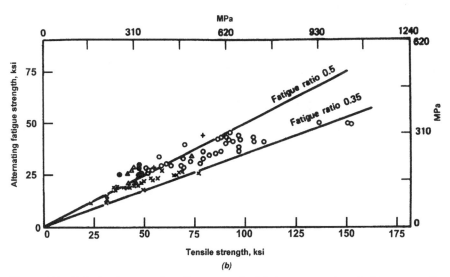

(b)

Figure 4.8 Relation between bending unnotched fatigue strength and ultimate tensile strength [8] (reprinted with permission of Pergamon Press Ltd.). (*a*) Carbon and alloy steels (10^7 to 10^8 cycles): (●) alloy steels, (x) carbon steels. (*b*) Wrought and cast irons (10^7 cycles): (x) flake-graphite cast iron, (○) nodular cast iron, (+) malleable cast iron, (△) ingot iron, (●) wrought iron. (*c*) Aluminum alloys (10^8 cycles): (x) wrought, (●) cast. (*d*) Wrought copper alloys.

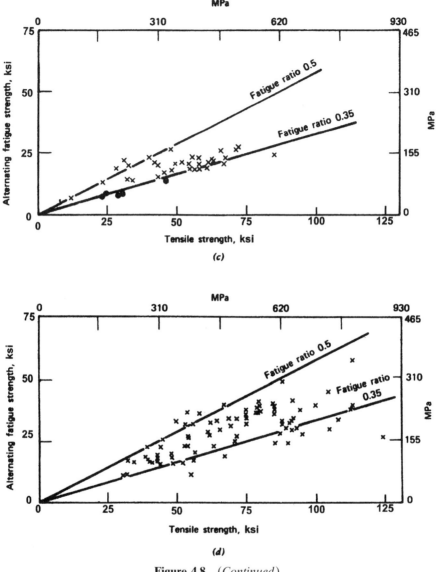

Figure 4.8   (*Continued*).

increases linearly with S_u. A careful examination of Fig. 4.8 shows that this is incorrect and that data bands tend to bend over at the higher ultimate strengths. This undesirable tendency, however, can be minimized by suitable mechanical or thermal surface treatment, as explained in Chapter 8. Also, for more recent cleaner high-strength steels (S_u > 1400 MPa), the fatigue limit is not as low as shown in Fig. 4.8*a* due to fewer inclusions and other impurities and hence fewer crack nucleation sites [9]. The fatigue resistance of these cleaner steels depends more on material strength than on inclusion/porosity content.

Figure 4.8*a* for steels indicates that substantial data are clustered near the fatigue ratio $S_f/S_u \approx 0.5$ for the low- and medium-strength steels. The data, however, actually fall between 0.35 and 0.6 for $S_u < 1400$ MPa (200 ksi). For $S_u > 1400$ MPa (200 ksi), S_f does not increase significantly. Thus, very common, loosely used estimates for unnotched, highly polished, small bending specimen fatigue limits for steels are

$$S_f \approx 0.5\, S_u \quad \text{for} \quad S_u \leq 1400 \text{ MPa (200 ksi)} \tag{4.3a}$$

$$S_f \approx 700 \text{ MPa (100 ksi)} \quad \text{for} \quad S_u \geq 1400 \text{ MPa (200 ksi)} \tag{4.3b}$$

For steels, S_u can be approximated from the Brinell hardness (HB) as

$$S_u \approx 3.45 HB \quad \text{for MPa units} \tag{4.4a}$$

$$S_u \approx 0.5 HB \quad \text{for ksi units} \tag{4.4b}$$

Other similar S_u–HB empirical formulas have been suggested, and data indicate that the above popular relations can be quite accurate and also can have a 10 to 20 percent error. *HB* values above 400 in Eq. 4.4 often give S_u values 10 to 15 percent lower than actual values [9]. By suitable mechanical or thermal surface treatments, fatigue limits of high-strength steels can be increased to agree more with Eq. 4.3a. Equations 4.3 are not unreasonable for the small, highly polished steel specimens tested, but empirical reduction factors for surface finish, size, stress concentration, temperature, and corrosion must also be considered [6–8], as discussed in Section 4.4 and Chapters 7, 8, and 11. We strongly warn against using a design fatigue limit equal to one-half of the ultimate strength of steels. Most data in Fig. 4.8 for irons, aluminum, and copper alloys fall below the 0.5 fatigue ratio. The aluminum and copper alloy data bands bend over at higher strengths, as do the bands for steels. Thus, high-strength steels, aluminum, and copper alloys generally do not exhibit correspondingly high unnotched fatigue limits.

Data for steels similar to those in Fig. 4.8*a* can be generalized into schematic scatter bands, as shown in Fig. 4.9 [10]. Here it is also seen that severely notched and/or corroding specimens have substantially lower fatigue limits than unnotched specimens. This behavior has too often been given insufficient importance in fatigue design. But again, this can be significantly altered in many cases by proper mechanical or thermal surface treatment, as described in Chapter 8.

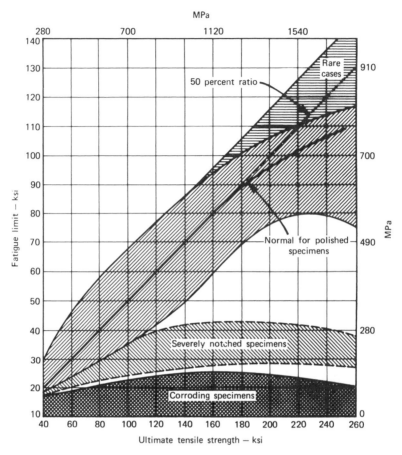

Figure 4.9   General relationship between fatigue limit and tensile strength for polished unnotched, notched, and corroding steel specimens [10] (reprinted with permission of John Wiley and Sons, Inc.).

4.3   MEAN STRESS EFFECTS ON *S–N* BEHAVIOR

The alternating stress, S_a, and the mean stress, S_m, are defined in Fig. 4.2 and Eq. 4.1. The mean stress, S_m, can have a substantial influence on fatigue behavior. This is shown in Fig. 4.10, where alternating stress, S_a, is plotted against the number of cycles to failure, N_f, for different mean stresses. It is seen that, in general, tensile mean stresses are detrimental and compressive mean stresses are beneficial. This is also shown by the three vertical lines indicating fatigue life: N_{ft}, N_{fo}, and N_{fc}, representing fatigue life for tensile, zero, and compressive mean stress, respectively, for a given alternating stress,

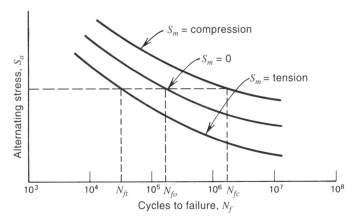

Figure 4.10 Effect of mean stress on fatigue life.

S_a. At intermediate or high stress levels under load control test conditions, substantial cyclic creep (also referred to as "cyclic ratcheting"), which increases the mean strain, can occur in the presence of mean stresses, as shown by Landgraf [11] in Fig. 4.11. This cyclic creep adds to the detrimental effects of tensile mean stress on fatigue life and results in additional undesirable excess deformation.

Substantial investigation of tensile mean stress influence on long-life fatigue strength has been made. Typical dimensionless plots are shown in Fig. 4.12 for steels and aluminum alloys, where S_a/S_f versus S_m/S_u is plotted. S_f is the

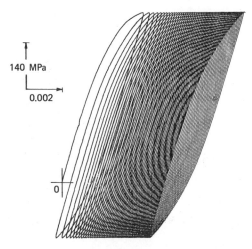

Figure 4.11 Cyclic creep under load control constant amplitude testing, $S_m > 0$ [11] (reprinted by permission of the American Society for Testing and Materials).

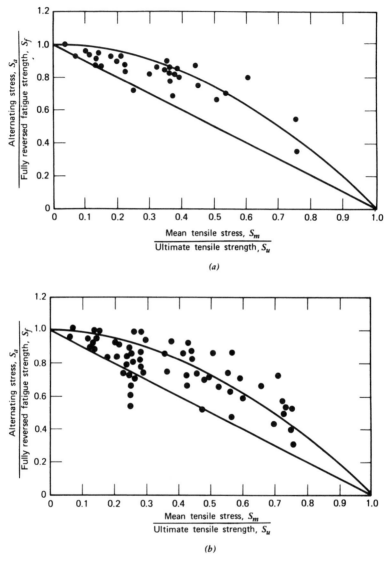

Figure 4.12 Effect of mean stress on alternating fatigue strength at long life [8] (reprinted with permission of Pergamon Press Ltd.). (a) Steels based on ~10^7 cycles. (b) Aluminum alloys based on ~5×10^7 cycles.

fully reversed, $(S_m = 0, R = -1)$, fatigue limit of smooth specimens, and S_u is the ultimate tensile strength. Similar behavior exists for other alloys. Substantial scatter exists, but the general trend indicating that tensile mean stresses are detrimental is quite evident. Small tensile mean stresses, however, often have only a small effect. It appears that much of the data fall between the straight and curved lines. The straight line is the modified Goodman line, and the curve is the Gerber parabola. An additional popular relationship has been formulated by replacing S_u with σ_f (Morrow line), where σ_f is the true fracture strength. The following equations represent these tensile mean stress effects for uniaxial state of stress:

Modified Goodman:
$$\frac{S_a}{S_f} + \frac{S_m}{S_u} = 1 \qquad (4.5a)$$

Gerber:
$$\frac{S_a}{S_f} + \left(\frac{S_m}{S_u}\right)^2 = 1 \qquad (4.5b)$$

Morrow:
$$\frac{S_a}{S_f} + \frac{S_m}{\sigma_f} = 1 \qquad (4.5c)$$

All three expressions have been used in fatigue design when modified for notches, size, surface finish, environmental effects, and finite life. A yield criterion has also been used in conjunction with these expressions.

Figure 4.12 does not provide information on compressive mean stress effects, as shown by Sines [12] in Fig. 4.13 for several steels and aluminum

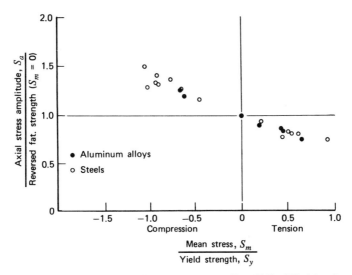

Figure 4.13 Compressive and tensile mean stress effect [12]. (●) Aluminum alloys, (○) steels.

alloys. It is seen that these compressive mean stresses cause increases of up to 50 percent in the alternating fatigue strength. Even higher increases have been shown [8]. This increase is too often overlooked, since compressive residual stresses can cause similar beneficial behavior. Based on the fact that compressive mean stresses at long lives are beneficial, the modified Goodman or Morrow equations can be conveniently extrapolated to the compressive mean stress region. The Gerber equation incorrectly predicts a detrimental effect of compressive mean stresses and does not properly represent notched component tensile mean stress fatigue behavior. Thus, we deemphasize the use of the Gerber equation for design. The modified Goodman and Morrow equations are shown in Fig. 4.14 for a given long life (e.g., 10^6, 10^7, 10^8 cycles) along with the criterion for uniaxial yielding:

$$\frac{S_a}{S_y'} + \frac{S_m}{S_y} = 1 \tag{4.6}$$

where S_y is the monotonic tensile yield strength and S_y' is the cyclic yield strength. S_y is used along the S_m axis since this axis represents monotonic loading only, and S_y' is used along the S_a axis since this axis represents the cyclic conditions used to obtain S_y', as described in Section 5.3. If S_y' is unknown, it can be replaced by S_y as an approximation.

In fatigue design with constant amplitude loading and unnotched parts, if the coordinates of the applied alternating and mean stresses fall within the modified Goodman or Morrow lines shown in Fig. 4.14, then fatigue failure should not occur prior to the given life. Note that the difference between the modified Goodman and Morrow equations is often rather small; thus, either model often provides similar results. If yielding is not to occur, then the applied alternating and mean stresses must fall within the two yield lines connecting $\pm S_y$ to S_y'. If both fatigue failure and yielding are not to occur, then neither criterion, as indicated by the three bold lines in Fig. 4.14, should be exceeded. In Fig. 4.14 the modified Goodman line was used for the bold

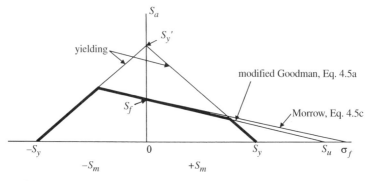

Figure 4.14   Fatigue and yielding criteria for constant life of unnotched parts.

fatigue line, but the Morrow line could also have been used. Fuchs [13] developed more general quantitative criteria for long-life fatigue design involving tensile and compressive mean or residual stresses and notches. These are discussed in later chapters.

4.4 FACTORS INFLUENCING *S–N* BEHAVIOR

The reference fatigue condition for $S–N$ behavior is usually fully reversed $R = -1$ bending or axial loading using small, unnotched specimens. Section 4.3 showed how mean stresses affect this reference fatigue condition at long life. Many other factors also affect the reference fatigue condition. Some of these are complex, and we have included specific chapters or sections to discuss them adequately. These topics, and where they are covered in later chapters or sections for the $S–N$ model, are as follows: notches and stress concentrations (Section 7.2), residual stress and surface treatment (Chapter 8), variable amplitude loading (Section 9.6), multiaxial and torsion loading (Section 10.3), corrosion (Section 11.1.2), fretting (Section 11.2), low temperature (Section 11.3.2), high temperature (Section 11.4.3), welds (Chapter 12), and statistical aspects (Chapter 13). Additional factors that influence $S–N$ behavior are discussed in this section.

4.4.1 Microstructure

In solid mechanics we often model metals as homogeneous, isotropic, and linearly elastic. At the microscopic level, none of these assumptions may exist and metal fatigue is significantly influenced by microstructure. This includes chemistry, heat treatment, cold working, grain size, anisotropy, inclusions, voids/porosity, delaminations, and other discontinuities or imperfections. If the actual $S–N$ data are available, microstructural effects are inherently accounted for and therefore do not have to be accounted for again. Chemistry, heat treatment, and cold working have a broadly significant influence on ultimate tensile strength, and the effects of ultimate tensile strength on fatigue limits for many metals were shown in Fig. 4.8. These three items have an enormous number of synergistic variations, and generalities concerning their effects on fatigue behavior are not practical here. However, some generalities can be formulated for the other microstructural aspects. Fine grains generally provide better $S–N$ fatigue resistance than coarse grains except at elevated temperatures, where creep/fatigue interaction exists. Fine grains reduce localized strains along slip bands, decreasing the amount of irreversible slip, and provide more grain boundaries to aid in transcrystalline crack arrest and deflection, thus reducing fatigue crack growth rates. Anisotropy caused by cold working gives increased $S–N$ fatigue resistance when loaded in the direction of the working than when loaded in the transverse direction. This is due to the elongated grain structure in the direction of the original cold working.

Inclusions, voids/porosity, and laminations act as stress concentrations and thus are common locations for microcracks to nucleate under cyclic loading, or to form during heat treatment or cold working prior to cyclic loading. Under either condition, fatigue resistance is reduced by these discontinuities. Minimizing inclusions, voids/porosity, laminations, and other discontinuities through carefully controlled production and manufacturing procedures is a key to good fatigue resistance.

4.4.2 Size Effects

Under unnotched bending conditions, if the diameter or thickness of the specimen is <10 mm (0.4 in.), then the *S–N* fatigue behavior is reasonably independent of the diameter or thickness. For larger size, *S–N* fatigue resistance is decreased. The decreases may differ slightly for rotating bending compared to nonrotating bending specimens or components. As the diameter or thickness increases to 50 mm (2 in.), the fatigue limit decreases to a limiting factor of about 0.7 to 0.8 of the fatigue limit for specimens less than 10 mm (0.4 in.) in diameter or thickness. Additional decreases can occur for larger specimens or components. Under unnotched axial conditions, *S–N* fatigue resistance is lower than for most bending conditions. The fatigue limit for axial loading can range from 0.75 to 0.9 of the bending fatigue limits for small specimens. Several factors are involved in the above size and axial loading effects. In bending, for a given nominal stress, the stress gradient depends upon the specimen's diameter or thickness. The larger the diameter or thickness, the smaller the bending stress gradient and hence the larger the average stress in a local region on the surface. The average stress in the local region, rather than the maximum stress, may be the governing stress for fatigue. For axial loaded unnotched specimens, a nominal stress gradient does not exist, and the average and maximum nominal stresses have the same magnitude, resulting in less size effect than in bending. In bending and axial loading, larger specimens have a higher probability of microstructural discontinuities in the highly stressed surface regions that contribute to the decrease in fatigue resistance. Another reason axial fatigue resistance is lower than bending fatigue resistance is possible eccentricity or alignment difficulties that superimpose bending stresses on the axial stresses. Also, with axial unnotched specimens, since the whole specimen is subjected to a uniform stress, hysteresis energy may not dissipate adequately and the specimen's temperature can rise, which may decrease fatigue resistance.

4.4.3 Surface Finish

Since most fatigue failures originate at the surface, the surface will have a substantial influence on fatigue behavior. Surface effects are caused by differences in surface roughness, microstructure, chemical composition, and residual stress. This influence will be more pronounced at long lives where a greater percentage of the cycles is usually involved with crack nucleation, as shown in Fig. 4.7. The reference fatigue strengths shown in Fig. 4.8 were for highly polished, smooth specimens. Most engineering parts, however, are not highly polished, and grinding or machining, even if done carefully, will cause degradation in fatigue strength relative to that shown in Fig. 4.8. Generalizations for grinding and machining effects are difficult because so many variables are involved. Noll and Lipson [14], however, provided data on these two operations for steels of varying ultimate strength, along with surface effects for hot-rolled and as-forged specimens. Their results have been further interpreted and included in machine design textbooks [6,7]. Figure 4.15 shows surface factors, k_s, applied to highly polished fatigue limits as a function of ultimate tensile strength for steels involving grinding, machining, hot-rolling, and as-forged surface conditions [6]. Reemsnyder [15] has evaluated these factors through more than 20 years of testing rotating beam and axial specimens and components. For constant amplitude load conditions, using finite life calculation models and these fatigue limits, he found that the surface factors are conservative. These values should thus be considered reasonable estimates of surface finish effects for steels. It is seen that, based on these surface factors, the higher the ultimate tensile strength and hardness, the greater the

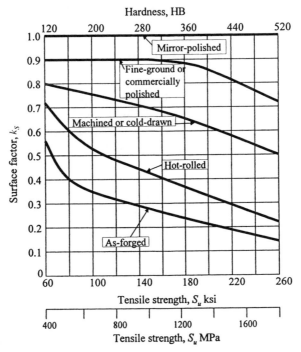

Figure 4.15 Effects of surface finish on the fatigue limit of steel [6] (reprinted by permission of John Wiley & Sons, Inc.).

degradation of fatigue limits. Surface factor reductions for grinding range from 0.9 to 0.7 and for machining from 0.8 to 0.5 as S_u is increased. These are significant reductions in the fatigue limit. However, the hot-rolled and as-forged conditions show even greater decreases, ranging from 0.75 to 0.15. The decreases caused by grinding and machining are more closely related to surface roughness and residual stresses, while hot-rolled and as-forged behavior include these two important factors along with significant surface microstructural and chemical composition changes such as decarburization and hence surface hardenability. One obvious conclusion here is to avoid the hot-rolled and as-forged surface conditions at local fatigue-sensitive locations. This can be accomplished by removing the undesirable surface by grinding or machining. Chapter 8 discusses the importance of inducing compressive residual stresses at the surface to enhance fatigue resistance. The estimates given in Fig. 4.15 are not applicable to nonsteels, and unfortunately, significant data for other metals are not generally available. Thus, surface effects for most metals have not been adequately quantified.

4.4.4 Frequency

The influence of the frequency on *S–N* behavior of metals is complicated because of the synergistic effects of test temperature, corrosive environment, stress–strain sensitivity to strain rate, and frequency. Independently, both elevated temperature and corrosive environments are usually detrimental to fatigue resistance. Specimen heating at higher test frequency due to internal hysteresis damping can increase the specimen's temperature and thus disguise the true fatigue behavior at ambient temperature. This is particularly important for lower-strength metals. Generation of heat due to cyclic loading depends on the volume of highly stressed material. Thus, axial loading and large specimens will produce more heat than small bending specimens or notched specimens, and therefore frequency effects could be different in these situations. If heating effects are negligible due to various cooling techniques and/or low stress amplitude during testing, along with negligible corrosion effects, then frequency effects can be evaluated. Under these conditions, using axial or bending specimens, frequencies ranging from less than 1 Hz to 200 Hz have had only a small effect on *S–N* behavior for most structural metals. At higher frequencies (still <200 Hz), fatigue strengths at 10^6 to 10^8 cycles have shown increases from zero to 10 percent [8,16]. At higher stress amplitudes, small increases in fatigue life have also occurred; however, many of these life changes are similar to typical scatter at a given frequency. In a few exceptional cases, fatigue resistance has decreased at higher frequencies. Based on the above, with known absence of corrosion and temperature effects and other aggressive environments, frequency effects of up to about 200 Hz have often been neglected in fatigue design and testing. Other effects may be more important. The key, however, is the absence of corrosion and an increase in temperature.

In the kHz frequency range, greater changes in fatigue resistance have occurred compared to those at less than 200 Hz. Temperature increases at these frequencies (1 to 25 kHz) are more difficult to control, and specimens have been cooled by air, water, water plus inhibitors, or oil that make contact with the specimen surface. Intermittent testing has also been used to maintain isothermal conditions. Test machines for kHz testing have been pneumatic, piezoelectric, and magnetostrictive, and specimens have been smooth, notched, or cracked. Fatigue limits have been obtained from 10^6 to 10^{10} cycles. Changes in fatigue limits at kHz frequency compared to those obtained below 200 Hz have ranged from a factor of less than 1 up to 2.6 [16,17], with the significant majority above 1. This more prevalent increase in fatigue limit has been accompanied by a beneficial shift in the finite life region ($>10^5$ cycles at these frequencies). In most cases, fatigue crack growth resistance has also increased at the kHz frequencies. However, despite the many trends toward increased fatigue resistance at kHz frequencies, it is still difficult to make this generalization due to the large number of complex test/material variables involved.

4.5 *S–N* CURVE REPRESENTATION AND APPROXIMATIONS

Actual fatigue data from either specimens or parts should be used in design if possible. This information may be available through data handbooks, design codes, industry standards, company data files, and previous tests. Often this information is not available and must be generated or approximations of *S–N* behavior must be made. There are many models depicting *S–N* curves, and these usually imply a median fatigue life. Figure 4.16 shows common reasonable *S–N* median fatigue life curves based upon straight-line log-log approxi-

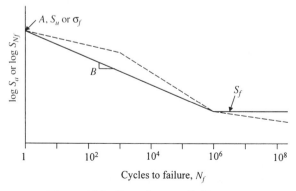

Figure 4.16 Basquin type *S–N* curves.

mations. Basquin in 1910 [18] suggested a log-log straight line S–N relationship such that

$$S_a \text{ or } S_{Nf} = A(N_f)^B \qquad (4.7)$$

where S_a is an applied alternating stress and S_{Nf} is the fully reversed, $R = -1$, fatigue strength at N_f cycles, A is the coefficient and represents the value of S_a or S_{Nf} at one cycle, and B is the exponent or slope of the log-log S–N curve. One approximate representation of the median S–N curve is a tri-slope model with one slope between 1 cycle and 10^3 cycles, one slope between 10^3 and 10^6 or 10^8 cycles, and another slope after 10^6 or 10^8 cycles. This model is represented by the dashed lines in Fig. 4.16. The tri-slope model indicates that a fatigue limit does not exist, which may be the case for in-service variable amplitude loading. The tri-slope model exists in some design codes, such as those for gears and welds. In the tri-slope model, the third, or long-life, slope could also be horizontal after 10^6 or 10^8 cycles. Other approximation models assume one sloping straight line from 1 cycle to 10^6, 10^7, or 10^8 cycles followed by a horizontal line or another sloped line. The intercept, A, at $N_f = 1$ could be chosen as the ultimate tensile strength, S_u, the true fracture strength, σ_f, or an intercept found from regression of the fatigue data. This regression is done in Chapter 5 for ε–N behavior. However, Basquin's equation there is based on reversals, $2N_f$, rather than cycles, N_f, and the coefficient, σ_f', is based on 1 reversal rather than 1 cycle. Values of σ_f', and b, defined as the fatigue strength coefficient and the fatigue strength exponent respectively, are given in Table A.2 for a few selected engineering alloys. However, the values of b in Table A.2 must be used with reversals, $2N_f$, not cycles, N_f, in Basquin's equation. Sometimes the difference between S_u, σ_f, and σ_f' can be small, and the three values have been used interchangeably for the coefficient A in Basquin's Eq. 4.7. It is best to use the actual values from fatigue tests for A, B, or b rather than estimates. The slope, B, depends upon many factors and for unnotched parts could vary from about -0.05 to -0.2. For example, the varying slope can be indicated by using surface effects with k_s taken from Fig. 4.15. The surface effects are dominant at long fatigue lives and less significant at short lives, with convergence of the S–N curves at the monotonic ultimate tensile strength at $N_f = 1$. This will give different values of the slope, B, for each surface condition. The convergence at $N_f = 1$ is reasonable because surface finish does not have an appreciable effect on monotonic properties for most smooth metal specimens.

Often the slope, B, for smooth, unnotched specimens is about -0.1. This suggests that for unnotched specimens the fatigue life is approximately inversely proportional to the 10th power of alternating stress. Thus, a 10 percent increase or decrease in alternating stress will cause a decrease or increase, respectively, of about a factor of 3 in fatigue life. For notched parts the slope of the S–N curve on logarithmic scales is steeper, yielding more extreme changes. Thus, even small changes in applied alternating stress can have a significant effect on fatigue life.

Constant fatigue life diagrams relating S_a and S_m are often modeled, as shown schematically in Fig. 4.17, by using one of the *S–N* models from Fig. 4.16 and the modified Goodman or Morrow equations for mean stress (Eqs. 4.5a and 4.5c). In Fig. 4.17, the intercepts, S_{Nf}, at $S_m = 0$ for a given life are found from the fully reversed, $R = -1$, modeled *S–N* curves. Modified Goodman straight lines are shown passing through these intercepts and the ultimate tensile strength, S_u, in Fig. 4.17. For these finite fatigue lives, the modified Goodman or Morrow equations (4.5a and 4.5c) should have S_f replaced with S_{Nf}, for a given life. Using these equations and Basquin's equation for the fully reversed, $R = -1$, finite life region, gives the following equations:

$$S_{Nf} = A(N_f)^B \tag{4.7}$$

and

$$\frac{S_a}{S_{Nf}} + \frac{S_m}{S_u} = 1 \quad \text{or} \quad \frac{S_a}{S_{Nf}} + \frac{S_m}{\sigma_f} = 1 \tag{4.8}$$

Either of Eqs. 4.8 along with Eq. 4.7 provides information to determine estimates of allowable S_a and S_m for a given fatigue life of unnotched parts. When Eqs. 4.7 and 4.8 are used, the *S–N* curve for a given mean stress is parallel to the fully reversed, $R = -1$, *S–N* curve, i.e., the slope, B, remains unchanged. Thus, the mean stress effect is handled in the same manner at both short and long lives. This may not represent the actual behavior in situations where mean stress relaxation occurs at short lives due to plastic

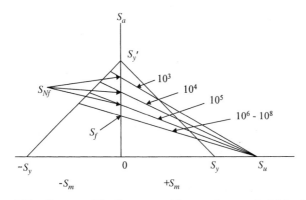

Figure 4.17 Constant life diagrams with superimposed yield criterion.

deformation, such as when the mean stress represents a residual stress. For such cases, it may be more realistic to apply the mean stress correction to the fatigue strength at long life, i.e., the fatigue limit, and connect it by a straight line to S_u or σ_f at one cycle. This assumes a full effect of mean stress at long life, with gradually decreasing effect, to no effect at 1 cycle. This method is used in the example problem in Section 7.2.4. The yield criterion

$$\frac{S_a}{S_y'} + \frac{S_m}{S_y} = 1 \qquad (4.6)$$

is also superimposed in Fig. 4.17. Negative (compressive) mean stress is beneficial even in the finite life regions, but as seen in Fig. 4.17, to avoid yielding, this region becomes very small at shorter lives. Constant life diagrams for notched parts are presented and discussed in Section 7.2.

4.6   EXAMPLE OF LIFE ESTIMATION USING THE *S–N* APPROACH

An unnotched circular rod with a diameter of 10 mm is subjected to constant amplitude bending at room temperature, with $S_m = 200$ MPa. The material is 4340 quenched and tempered alloy steel with $S_u = 1240$ MPa, $S_y = 1170$ MPa, and $S_y' = 1000$ MPa. If the rod is commercially polished, estimate the values of S_a, S_{max}, S_{min}, and R for a median fatigue life of 50 000 cycles and no yielding.

Since a mean stress other than zero is involved, a constant life diagram similar to Fig. 4.17 is needed. This diagram is also represented by Eqs. 4.8 and 4.6. We could use either the modified Goodman or the Morrow mean stress equation, but since σ_f is not given, we choose the finite life modified Goodman equation:

$$\frac{S_a}{S_{Nf}} + \frac{S_m}{S_u} = 1 \qquad (4.8)$$

We also assume Basquin's equation and a noncorrosive environment. Therefore,

$$S_{Nf} = A(N_f)^B \qquad (4.7)$$

where S_{Nf} is the fully reversed, $R = -1$, fatigue strength at N_f cycles to failure. For yielding we use Eq. 4.6:

$$\frac{S_a}{S_y'} + \frac{S_m}{S_y} = 1 \qquad (4.6)$$

Since S_u, S_y, S_y', S_m, and $N_f = 50\,000$ cycles are known, we need to determine S_{Nf} at 50 000 cycles, along with A and B, before we can solve for S_a, S_{max}, S_{min}, and R. We assume that the line represented by Basquin's equation passes through $S_u = 1240$ MPa at 1 cycle and S_f at 10^6 cycles (this could also be 10^7 cycles) for steels. From the first condition, $A = S_u = 1240$ MPa. From the second condition

$$S_{Nf} = A(N_f)^B = S_u\,(N_f)^B \quad \text{or} \quad S_f = S_u(10^6)^B$$

then

$$\log\,(S_f/S_u) = B\,\log\,10^6 = 6B \quad \text{or} \quad B = (1/6)\,\log\,(S_f/S_u)$$

For highly polished, small rotating beam steel specimens

$$S_f \approx 0.5\widehat{S}_u \quad \text{if} \quad S_u = \leq 1400 \text{ MPa}$$

and therefore

$$S_f \approx 0.5 \times 1240 = 620 \text{ MPa}$$

For commercial polishing, a correction factor for S_f from Fig. 4.15 is needed. With $S_u = 1240$ MPa (180 ksi), this correction factor is 0.87. For the diameter equal to 10 mm, no size effect adjustment is needed and therefore

$$S_f \approx 0.87 \times 620 = 540 \text{ MPa} \quad \text{and} \quad B = (1/6)\,\log\,(540/1240) = -0.06$$

At 50 000 cycles and for $R = -1$

$$S_{Nf} = 1240 \times (50\,000)^{-0.06} = 648 \text{ MPa}$$

and

$$\frac{S_a}{648} + \frac{200}{1240} = 1$$

resulting in

$$S_a = \underline{543 \text{ MPa}}$$

$$\frac{S_a}{S_y'} + \frac{S_m}{S_y} = \frac{543}{1000} + \frac{200}{1170} = 0.714$$

which is <1, and therefore yielding does not occur.

$$S_{min} = S_m - S_a = 200 - 543 = \underline{-343 \text{ MPa}}$$

$$S_{max} = S_m + S_a = 200 + 543 = \underline{743 \text{ MPa}}$$

$$R = \frac{S_{min}}{S_{max}} = -343/743 = \underline{-0.46}$$

If a few experimental tests were performed using the calculated value of S_a = 543 MPa and the given value of S_m = 200 MPa, failures would most likely not be 50 000 cycles due to the several approximations assumed and the scatter inherent in fatigue tests. However, this is our best median estimate.

4.7 SUMMARY

Test systems are available to perform fatigue and durability tests for almost every conceivable situation, ranging from a small, highly polished laboratory specimen to a wheel, suspension system, jet engine, automobile, tractor, or aircraft structure. ASTM standards volume 03.01 contains many different standards (Table 4.1) dealing with fatigue testing and data evaluation. ISO draft standards are also available on fatigue testing and data evaluation (Table 4.2). These ASTM and ISO standards address variability in fatigue life that is inherent, where for a given test condition it can vary from a factor of nearly 1 to several orders of magnitude, depending on the stress levels and the part tested. However, quite often the engineer will not have sufficient data, and median S–N data must be used in design.

The fatigue limit under constant amplitude loading conditions occurs for a few metals (notably low- and medium-strength steels), but under in-service, variable amplitude loading with corrosive, temperature, or other environmental conditions, the fatigue limit is rare. The fully reversed, rotating beam, smooth specimen fatigue strength, S_f, at 10^6 to 10^8 cycles ranges from about 25 to 65 percent of the ultimate tensile strength, S_u. For real parts, this can vary from about 0.01 to 0.7. By proper mechanical or thermal treatments, values of S_f for real parts can be raised significantly by altering the surface finish, microstructure, chemistry, and residual stresses. Quantitative determination of the many variables that affect fatigue resistance is complex, and many of these variables are treated in specific sections of this book. In some cases, excellent quantitative information is available; in other cases, estimates or approximations must be used. This obviously relies on cumulative experience, testing, and in-service inspection. Yielding, finite or long life, and mean stress effects can be approximated with the following equations:

Yielding:
$$\frac{S_a}{S_y'} + \frac{S_m}{S_y} = 1 \qquad (4.6)$$

Basquin equation:
$$S_{Nf} = A(N_f)^B \qquad (4.7)$$

Mean stress finite life:
$$\frac{S_a}{S_{Nf}} + \frac{S_m}{S_u} = 1 \quad \text{or} \quad \frac{S_a}{S_{Nf}} + \frac{S_m}{\sigma_f} = 1 \qquad (4.8)$$

Mean stress long life
$$\frac{S_a}{S_f} + \frac{S_m}{S_u} = 1 \quad \text{or} \quad \frac{S_a}{S_f} + \frac{S_m}{\sigma_f} = 1 \qquad (4.5)$$

4.8 DOS AND DON'TS IN DESIGN

1. Do consider the wide range of test systems and specimens available for fatigue testing. Tests can range from those performed on small, highly polished specimens for material characterization to full-scale durability tests of large structures.

2. Don't neglect to refer to ASTM, ISO, or similar standards on fatigue testing and data reduction techniques.

3. Do consider that the fully reversed fatigue strength, S_f, at 10^6 to 10^8 cycles for components can vary from about 1 to 70 percent of the ultimate tensile strength and that the engineer can substantially influence this value by proper design and manufacturing decisions.

4. Do note that cleaner metals, and generally smaller grain size for ambient temperature, have better fatigue resistance.

5. Do recognize that frequency effects are generally small only when corrosion, temperature, or other aggressive environmental effects are absent.

6. Don't use the median constant amplitude fatigue material properties given in Tables A.1 and A.2 for design unless they are modified for lower probabilities of failure and various conditions, such as corrosive and temperature environmental effects, surface finish and size effects, residual stresses, and notches. This also applies to most other published fatigue material properties.

7. Do consider that surface finish can have a substantial influence on fatigue resistance, particularly at longer lives.

8. Don't neglect the advantages of compressive mean or compressive residual stresses in improving fatigue life, and the detrimental effect of tensile mean or tensile residual stresses in decreasing fatigue life, and that models are available to account for these effects.

9. Do attempt to use actual fatigue data in design; however, if this is not possible or reasonable, approximate estimates of median fatigue behavior can be made.

REFERENCES

1. W. Schütz, "Standardized Stress-Time Histories—An Overview," *Development of Fatigue Loading Spectra,*" J. M. Potter and R. T. Watanabe, eds., ASTM STP 1006, ASTM, West Conshohocken, PA, 1989, p. 3.

2. R. M. Wetzel, ed., *Fatigue Under Complex Loading: Analysis and Experiments,* SAE, Warrendale, PA, 1977.

3. J. M. Barsom, "Fatigue-Crack Growth Under Variable-Amplitude Loading in ASTM A514-B Steel," *Progress in Flaw Growth and Fracture Toughness Testing,* ASTM STP 536, ASTM, West Conshohocken, PA, 1973, p. 147.

4. *Annual Book of ASTM Standards, Metals Test Methods and Analytical Procedures,* Vol. 03.01, ASTM, West Conshohocken, PA, 2000.

5. *Aluminum Standards and Data 1997,* The Aluminum Association, Washington, DC, 1997.

6. R. C. Juvinall and K. M. Marshek, *Fundamentals of Machine Component Design,* 2nd ed., John Wiley and Sons, New York, 1991.

7. J. E. Shigley and C. R. Mischke, *Mechanical Engineering Design,* 5th ed., McGraw-Hill Book Co., New York, 1989.

8. P. G. Forrest, *Fatigue of Metals,* Pergamon Press, London, 1962.

9. M. L. Roessle, and A. Fatemi, "Strain-Controlled Fatigue Properties of Steels and Some Simple Approximations," *Int. J. Fatigue,* Vol. 22, No. 6, 2000, p. 495.

10. D. K. Bullens, *Steel and Its Heat Treatment,* Vol. 1, John Wiley and Sons, New York, 1938, p. 37.

11. R. W. Landgraf, "The Resistance of Metals to Cyclic Deformation," *Achievement of High Fatigue Resistance in Metals and Alloys,* ASTM STP 467, ASTM, West Conshohocken, PA, 1970, p. 3.

12. G. Sines, "Failure of Materials Under Combined Repeated Stresses with Superimposed Static Stresses," NACA TN 3495, 1955.

13. H. O. Fuchs, "A Set of Fatigue Failure Criteria," *Trans. ASME, J. Basic Eng.,* Vol. 87, 1965, p. 333.

14. C. G. Noll and C. Lipson, "Allowable Working Stresses," *Proc. Soci. Exp. Stress Analysis,* Vol. III, No. 2, 1946, p. 49.

15. H. S. Reemsnyder, "Simplified Stress-Life Model," Bethleham Steel Corporation Report, Bethleham, PA, 1985.

16. T. B. Adams, *Cumulative Damage of a Repeated High Frequency Vibratory Loading,* M.S. thesis, The University of Iowa, 1991.

17. A. Puskar, *The Use of High Intensity Ultrasonics,* Elsevier Science, New York, 1982.

18. O. H. Basquin, "The Exponential Law of Endurance Tests," *Proc. ASTM,* Vol. 10, Part II, ASTM, West Conshochoken, PA, 1910, p. 625.

PROBLEMS

1. If a structural component is subjected to repeated stress cycles where $S_{max} = 400$ MPa and $S_{min} = -600$ MPa, determine the following: (*a*) S_m, (*b*) S_a, (*c*) ΔS, (*d*) R, and (*e*) A.

2. Fatigue testing can take an appreciable amount of time. Calculate the number of hours, days, or weeks it would take to apply 10^6, 10^7, and 10^8 cycles for test frequencies of (*a*) 1 Hz (approximate speed of Wöhler's

original work). One of Wöhler's tests ran for 1.3×10^8 cycles. How long did the test run? (b) 30 Hz (speed of many common test machines). (c) 150 Hz (speed of some rotating beam test machines). (d) 20 kHz, but calculate the time for 10^8, 10^9, and 10^{10} cycles.

3. Estimate the number of cycles the following items must endure during their expected lifetime: bicycle pedal shaft, truck engine valve spring, home light switch, automobile brake pedal.

4. Show that the stress in a rotating beam fatigue specimen is sinusoidal as a function of time when rotated at a constant angular velocity.

5. For the following $R = -1$ AISI 1090 steel test data, plot two S–N curves, one using rectangular coordinates and one using log-log coordinates. Draw median and lower bound S–N curves. What advantages or disadvantages exist in the two coordinate systems? From the log-log S–N curve determine the median: (a) fatigue limit, (b) fatigue strength at 5×10^5 cycles, and (c) fatigue life at $S_a = 260$ MPa. Comment on the scatter and how you handled it for parts a through c.

S_a (MPa)	Cycles to Failure	S_a (MPa)	Cycles to Failure
340	15×10^3	250	301×10^3
300	24×10^3	235	290×10^3
290	36×10^3	230	361×10^3
275	80×10^3	220	881×10^3
260	177×10^3	215	1.3×10^6
255	162×10^3	210	2.5×10^6

In addition the following stress levels had $>10^7$ cycles without failure: 210, 210, 205, 205, and 205 MPa.

6. Why should unnotched axial fatigue limits be 10 to 25 percent less than unnotched fatigue limits obtained from rotating bending? List several contributing factors.

7. Obtain a copy of the ASTM annual book of standards (Vol. 03.01) and write a one-page summary/critique of one of the following fatigue test standards: E466, E467, E468, E739, or E1012.

8. Based on the data in Fig. 4.8a, how often will Eqs. 4.3a and 4.3b be conservative and how often will they be nonconservative? Comment on how this affects fatigue design accuracy if Eqs. 4.3a and 4.3b are used.

9. A smooth uniaxial rod with a cross-sectional area of 0.003 m² is made from a material with $S_f = 300$ MPa under $R = -1$ conditions, $S_u = 650$ MPa, and $\sigma_f = 700$ MPa. If the rod is subjected to a mean force of 180 kN, what is the allowable alternating force that will not cause failure in 10^6 cycles according to the Morrow criterion?

10. A smooth uniaxial bar is subjected to a minimum stress of 35 MPa in compression. S_f for $R = -1$ was 220 MPa, S_u was 500 MPa, and σ_f was 600 MPa. Using the modified Goodman criterion, determine the maximum tensile stress the bar will withstand without failure in 10^6 cycles. Repeat the calculation using the Morrow criterion. Comment on the difference between the two calculations and its significance.

11. A 2024-T3 aluminum alloy smooth bar has a diameter of 15 mm and is subjected to axial stresses. Determine the following using reasonable approximate fatigue models: (*a*) fully reversed fatigue strength at 5×10^8 cycles and (*b*) S_{max}, S_{min}, S_a, and S_m, for fatigue strength at 5×10^8 cycles with $R = -0.2$. (*c*) Repeat part (*b*) for 10^5 cycles.

12. An as-forged 2 in. diameter 1040 steel rod has an ultimate tensile strength of 100 ksi and a yield strength of 75 ksi and is subjected to constant amplitude cyclic bending. Determine the following using appropriate approximation models: (*a*) the fully reversed fatigue strength at 10^6 cycles, (*b*) S_a and S_m for 10^6 cycles if $R = 0$, and (*c*) S_a and S_m for 10^4 cycles if $R = 0$.

13. Repeat Problem 12 but with the small as-forged surface thickness machined off. Comment on the effect of removing the as-forged surface thickness on fatigue resistance.

CHAPTER 5

CYCLIC DEFORMATION AND THE STRAIN–LIFE (ε–N) APPROACH

5.1 MONOTONIC TENSION TEST AND STRESS–STRAIN BEHAVIOR

Monotonic tension or compression stress–strain properties are usually reported in handbooks and are called for in many specifications. These tests are easy to perform and provide information that has become conventionally accepted. However, their relation to fatigue behavior may be rather remote, as noted in the following excerpt from the preface of a small book on fatigue written by Spangenburg in 1876 [1]:

> Spangenburg's experiments, given in the following treatise, were, as will be seen, in continuation of Wöhler's. The results of these very important experiments have been before the profession for some years, but strange to say, seem to have attracted no attention; and tests of iron and steel still go on for the purpose of determining their elasticity, their elongation under strain, their ultimate strength and other qualities, while Wöhler and Spangenburg's experiments show that it is very doubtful that these bear any proportion to the durability of the metals.

This excerpt is still pertinent to modern engineers, who sometimes like to think that fatigue behavior can be accurately predicted from simple tests such as the monotonic tension test. However, before discussing strain-controlled fatigue behavior, an understanding of the material stress–strain response under static or monotonic uniaxial loading is necessary. Monotonic behavior is obtained from a tension test in which a specimen with a circular or rectangular cross section within the uniform gage length is subjected to a monotonically increasing tensile force until it fractures. Details of tension testing for metallic materials are provided in ASTM Standard E8 or E8M [2].

Monotonic uniaxial stress–strain behavior can be based on "engineering" stress–strain or "true" stress–strain relationships. The nominal engineering stress, S, in a uniaxial test specimen is defined by

$$S = \frac{P}{A_0} \tag{5.1}$$

where P is the axial force and A_0 is the original cross-sectional area. The true stress, σ, in this same test specimen is given by

$$\sigma = \frac{P}{A} \tag{5.2}$$

where A is the instantaneous cross-sectional area. The true stress in tension is larger than the engineering stress since the cross-sectional area decreases during loading. The engineering strain, e, is based on the original gage length and is given by

$$e = \frac{(l - l_0)}{l_0} = \frac{\Delta l}{l_0} \tag{5.3}$$

where l is the instantaneous gage length, and Δl is the change in the original gage length, l_0. The true or natural strain, ε, is based on the instantaneous gage length and is given by

$$d\varepsilon = \frac{dl}{l} \quad \text{or} \quad \varepsilon = \int \frac{dl}{l} = \ln\left(\frac{l}{l_0}\right) \tag{5.4}$$

For small strains, less than about 2 percent, the engineering stress, S, is approximately equal to the true stress, σ, and the engineering strain, e, is approximately equal to the true strain, ε. No distinction between engineering and true values is needed for these small strains. However, for larger strains, the differences become appreciable. With large, inelastic deformations, a constant volume condition can be assumed such that up to necking $Al = A_0 l_0$. This is a reasonable assumption since plastic strain does not usually contribute to volume change. The following relationships between S, σ, e, and ε can then be derived:

$$\sigma = S\,(1 + e) \tag{5.5}$$

$$\varepsilon = \ln\frac{A_0}{A} = \ln(1 + e) \tag{5.6}$$

Equations 5.5 and 5.6 are valid up to necking which takes place when the ultimate tensile strength is reached, since after necking, plastic deformation becomes localized and strain is no longer uniform throughout the gage section.

Representative engineering and true stress–strain behavior is shown in Fig. 5.1. The elastic region has also been plotted on an extended scale to better indicate the yield strength and modulus of elasticity. The following properties are usually obtained from monotonic tension tests, some of which have been and still are used in fatigue design:

E = modulus of elasticity, MPa (ksi)

S_y = yield strength, MPa (ksi)

S_u = ultimate tensile strength, MPa (ksi) = P_{max}/A_0

σ_f = true fracture strength, MPa (ksi)

$\%RA$ = percent reduction in area = $100 (A_0 - A_f)/A_0$

ε_f = true fracture strain or ductility = $\ln (A_0/A_f) = \ln [100/(100 - \%RA)]$

$\%EL$ = percent elongation = $100 (l_f - l_0)/l_0$

The subscript f represents fracture. The true fracture strength, σ_f, can be calculated from P_f/A_f but is usually corrected for necking, which causes a biaxial state of stress at the neck surface and a triaxial state of stress at the neck interior. The Bridgman correction factor is used to compensate for this

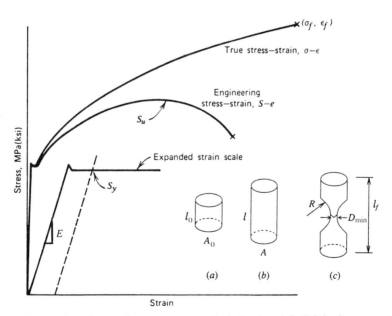

Figure 5.1 Engineering and true stress–strain behavior. (*a*) Original gage section. (*b*) Uniformly elongated gage section. (*c*) Gage section at fracture with necking.

triaxial state of stress and applies to cylindrical specimens [3]. The corrected true fracture strength is then given by

$$\sigma_f = \frac{P_f/A_f}{(1+4\ R/D_{min})\ \ln(1+D_{min}/4R)} \tag{5.7}$$

where R is the radius of curvature of the neck and D_{min} is the diameter of the cross section in the thinnest part of the neck. Materials with brittle tensile behavior do not exhibit necking and therefore do not require this correction factor.

Representative values of these monotonic tensile material properties of unnotched specimens, along with accompanying fatigue properties, are given in Tables A.1 and A.2 for selected engineering alloys. Values of S_y, S_u, and σ_f are indicators of material strength, and values of %RA, %EL, and ε_f are indicators of material ductility. Indicators of the energy absorption capacity of a material are resilience and tensile toughness. Resilience is the elastic energy absorbed by the specimen and is equal to the area under the elastic portion of the stress–strain curve. Tensile toughness is the total energy density or energy per unit volume absorbed during deformation (up to fracture) and is equal to the total area under the engineering stress–strain curve. A material with high tensile toughness and with a good combination of both high strength and ductility is often desirable.

Inelastic or plastic strain results in permanent deformation which is not recovered upon unloading. The unloading curve is elastic and parallel to the initial elastic loading line, as shown in Fig. 5.2. The total strain, ε, is composed of two components: an elastic strain, $\varepsilon_e = \sigma/E$, and a plastic component, ε_p. Even though true stress and true strain symbols are used in Fig. 5.2, engineering values could also have been used due to small differences between the true

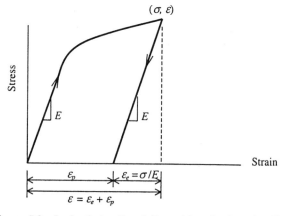

Figure 5.2   Inelastic loading followed by elastic unloading.

and engineering values for the small strains shown. For many metals, a plot of true stress versus true plastic strain in log-log coordinates results in a linear curve. An example of such a plot is shown in Fig. 5.3 for AISI 11V41 steel. To avoid necking influence, only data between the yield strength and the ultimate tensile strength portions of the stress–strain curve are used to generate this plot [4]. This curve is represented by the power function

$$\sigma = K \, (\varepsilon_p)^n \tag{5.8}$$

where K is the strength coefficient (stress intercept at $\varepsilon_p = 1$) and n is the strain hardening exponent (slope of the line). The total true strain is then given by

$$\varepsilon = \varepsilon_e + \varepsilon_p = \frac{\sigma}{E} + \left(\frac{\sigma}{K}\right)^{1/n} \tag{5.9}$$

This type of true stress–true strain relationship is often referred to as the "Ramberg-Osgood relationship." The value of n gives a measure of the material's work hardening behavior and is usually between 0 and 0.5. Values of K and n for some engineering alloys are also given in Table A.2.

The monotonic tensile stress–strain behavior of notched specimens may be similar to or quite different from that of unnotched specimens. For most notched tension tests, the ductility is reduced. In low- and intermediate-strength metals that behave in a ductile manner, the ultimate tensile strength based upon net section dimensions is usually increased in notched specimens, while for high-strength metals that behave in a brittle manner, the ultimate tensile strength is usually reduced in notched specimens. Notch strength is commonly

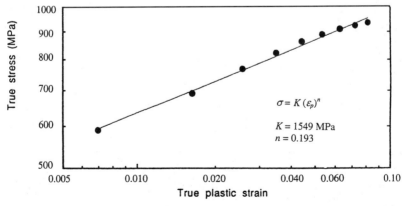

Figure 5.3 True stress versus true plastic strain behavior for AISI 11V41 steel (courtesy of A. Fatemi).

unavailable in material property handbooks due to its dependence on notch geometry; however, the above differences do exist. Also, the stress–strain behavior of a material can be sensitive to the strain rate, particularly at elevated temperatures. A significant increase in the strain rate generally increases the strength but reduces the ductility of the material. For metals and alloys, however, the strain rate effect can be small at room temperature.

5.2   STRAIN-CONTROLLED TEST METHODS

An important aspect of the fatigue process is plastic deformation. Fatigue cracks usually nucleate from plastic straining in localized regions. Therefore, cyclic strain-controlled tests can better characterize the fatigue behavior of a material than cyclic stress-controlled tests can, particularly in the low-cycle fatigue region and/or in notched members, where significant localized plastic deformation is often present. As a result, strain-controlled fatigue testing has become very common, even though the testing equipment and control are more complicated than in traditional load- or stress-controlled testing.

Strain-controlled testing is usually conducted on a servo-controlled closed-loop testing machine similar to that shown in Figs. 4.3d and 4.4. An unnotched specimen with a uniform gage section such as that shown in Fig. 4.5b is subjected to axial straining. An extensometer is attached to the uniform gage length to control and measure strain over the gage section, as shown in Fig. 4.4. A standard strain-controlled test consists of constant amplitude, completely reversed straining at a constant or nearly constant strain rate. The most common strain-time control signals used are triangular (sawtooth) and sinusoidal waveforms. Since strain limits are controlled at constant values, the stress response generally changes with continued cycling. Stress and plastic strain variations are, therefore, usually recorded periodically throughout the test, and cycling is continued until fatigue failure occurs. Several definitions of fatigue failure can be used, and these are discussed in Section 5.4.

An important consideration in axial fatigue testing is uniformity of stress and strains in the specimen gage section. A major source of nonuniformity of gage section stress and strains is a bending moment resulting from specimen misalignment that can significantly shorten the fatigue life. Specimen misalignment can result from eccentricity and/or tilt in the load-train components (including load cell, grips, and load actuator), from improper specimen gripping, or from lateral movement of the load-train components during the test due to their inadequate stiffness. The details of strain-controlled cyclic testing are described in ASTM Standard E606 [5] and SAE Standard J1099 [6].

5.3   CYCLE-DEPENDENT MATERIAL DEFORMATION AND CYCLIC STRESS–STRAIN BEHAVIOR

The stress–strain behavior obtained from a monotonic tension or compression test can be quite different from that obtained under cyclic loading. This was first observed during the late nineteenth century by Bauschinger [7]. His experiments indicated that the yield strength in tension or compression was reduced after applying a load of the opposite sign that caused inelastic deformation. This can be clearly seen in Fig. 5.4, where the yield strength in compression is significantly reduced by prior yielding in tension. Thus, one single reversal of inelastic strain can change the stress–strain behavior of metals.

Continuous monitoring of the stress–strain curve by Morrow [8] during cyclic strain-controlled testing of copper in three initial conditions is shown in Fig. 5.5. These tests were performed on axially loaded specimens in the (*a*) fully annealed condition, (*b*) partially annealed condition, and (*c*) cold-worked condition. The number of applied reversals at different positions of the hysteresis loops are indicated. The area within a hysteresis loop is energy per unit volume dissipated during a cycle, usually in the form of heating. This energy represents the plastic work resulting from the loading cycle and is used in some energy-based fatigue theories. The solid curve from the origin to the first reversal represents the monotonic tensile stress–strain behavior. The last hysteresis loop is also shown solid for ease of comparison. The remaining curves show the appreciable progressive changes in the stress–strain behavior during inelastic cyclic straining. The fully annealed soft copper (*a*) cyclic hardened, as indicated by an increase in stress range to reach the constant amplitude strain range, while the cold-worked copper (*c*) cyclic softened, as indicated by the decrease in stress to reach the prescribed strain. The partially annealed copper (*b*) showed initial cyclic hardening followed by cyclic softening. These changes, however, were much less than those of the fully annealed or cold-worked copper. The change in stress amplitude for the constant strain amplitude tests was about 400 percent for the fully annealed copper, 33 percent for the cold-worked copper, and 15 percent for the partially annealed copper. The mechanisms of hardening and softening were briefly described in terms

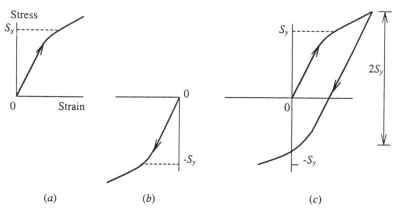

Figure 5.4 Bauschinger effect. (*a*) Tension loading. (*b*) Compression loading. (*c*) Tension loading followed by compression loading.

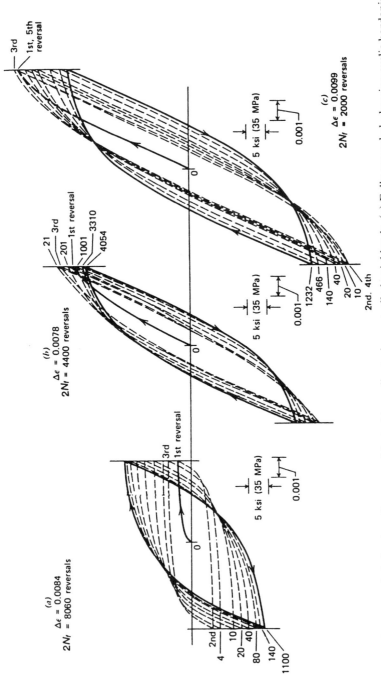

Figure 5.5 Stress–strain behavior of copper subjected to cyclic strain-controlled axial loads. (*a*) Fully annealed, showing cyclic hardening. (*b*) Partially annealed, showing small cyclic hardening and softening. (*c*) Cold-worked, showing cyclic softening [8] (reprinted by permission of the American Society for Testing and Materials).

of dislocation substructure and motion in Section 3.2. A more extensive discussion can be found in [9].

The extent and rate of cyclic hardening or softening under strain-controlled testing conditions can be evaluated by recording stress variation as a function of cycles, as shown in Fig. 5.6. Cyclic hardening (Fig. 5.6*b*) indicates increased resistance to deformation, whereas cyclic softening (Fig. 5.6*c*) indicates the opposite effect. Changes in cyclic deformation behavior are more pronounced at the beginning of cyclic loading, but the material usually stabilizes gradually with continued cycling. Therefore, such cyclic deformation behavior is referred to as "cyclic transient behavior." Cyclic stabilization is usually reasonably complete within 10 to 40 percent of the total fatigue life. A hysteresis loop from about half of the fatigue life is often used to represent the stable or steady-state cyclic stress-strain behavior of the material. A stable stress-strain loop is shown in Fig. 5.7. The total true strain range is denoted by $\Delta\varepsilon$, and $\Delta\sigma$ is the true stress range. The true elastic strain range, $\Delta\varepsilon_e$, can be calculated from $\Delta\sigma/E$. By definition

$$\Delta\varepsilon = \Delta\varepsilon_e + \Delta\varepsilon_p = \frac{\Delta\sigma}{E} + \Delta\varepsilon_p \qquad (5.10)$$

where $\Delta\varepsilon_p$ is the true plastic strain range. Even though true stress and strains are used in Figs. 5.6 and 5.7, no distinction is usually made between true values and engineering values. This is because the differences between true and engineering values during the tension and compression parts of the cycle are opposite to each other and therefore cancel out. In addition, strain levels in cyclic loading applications are often small (typically less than 2 percent), compared to strain levels in monotonic loading.

A family of stabilized hysteresis loops at different strain amplitudes can be used to obtain the cyclic stress–strain curve for a given material. The tips from the family of multiple loops are connected, as shown in Fig. 5.8 for Man-Ten steel, to form the cyclic stress–strain curve [10]. This curve does not

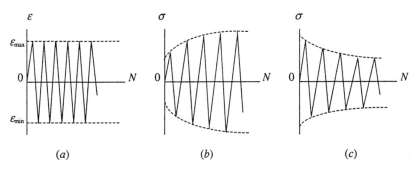

Figure 5.6 Stress response under constant strain amplitude cycling. (*a*) Constant strain amplitude. (*b*) Cyclic hardening. (*c*) Cyclic softening.

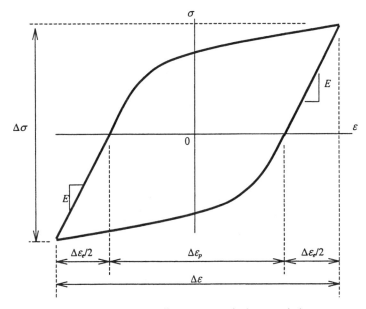

Figure 5.7 Stable cyclic stress–strain hysteresis loop.

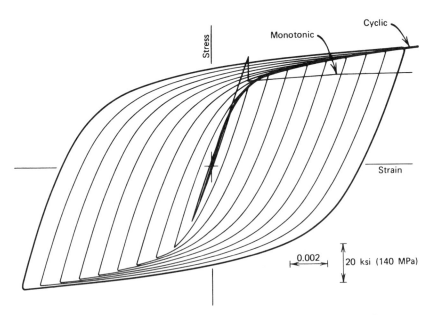

Figure 5.8 Stable hysteresis loops for determining the cyclic stress–strain curve and comparison with the monotonic stress–strain curve for Man-Ten steel [10] (reprinted with permission of the Society of Automotive Engineers).

contain the monotonic upper and lower yield points. Three methods commonly used to obtain the cyclic stress–strain curve are the companion, incremental-step, and multiple-step test methods. The companion method requires a series of test specimens, each of which is subjected to a constant strain amplitude until failure occurs. Half-life or near half-life hysteresis loops from each specimen and strain amplitude are then used to obtain the cyclic stress–strain curve. If the experimental program includes strain-controlled fatigue tests, the cyclic stress–strain curve can be obtained from the same fatigue data using the companion method. In the incremental-step method, a single specimen is subjected to repeated blocks of incrementally increasing and decreasing strains, as shown in Fig. 5.9. After the material has stabilized (usually after several strain blocks), the hysteresis loops from half of a stable block are then used to obtain the cyclic stress–strain curve. The multiple-step test method is similar to the incremental-step test method, except that rather than incrementally increasing and decreasing the strain in each block, the strain amplitude is kept constant. Once cyclic stability is reached at the constant strain amplitude, the stable hysteresis loop is recorded and strain amplitude is increased to a higher level. This process is repeated until a sufficient number of stable hysteresis loops is recorded to construct the cyclic stress–strain curve. Even though some differences exist between the results of the three methods, they are small in most cases [11].

Landgraf, Morrow, and Endo [12] obtained both cyclic and monotonic stress–strain curves for several metals shown superimposed in Fig. 5.10. Cyclic softening exists if the cyclic stress–strain curve is below the monotonic curve, and cyclic hardening is present if it lies above the curve. The difference between the two curves is small in some cases and substantial in others. Low-strength, soft metals tend to cyclic harden and high-strength, hard metals tend to cyclic soften. Using monotonic properties in a cyclic loading application

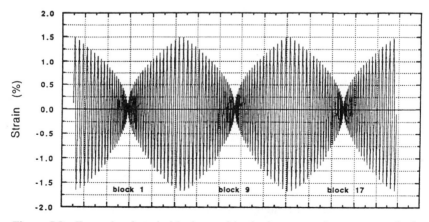

Figure 5.9 Example of strain blocks used in the incremental step-test method.

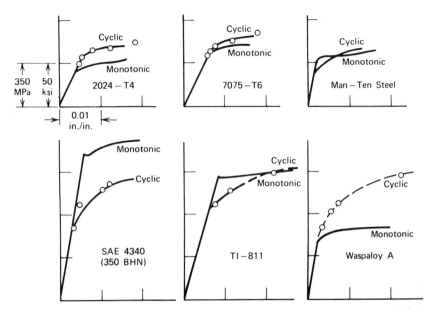

Figure 5.10   Monotonic and cyclic stress–strain curves for several materials [12] (reprinted by permission of the American Society for Testing and Materials).

for a cyclic softening material, such as the SAE 4340 steel in Fig. 5.10, can significantly underestimate the extent of plastic strain which may be present.

Similar to the monotonic deformation in a tension test, a plot of true stress amplitude, σ_a, versus true plastic strain amplitude, $\Delta\varepsilon_p/2$, in log-log coordinates for most metals results in a linear curve represented by the power function

$$\sigma_a = K'\left(\frac{\Delta\varepsilon_p}{2}\right)^{n'} \tag{5.11}$$

where K' and n' are the cyclic strength coefficient and cyclic strain hardening exponent, respectively. Substituting plastic strain amplitude obtained from Eq. 5.11 into Eq. 5.10 results in the cyclic stress–strain equation represented by a Ramberg-Osgood-type relationship:

$$\varepsilon_a = \frac{\Delta\varepsilon}{2} = \frac{\Delta\varepsilon_e}{2} + \frac{\Delta\varepsilon_p}{2} = \frac{\Delta\sigma}{2E} + \left(\frac{\Delta\sigma}{2K'}\right)^{1/n'} = \frac{\sigma_a}{E} + \left(\frac{\sigma_a}{K'}\right)^{1/n'} \tag{5.12}$$

Values of K' and n' for selected engineering alloys are given in Table A.2. The value of n' is found to be between about 0.05 and 0.25 for most metals. This range is smaller than the monotonic range. The cyclic yield strength, S_y', is defined at 0.2 percent strain offset which corresponds to a plastic strain amplitude of 0.002 on the cyclic stress–strain curve. It can be estimated by substituting $\Delta\varepsilon_p/2 = 0.002$ in Eq. 5.11 or obtained graphically from the cyclic stress–strain curve.

The stabilized hysteresis loop curve shown in Fig. 5.7 can be obtained by doubling the size of the cyclic stress–strain curve. The equation for the stable hysteresis loop curve can then be written as

$$\Delta\varepsilon = \frac{\Delta\sigma}{E} + 2\left(\frac{\Delta\sigma}{2K'}\right)^{1/n'} \tag{5.13}$$

This equation represents the stable hysteresis loop curve of metals with symmetric deformation behavior in tension and compression very well. Materials for which the hysteresis loop can be described by magnifying the cyclic stress-strain curve by a factor of 2 are said to exhibit "Masing-type" behavior. This type of behavior is common for many metals. More detailed discussion of Masing and non-Masing behaviors can be found in [13].

The macroscopic material behaviors given in this section were obtained from uniaxial test specimens with cross-sectional dimensions of about 3 to 10 mm (about 1/8 to 3/8 in.). In most cases, gross plastic deformation was involved. We should ask the question, how does all this gross plasticity apply to most service fatigue failures where gross plastic deformation does not exist? The answer is quite simple and extremely important. Most fatigue failures begin at local discontinuities where local plasticity exists and crack nucleation and growth are governed by local plasticity at the crack tip. The type of behavior shown for gross plastic deformation in Figs. 5.5 to 5.8 and 5.10 is similar to that which occurs locally at notches and crack tips, and thus is extremely important in fatigue of components and structures.

5.4 STRAIN-BASED (ε–N) APPROACH TO LIFE ESTIMATION

The strain-based approach to fatigue problems is widely used at present. Strain can be measured and has been shown to be an excellent quantity for correlating with low-cycle fatigue. For example, gas turbines operate at fairly steady stresses, but when they are started or stopped, they are subjected to a very high stress range. The local strains can be well above the yield strain, and the stresses are more difficult to measure or estimate than the strains. The most common application of the strain-based approach, however, is in fatigue of notched members. In a notched component or specimen subjected to cyclic external loads, the behavior of material at the root of the notch is best considered in terms of strain. As long as there is elastic constraint surrounding a local plastic zone at the notch, the strains can be calculated more easily than the stress. This concept has motivated the strain–life design method based on relating the fatigue life of notched parts to the life of small, unnotched specimens that are cycled to the same strains as the material at the notch root. Since fatigue damage is assessed directly in terms of local strain, this approach is called the "local strain approach." A reasonable expected fatigue life, based on the nucleation or formation of small macrocracks, can then be determined if one knows the local strain–time history at a notch in a component and the unnotched strain–life fatigue properties of the material. The remaining fatigue crack growth life of a component can be analyzed using fracture mechanics concepts, which are discussed in Chapter 6.

Substantial strain–life fatigue data needed for the above procedure have been accumulated and published in the form of simplified fatigue material properties [6]. Some of these data are included in Table A.2 for selected engineering alloys. These properties are obtained from small, polished, unnotched axial fatigue specimens similar to those in Figs. 4.5b and 4.5c. Tests are performed under constant amplitude, fully reversed cycles of strain, as shown in Figs. 5.5 and 5.6. Steady-state hysteresis loops can predominate throughout most of the fatigue life, and these loops can be reduced to elastic and plastic strain ranges or amplitudes. Cycles to failure can range from about 10 to 10^6 cycles, and frequencies can range from about 0.1 to 10 Hz. Beyond 10^6 cycles, load or stress-controlled tests at higher frequencies can often be performed because of the small or lack of plastic strains and the longer time to failure. The strain–life curves are often called "low-cycle fatigue data" because much of the data are for fewer than about 10^5 cycles.

Strain–life fatigue curves plotted on log-log scales are shown schematically in Fig. 5.11, where N_f or $2N_f$ is the number of cycles or reversals to failure, respectively. Failure criteria for strain–life curves have not been consistently defined in that failure may be the life to a small detectable crack, life to a certain percentage decrease in tensile load, life to a certain decrease in the ratio of unloading to loading moduli, or life to fracture. Differences in fatigue life depending on these four criteria may be small or appreciable. Crack lengths associated with these failure criteria are discussed later in this section. When tensile load drop is used as the failure criterion, a 50 percent drop level is recommended by ASTM Standard E606 [5].

The total strain amplitude shown in Fig. 5.11 has been resolved into elastic and plastic strain components from the steady-state hysteresis loops. At a given life, N_f, the total strain is the sum of the elastic and plastic strains. Both the elastic and plastic curves can be approximated as straight lines. At large strains or short lives the plastic strain component is predominant, and at small strains or longer lives the elastic strain component is predominant. This is indicated by the straight-line curves and the sizes of the hysteresis loop in Fig. 5.11. The intercepts of the two straight lines at $2N_f = 1$ are σ_f'/E for the elastic component and ε_f' for the plastic component. The slopes of the elastic and plastic lines are b and c, respectively. This provides the following equation for strain–life data of small smooth axial specimens:

$$\frac{\Delta\varepsilon}{2} = \varepsilon_a = \frac{\Delta\varepsilon_e}{2} + \frac{\Delta\varepsilon_p}{2} = \frac{\sigma_f'}{E}(2N_f)^b + \varepsilon_f'(2N_f)^c \qquad (5.14)$$

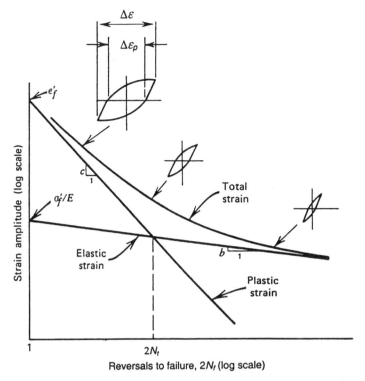

Figure 5.11 Strain–life curves showing total, elastic, and plastic strain components.

where $\Delta\varepsilon/2$ = total strain amplitude = ε_a

$\Delta\varepsilon_e/2$ = elastic strain amplitude = $\Delta\sigma/2E = \sigma_a/E$

$\Delta\varepsilon_p/2$ = plastic strain amplitude = $\Delta\varepsilon/2 - \Delta\varepsilon_e/2$

ε_f' = fatigue ductility coefficient

c = fatigue ductility exponent

σ_f' = fatigue strength coefficient

b = fatigue strength exponent

E = modulus of elasticity

$\Delta\sigma/2$ = stress amplitude = σ_a

To solve this equation for N_f for a given strain amplitude requires iterations, using numerical or graphical solutions. The straight-line elastic behavior can be transformed into

$$\frac{\Delta\sigma}{2} = \sigma_a = \sigma_f' \, (2N_f)^b \qquad (5.15)$$

which is Basquin's equation, analogous to Eq. 4.7, proposed in 1910 [14]. The fitting constants in Eqs. 4.7 and 5.15 are related by $B = b$ and $A = (2)^b \sigma_f'$. The relation between plastic strain and life is

$$\frac{\Delta \varepsilon_p}{2} = \varepsilon_f'(2N_f)^c \tag{5.16}$$

which is the Manson-Coffin relationship proposed in the early 1960s [15,16]. A typical complete strain–life curve with data points is shown in Fig. 5.12 for 4340 steel [17]. Eleven test specimens were used to form these strain–life curves. Representation of the strain-controlled fatigue behavior by the strain–life curves of Eqs. 5.14, 5.15, and 5.16 has been confirmed experimentally for many metals.

The life at which elastic and plastic components of strain are equal is called the "transition fatigue life," $2N_t$. This is the life at which the elastic and plastic strain–life curves intersect, as shown in Fig. 5.11. The equation for transition fatigue life can be derived by equating the elastic and plastic strains in Eqs. 5.15 and 5.16, respectively, resulting in

$$2N_t = \left(\frac{\varepsilon_f' E}{\sigma_f'}\right)^{\frac{1}{b-c}} \tag{5.17}$$

For lives less than $2N_t$ the deformation is mainly plastic, whereas for lives larger than $2N_t$ the deformation is mainly elastic. The transition fatigue life decreases with increasing hardness for steels and can be just a few cycles for high-strength metals and on the order of 10^5 cycles for metals with ductile behavior. This indicates that even at relatively long lives of more than 10^5 cycles, significant plastic strain can be present; therefore, the strain-based approach is an appropriate approach to use.

The general differences between metals under strain-controlled tests are shown schematically in Fig. 5.13. Some metals have similar life at a total strain amplitude of about 0.01. At large strains, increased life is dependent more on ductility, while at small strains longer life is obtained from higher-strength materials. The optimum overall strain–life behavior is for tough metals, which have good combinations of strength and ductility. "Fatigue life" here refers to the nucleation or formation of a small detectable crack, a percentage decrease in the tensile load caused by crack nucleation and growth, a decrease in the ratio of unloading to loading moduli due to the presence of a crack or cracks, or final fracture. For the final fracture criterion, the crack would grow to about 10 to 50 percent of the test specimen's cross section. Since strain–life test specimens are usually between about 3 and 10 mm (1/8 and 3/8 in.) in diameter, this implies that the strain–life fracture criteria are based on cracks growing to a depth of about 0.25 to 5 mm (0.01 to 0.2 in.). The actual value depends on the strain amplitude, the modulus of elasticity, and the material's

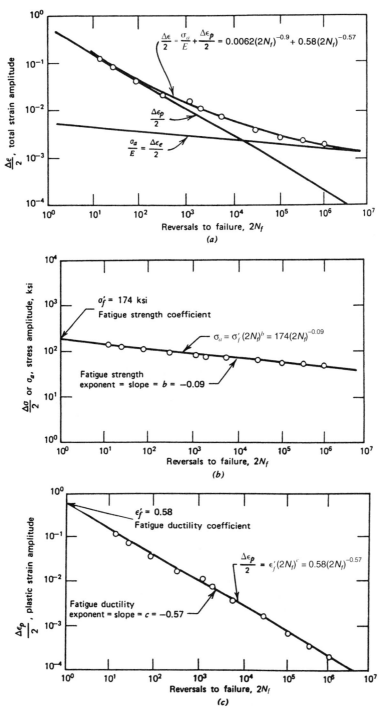

Figure 5.12 Low-cycle fatigue behavior of annealed 4340 steel [17] (reprinted with permission of the Society of Automotive Engineers). (*a*) Total strain amplitude. (*b*) Elastic strain amplitude multiplied by *E*. (*c*) Plastic strain amplitude.

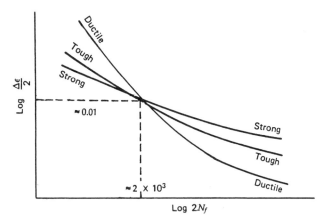

Figure 5.13   Schematic of strain–life curves for different materials.

fracture toughness. The other three criteria are based on life to cracks, which are usually smaller than those at fracture. In general, cracks less than 0.25 mm (0.01 in.) would not be readily observed in these tests and would probably not cause sufficient decrease in the tensile load or unloading to loading moduli to terminate a test. Thus, a reasonably important conclusion concerning failure criteria in strain-life testing of unnotched smooth specimens is that the life to failure means life to fatigue crack lengths of 0.25 to 5 mm (0.01 to 0.2 in.). This range is quite large; thus, we shall refer to this fatigue crack length as a length "on the order of 1 mm."

Low-cycle strain–life fatigue data in Table A.2 and in [6] were obtained under the above conditions. Both monotonic tensile properties and strain–life material properties are included for completeness in both SI and American/ British units. The terms in Table A.2 were described earlier. Fatigue limit, S_f, can also be estimated by substituting the proper material constants and $2N_f = 2 \times 10^6$ to 10^8 cycles, as appropriate for the type of material, into Eq. 5.15. This value is an approximation of reasonable long-life, unnotched, smooth axial fatigue strength and should be somewhat compatible with the S_f values given in Table A.1. Fatigue limits from axial fatigue tests should be about 10 to 25 percent lower than those given in Table A.1 from rotating bending tests for a given material because of the difference between bending and axial long-life fatigue strengths (see Section 4.4.2 and Problem 6 in Chapter 4). The material properties given in Table A.2 also omit influences of surface finish, size, stress concentration, temperature, and corrosion. These must be included in fatigue design; thus, the values in Table A.2 do not represent final fatigue design properties.

The strain-based approach unifies the treatment of low-cycle and high-cycle fatigue. This approach also applies to long-life applications where small plastic strains may exist. In this case, the plastic strain term in Eq. 5.14 is negligible and the total strain–life equation reduces to Basquin's Eq. 5.15, which was also used for the stress–life $(S–N)$ approach. Therefore, the strain-based approach is a comprehensive approach which can be applied to both low-cycle and high-cycle fatigue regimes.

5.5 DETERMINATION OF STRAIN–LIFE FATIGUE PROPERTIES

Strain–life fatigue properties, which are also often referred to as "low-cycle fatigue properties," are obtained by curve fitting data from stable hysteresis loops, such as that shown in Fig. 5.7. The fatigue strength coefficient, σ_f', and the fatigue strength exponent, b, are the intercept and slope of the linear least squares fit to stress amplitude, $\Delta\sigma/2$, versus reversals to failure, $2N_f$, using a log-log scale. Similarly, the fatigue ductility coefficient, ε_f', and the fatigue ductility exponent, c, are the intercept and slope of the linear least squares fit to plastic strain amplitude, $\Delta\varepsilon_p/2$, versus reversals to failure, $2N_f$, using a log-log scale. Plastic strain amplitudes can either be measured directly from half of the width of stable hysteresis loops at $\sigma = 0$ or calculated from

$$\frac{\Delta\varepsilon_p}{2} = \frac{\Delta\varepsilon}{2} - \frac{\Delta\sigma}{2E} \qquad (5.18)$$

A difference usually exists between the measured and calculated values, which results from the difference between monotonic and cyclic moduli of elasticity, as well as from rounding of the hysteresis loops near the strain axis that many materials exhibit. This difference, however, is usually small [18], and Eq. 5.18 is often more conveniently used to obtain the plastic strain amplitude.

When fitting the data to obtain the four strain–life properties, stress and plastic strain amplitudes should be treated as independent variables and fatigue life as the dependent variable. This is because fatigue life cannot be controlled and is dependent upon the applied strain amplitude. Also, very short life (sometimes fewer than 10 cycles but usually fewer than 100 cycles) and very long life (typically more than 10^5 or 10^6 cycles for steels) data are not usually included in data fits. This is because buckling at high loads and fatigue limit effects and inaccuracies in measuring small plastic strains at low loads can influence the test results. Therefore, the strain–life equation based on the obtained properties is valid only for the same life regime as the data used, even though extrapolations are often made to both shorter and longer lives. More details on statistical aspects of equation fitting can be found in [19] and in Chapter 13.

As discussed in Section 5.3, the cyclic strength coefficient, K', and the cyclic strain hardening exponent, n', are obtained by fitting stable stress amplitude versus plastic strain amplitude data. Rough estimates of K' and n' can also be calculated from the low-cycle fatigue properties using

$$K' = \frac{\sigma_f'}{(\varepsilon_f')^{\frac{b}{c}}} \quad \text{and} \quad n' = \frac{b}{c} \tag{5.19}$$

These equations are derived from compatibility between Eqs. 5.11, 5.15, and 5.16. Values of K' and n' obtained from direct fitting of the experimental data and calculated from the relations in Eq. 5.19 can be very similar or very different, depending on the goodness of linearized fits represented by Eqs. 5.11, 5.15, and 5.16. A large difference for a material can indicate that the elastic and plastic strain–life behavior are not well represented by log-log linearized fits. In this case, Eq. 5.14 may not be representative of the strain–life behavior for the material. It is recommended that the values of K' and n' obtained from direct fitting of the experimental data be used in fatigue design rather than the values calculated from Eq. 5.19.

The exponent b ranges from about -0.06 to -0.14, with -0.09 or -0.1 as representative values. The exponent c ranges from about -0.4 to -0.7, with -0.6 as a representative value. The term ε_f' is somewhat related to the true fracture strain, ε_f, in a monotonic tension test and in most cases ranges from about 0.35 to 1.0 times ε_f. The coefficient σ_f' is somewhat related to the true fracture strength, σ_f, in a monotonic tension test, which for steels can be approximated from

$$\sigma_f \approx S_u + 345 \quad \text{and} \quad S_u \approx 3.45HB \tag{5.20}$$

where σ_f and S_u are in MPa and HB is Brinell hardness. Sometimes ε_f' and σ_f' may be taken as ε_f and σ_f, respectively, as a rough first approximation. However, this is not recommended.

Muralidharan and Manson [20] have approximated Eq. 5.14 with their method of Universal Slopes, where

$$\frac{\Delta\varepsilon}{2} = 0.623 \left(\frac{S_u}{E}\right)^{0.832} (2N_f)^{-0.09} + 0.0196 \, (\varepsilon_f)^{0.155} \left(\frac{S_u}{E}\right)^{-0.53} (2N_f)^{-0.56} \tag{5.21}$$

S_u, E, and ε_f are all obtained from a monotonic tension test. It is assumed that the two exponents are fixed for all metals and that only S_u, E, and ε_f control the fatigue behavior. Equation 5.21 was obtained based on data from 47 metals including steels, aluminum, and titanium alloys. Thus, the Universal Slopes method can be a first approximation for the fully reversed strain–life curve for unnotched, smooth specimens based on monotonic tension properties. Another approximation that uses only hardness and E has been shown to provide good agreement with experimental data from 69 steels and is given in [21].

5.6 MEAN STRESS EFFECTS

Strain-controlled deformation and fatigue behavior discussed in the previous sections were for completely reversed straining, $R = \varepsilon_{min}/\varepsilon_{max} = -1$. In many applications, however, a mean strain can be present. Strain-controlled cycling with a mean strain usually results in a mean stress which may relax fully or partially with continued cycling, as shown in Fig. 5.14. This relaxation is due to the presence of plastic deformation, and therefore, the rate or amount of relaxation depends on the magnitude of the plastic strain amplitude. As a result, there is more mean stress relaxation at larger strain amplitudes. A model for predicting the amount of mean stress relaxation as a function of cycles was proposed in [22]. Stress relaxation is different from cyclic softening and can occur in a cyclically stable material.

Mean strain does not usually affect fatigue behavior unless it results in a non-fully relaxed mean stress. Since there is more mean stress relaxation at higher strain amplitudes due to larger plastic strains, mean stress effect on fatigue life is smaller in the low-cycle fatigue region and larger in the high-cycle fatigue region. Such behavior can be seen from Fig. 5.15 for SAE 1045 hardened steel [23]. The general effect of mean stress on fatigue life was discussed in Section 4.3, and several equations were presented to account for mean stress effects using the S–N approach. The inclusion of mean stress effects in fatigue life prediction methods involving strain-life data is more complex. Several models dealing with mean stress effects on strain–life fatigue behavior are discussed in [24]. One method, often referred to as "Morrow's mean stress method," replaces σ_f' with $\sigma_f' - \sigma_m$ in Eq. 5.14 [17], where σ_m is the mean stress, such that

$$\frac{\Delta \varepsilon}{2} = \varepsilon_a = \frac{\sigma_f' - \sigma_m}{E} (2N_f)^b + \varepsilon_f'(2N_f)^c \qquad (5.22)$$

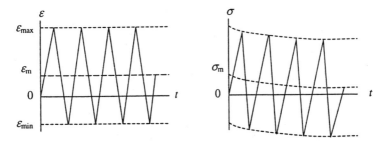

Figure 5.14 Mean stress relaxation under strain-controlled cycling with a mean strain.

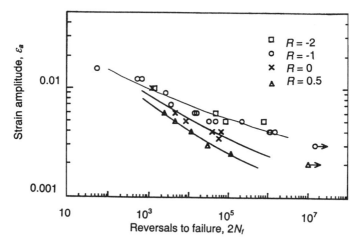

Figure 5.15   Effect of mean strain on fatigue life for SAE 1045 hardened steel [23] (reprinted with permission of Elsevier Science).

Here σ_m is taken to be positive for tensile values and negative for compressive values. This equation predicts that tensile mean stress is detrimental and compressive mean stress is beneficial, as indicated in Chapter 4. Equation 5.22 also predicts more effect of mean stress at long lives, as shown in Fig. 5.15. An alternative version of Morrow's mean stress parameter where both the elastic and plastic terms are affected by the mean stress is given by [25]

$$\frac{\Delta\varepsilon}{2} = \varepsilon_a = \frac{\sigma_f' - \sigma_m}{E}(2N_f)^b + \varepsilon_f'\left(\frac{\sigma_f' - \sigma_m}{\sigma_f'}\right)^{c/b}(2N_f)^c \qquad (5.23)$$

Another equation suggested by Smith, Watson, and Topper [26] (often called the "SWT parameter"), based on strain–life test data obtained with various mean stresses, is

$$\sigma_{max}\varepsilon_a E = (\sigma_f')^2 (2N_f)^{2b} + \sigma_f'\varepsilon_f' E(2N_f)^{b+c} \qquad (5.24)$$

where $\sigma_{max} = \sigma_m + \sigma_a$ and ε_a is the alternating strain. This equation is based on the assumption that for different combinations of strain amplitude, ε_a, and mean stress, σ_m, the product $\sigma_{max}\,\varepsilon_a$ remains constant for a given life. If σ_{max} is zero, Eq. 5.24 predicts infinite life, which implies that tension must be present for fatigue fractures to occur. All three equations have been used to handle mean stress effects. Equation 5.24 has been shown to correlate mean stress data better for a wider range of materials and is therefore regarded as more promising for general use. Derivations of Eqs. 5.23 and 5.24 are left as problems at the end of this chapter.

5.7 SURFACE FINISH AND OTHER FACTORS INFLUENCING STRAIN–LIFE BEHAVIOR

Similar to the S–N approach, in addition to the mean stress, many other factors can influence the strain–life fatigue behavior of a material. These include stress concentrations, residual stresses, multiaxial stress states, environmental effects, size, and surface finish effects. The effects of many of these factors are similar to those on S–N behavior discussed in Section 4.4.

As mentioned previously, an advantage of the strain–life approach to fatigue life prediction is its ability to account directly for the plastic strains often present at stress concentrations. Notch strain analysis and life predictions using the strain–life approach are discussed in Chapter 7. Residual stress effects on fatigue life are similar to mean stress effects. Therefore, there is little or no influence at short lives due to stress relaxation resulting from plastic deformation and more influence at long lives, where strains are mainly elastic. Residual stress effects on fatigue life are discussed in Chapter 8. The discussion of strain-controlled deformation and fatigue behavior in this chapter is restricted to uniaxial stress states. This is extended to multiaxial stress states, discussed in Chapter 10. Environmental effects on strain–life behavior, including low and high temperatures and corrosion, are discussed in Chapter 11. Size effects on strain-life behavior are similar to those found using the S–N approach, which is discussed in Section 4.4.2.

Surface finish effects are also similar to those for the S–N approach. Since fatigue cracks often nucleate early in the low-cycle region due to large plastic strains, there is usually little influence of surface finish at short lives. Conversely, there is more influence in the high-cycle fatigue regime, where elastic strain is dominant. Therefore, only the elastic portion of the strain–life curve is modified to account for the surface finish effect. This is done by reducing the slope of the elastic strain–life curve, b, in an analogous manner to the modification of the S–N curve for surface finish discussed in Section 4.4.3 [27]. The procedure is shown schematically in Fig. 5.16. In this figure, S_f denotes a fatigue limit for a polished surface finish which can be estimated from Eq. 5.15 by substituting the appropriate number of cycles to the fatigue limit of the material for N_f. The slope of the elastic strain–life curve for the polished surface condition is then reduced from b to b' by lowering the fatigue limit with the correction factor, k_s, as appropriate for the surface finish. Values of the surface finish correction factor for steels are given in Fig. 4.15. The slope b' for steels with a fatigue limit assumed at 10^6 cycles can be calculated from $b' = b + 0.159 \log k_s$. This approach for surface finish correction can also be used for nonzero mean stress loading situations.

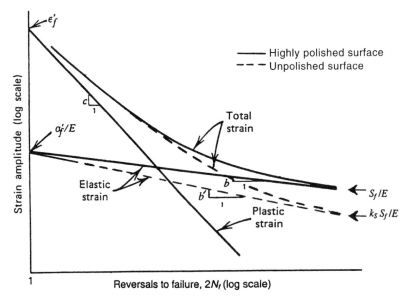

Figure 5.16 Schematic representation of surface finish effect on strain–life curves.

5.8 SUMMARY

Basic material mechanical properties such as strength and ductility can be obtained from simple monotonic tension tests. The stress–strain behavior obtained from such a test can, however, be quite different from that obtained under cyclic loading. Cyclic loading can cause hardening and/or softening of the material. Low-strength metals that behave in a ductile manner tend to cyclic harden, and those with high-strength, low-ductility behavior tend to cyclic soften. Changes in cyclic deformation behavior are more pronounced at the beginning of cyclic loading, as the material usually stabilizes gradually with continued cycling. The stable cyclic stress–strain curve can be obtained by connecting the tips of a family of stabilized hysteresis loops at different strain amplitudes. This curve can be represented by the following equation for many metals

$$\varepsilon_a = \frac{\Delta \varepsilon}{2} = \frac{\Delta \sigma}{2E} + \left(\frac{\Delta \sigma}{2K'}\right)^{1/n'} = \frac{\sigma_a}{E} + \left(\frac{\sigma_a}{K'}\right)^{1/n'} \tag{5.12}$$

where K' and n' are material cyclic deformation properties. The hysteresis loop curve (Eq. 5.13) can be obtained by doubling the size of the cyclic stress–strain curve.

Fatigue cracks usually nucleate from plastic straining in localized regions. Strain-controlled fatigue testing has become common, and the strain-based approach to fatigue problems is widely used at present. The total strain amplitude can be resolved into elastic and plastic strain components, each of which has been shown to be correlated with fatigue life in a linear fashion using a log-log scale for most metals. The total strain–life equation is therefore expressed as

$$\varepsilon_a = \frac{\Delta\varepsilon}{2} = \frac{\sigma_f'}{E}(2N_f)^b + \varepsilon_f'(2N_f)^c \qquad (5.14)$$

The strain–life approach is a comprehensive approach which can be applied for the treatment of both low-cycle and high-cycle fatigue. In the low-cycle region the plastic strain component is dominant, whereas in the high-cycle region the elastic strain component is dominant. Therefore, at large strains, better fatigue resistance depends more on ductility, while at smaller strains it depends more on strength.

Strain-controlled cycling with a mean strain results in a mean stress that usually relaxes at large strain amplitudes due to the presence of plastic deformation. A nonrelaxing mean stress can significantly affect the fatigue life, with a tensile mean stress having a detrimental effect and a compressive mean stress having a beneficial effect. Three popular strain–life models have been used to account for the mean stress effect. Other synergistic effects of loading, environment, and component or material processing can also influence strain–life behavior.

5.9 DOS AND DON'TS IN DESIGN

1. Do consider that inelastic stress–strain behavior under repeated loading is not the same as that determined under monotonic tensile or compressive loading. Under repeated loading the difference between materials is less than that under monotonic loading.

2. Don't ignore the role of material hardening or softening in cyclic loading applications. Using a monotonic stress–strain curve of a cyclic softening material in a cyclic loading application can significantly underestimate the extent of plastic deformation present.

3. Do consider the importance of material ductility on low-cycle or plastic strain dominated fatigue resistance and the importance of material strength on the high-cycle or elastic strain dominated fatigue resistance.

4. Do recognize that strain–life fatigue data of smooth uniaxial specimens are based on cycles to failure, where failure represents the formation of cracks on the order of 1 mm in depth, which may or may not have caused fracture.

5. Do recognize that mean strains generally affect fatigue resistance only if they produce a nonrelaxing mean stress. The greatest effect of mean stress is in the high-cycle fatigue regime.

6. Don't ignore the influence of synergistic effects of loading, environment, and component geometry and processing on strain–life fatigue behavior.

REFERENCES

1. L. Spangenburg, *The Fatigue of Metals Under Repeated Strains,* D. Van Nostrand Co., Princeton, NJ, 1876.

2. "Tension Testing of Metallic Materials," ASTM Standard E8 or E8M, *Annual Book of ASTM Standards,* Vol. 03.01, ASTM, West Conshohocken, PA, 1998.

3. P. W. Bridgman, "The Stress Distribution at the Neck of a Tension Specimen," *Trans. ASM,* Vol. 32, 1944, p. 553.

4. "Tensile Strain Hardening of Metallic Sheet Materials," ASTM Standard E646, *Annual Book of ASTM Standards,* Vol. 03.01, ASTM, West Conshohocken, PA, 1998, p. 555.

5. "Constant Amplitude Low Cycle Fatigue Testing," ASTM Standard E606, *Annual Book of ASTM Standards,* Vol. 03.01, ASTM, West Conshohocken, PA, 1998, p. 528.

6. "Technical Report on Low Cycle Fatigue Properties: Ferrous and Nonferrous Metals," SAE Standard J1099, SAE, Warrendale, PA, 1998.

7. J. Bauschinger, "On the Change of the Position of the Elastic Limit of Iron and Steel under Cyclic Variations of Stress," *Mitt. Mech.-Tech. Lab., Munich,* Vol. 13, No. 1, 1886.

8. J. Morrow, "Cyclic Plastic Strain Energy and Fatigue of Metals," *Internal Friction, Damping, and Cyclic Plasticity,* ASTM STP 378, ASTM, West Conshohocken, PA. 1965, p. 45.

9. M. Klensil and P. Lucas, *Fatigue of Metallic Materials,* 2nd ed., Elsevier Science, Amsterdam, 1992.

10. L. E. Tucker, "A Procedure for Designing Against Fatigue Failure of Notched Parts," SAE paper no. 720265, SAE, Warrendale, PA, 1972.

11. D. F. Socie and J. Morrow, "Review of Contemporary Approaches to Fatigue Damage Analysis," in *Risk and Failure Analysis for Improved Performance and Reliability,* J. J. Burke and V. Weiss, eds., Plenum Publishing, New York, 1980, p. 141.

12. R. W. Landgraf, J. Morrow, and T. Endo, "Determination of the Cyclic Stress-Strain Curve," *J. Mater.,* Vol. 4, No. 1, 1969, p. 176.

13. F. Ellyin, *Fatigue Damage, Crack Growth and Life Prediction,* Chapman and Hall, London, 1997.

14. O. H. Basquin, "The Exponential Law of Endurance Tests," *Proc. ASTM,* Vol. 10, Part 11, 1910, ASTM, West Conshohocken, PA, p. 625.

15. J. F. Tavernelli and L. F. Coffin, Jr., "Experimental Support for Generalized Equation Predicting Low Cycle Fatigue," *Trans. ASME, J. Basic Eng.,* Vol. 84, No. 4, 1962, p. 533.

16. S. S. Manson, discussion of reference 15, *Trans. ASME, J. Basic Eng.,* Vol. 84, No. 4, 1962, p. 537.

17. J. A. Graham, ed., *Fatigue Design Handbook, SAE,* Warrendale, PA, 1968.

18. M. L. Roessle, A. Fatemi, and A. K. Khosrovaneh, "Variation in Cyclic Deformation and Strain Controlled Fatigue Properties Using Different Curve Fitting and Measurement Techniques," SAE paper no. 1999-01-0364, SAE, Warrendale, PA, 1999.

19. "Standard Practice for Statistical Analysis of Linear or Linearized Stress-Life (*S–N*) and Strain-Life (*ε–N*) Fatigue Data," ASTM Standard E739, *Annual Book of ASTM Standards,* Vol. 03.01, ASTM, West Conshohocken, PA, 1998, p. 599.

20. U. Muralidharan and S. S. Manson, "Modified Universal Slopes Equation for Estimation of Fatigue Characteristics," *Trans. ASME, J. Eng. Mater. Tech.,* Vol. 110, 1988, p. 55.

21. M. L. Roessle and A. Fatemi, "Strain-Controlled Fatigue Properties of Steels and Some Simple Approximations," *Int. J. Fatigue,* Vol. 22, No. 6, 2000, p. 495.

22. R. W. Landgraf and R. A. Chernenkoff, "Residual Stress Effects on Fatigue of Surface Processed Steels," *Analytical and Experimental Methods for Residual Stress Effects in Fatigue,* ASTM STP 1004, R. L. Champoux, J. H. Underwood, and J. A. Kapp, eds., ASTM, West Conshohocken, PA, 1988, p. 1.

23. T. Wehner and A. Fatemi, "Effect of Mean Stress on Fatigue Behavior of a Hardened Carbon Steel," *Int. J. Fatigue,* Vol. 13, No. 3, 1991, p. 241.

24. M. Nihei, P. Heuler, C. Boller, and T. Seeger, "Evaluation of Mean Stress Effect on Fatigue Life by Use of Damage Parameters," *Int. J. Fatigue,* Vol. 8, 1986, p. 119.

25. S. S. Manson and G. R. Halford, "Practical Implementation of the Double Linear Damage Rule and Damage Curve Approach for Treating Cumulative Fatigue Damage, *Int. J. Fract.,* Vol. 17, No. 2, 1981, p. 169.

26. K. N. Smith, P. Watson, and T. H. Topper, "A Stress-Strain Function for the Fatigue of Metals," *J. Mater.,* Vol. 5, No. 4, 1970, p. 767.

27. N. E. Dowling, *Mechanical Behavior of Materials,* 2nd ed., Prentice-Hall, Upper Saddle River, NJ, 1998.

PROBLEMS

1. What difference exists between true stress, σ, and engineering stress, S, and between true strain, ε, and engineering strain, e, if engineering strain is 2 percent? Can this difference be neglected for engineering design purposes?

2. Derive Eqs. 5.5 and 5.6 for true stress and true strain, respectively.

3. The initial part of the load-displacement curve from the tension test of a 6061-T6 aluminum alloy is shown. A cylindrical specimen with an initial gage section diameter of 6.3 mm and an initial uniform gage section length of 12.7 mm was used. After fracture, which occurred at a load of 7.2 kN, the minimum diameter in the neck region, D_{min}, and the neck radius, R, were measured to be 4.2 mm and 3.3 mm, respectively. (*a*) Obtain and superimpose plots of engineering and true stress-strain curves. (*b*) Deter-

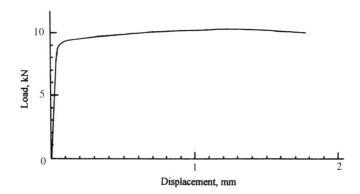

mine the following monotonic tensile properties: E, 0.2 percent offset S_y, S_u, σ_f, ε_f, and %RA. (c) Determine the strength coefficient, K, and the strain hardening exponent, n.

4. Using Table A.2, comment on any possible correlation between the monotonic and cyclic strain hardening exponents n and n'.

5. Why is it important to know if a material cyclic softens or hardens in fatigue?

6. Plot curves of the plastic strain range, $\Delta\varepsilon_p$, versus applied reversals, $2N$, for each of the materials in Fig. 5.5. How do these curves relate to Figs. 5.6b and 5.6c, and what significance can you ascertain from them?

7. Using data from Table A.2, construct the monotonic and cyclic stress-strain curves for RQC-100 hot-rolled steel for strains between 0 and 2 percent. Does RQC-100 cyclic harden or soften?

8. Using the data in Table A.2, determine the total, elastic, and plastic strain amplitudes for smooth uniaxial test specimens of Man-Ten and RQC-100 steels for life to (a) 10^3 cycles and (b) 10^5 cycles (note that one cycle equals two reversals).

9. If $\Delta\varepsilon/2 = 0.01$, find the number of cycles to failure for Man-Ten and RQC-100 steels using Eq. 5.14. How do these results compare with those obtained using Fig. 5.13?

10. Superimpose plots of elastic, plastic, and total strain versus life curves for normalized 1038 steel alloy in Table A.2. Determine the transition fatigue life, $2N_t$, and the total strain amplitude at this life for this alloy.

11. Derive the compatibility Eqs. 5.19 from interrelationships between Eqs. 5.11, 5.15, and 5.16.

12. Completely reversed, strain-controlled fatigue tests of a steel alloy with $E = 216$ GPa resulted in the following data:

Strain amplitude, ε_a	0.02	0.015	0.01	0.006	0.0035	0.002
Stress amplitude, σ_a (MPa) (at half-life)	650	625	555	480	395	330
Cycles to failure, N_f	200	350	1 100	4 600	26 000	560 000

(a) Determine the strain-life fatigue properties for this material. (b) Determine the cyclic strength coefficient, K', and the cyclic strain hardening exponent, n', from these data, plot the cyclic stress-strain curve, and compare the results with those from Eq. 5.19. (c) Estimate and plot the stable cyclic stress-strain hysteresis loop at a strain amplitude of 0.006.

13. Examine Table A.2 to see how ε_f' and ε_f compare and how σ_f' and σ_f compare by plotting ε_f' versus ε_f and σ_f' versus σ_f.

14. Repeat Problem 8 using Eq. 5.21 and comment on any differences from Problem 8 and their significance.

15. Derive Eq. 5.23 by using Eq. 5.15 in two different ways and equating the results as follows. Use Eq. 5.15 once by incorporating the mean stress effect in the fatigue strength coefficient as $\sigma_a = (\sigma_f' - \sigma_m) (2N_f)^b$. Then use this equation again but incorporate the mean stress effect in the fatigue life by changing the fatigue life without mean stress, N_f, to an equivalent, completely reversed life, N_f', in the presence of mean stress, $\sigma_a = \sigma_f' (2N_f')^b$. Obtain N_f' from this equality and substitute it in Eq. 5.14 with N_f' to obtain Eq. 5.23.

16. Derive Eq. 5.24, which is based on the assumption that $\sigma_{max} \varepsilon_a = $ constant for a given life.

17. If $\sigma_m = +\sigma_f'/4$, what approximate effect does this have on Man-Ten steel strain amplitude for 10^3 and 10^5 cycles using the three mean stress equations presented in Section 5.6? Repeat for $\sigma_m = -\sigma_f'/4$. Compare the results with those of Problem 8. Where does σ_m have its greatest influence?

18. Using the data in Table A.2, determine the completely reversed total, elastic, and plastic strain amplitudes for uniaxial test specimens of quenched and tempered 1090 steel alloy for life to 10^4 cycles with (a) a mirror-polished surface finish and (b) a machined surface finish.

19. An axially loaded component made of 1141 normalized steel is subjected to a total strain amplitude of 0.005 and a tensile mean stress of $\sigma_m = 0.25S_y'$. Determine the expected fatigue life if the component has a hot-rolled surface finish. Use the Smith-Watson-Topper parameter for mean stress correction.

CHAPTER 6

FUNDAMENTALS OF LEFM AND APPLICATIONS TO FATIGUE CRACK GROWTH

The presence of a crack can significantly reduce the life of a component or structure, as discussed in Chapter 3. This chapter introduces the concept and use of fracture mechanics for fatigue crack growth and damage-tolerant design. In order to make life estimations or predictions for fatigue crack growth and damage-tolerant design, several pieces of information are often needed. These include the following:

The stress intensity factor, K

The fracture toughness, K_c

The applicable fatigue crack growth rate expression

The initial crack size, a_i (or a_o)

The final or critical crack size, a_f (or a_c)

Information needed to attain, calculate, or estimate these values is presented in this chapter.

Perhaps of greatest significance in fatigue crack growth and damage-tolerant design in comparison to the fatigue design concepts introduced in Chapters 4 and 5 is the existence of a crack. Cracks can form due to fatigue, or they can exist as a consequence of manufacturing processes such as deep machining marks or voids in welds and metallurgical discontinuities such as foreign particle inclusions. Use of fracture mechanics in fatigue crack growth and damage-tolerant design requires knowledge of these preexisting cracks, either assumed or found using nondestructive flaw detection techniques, as reviewed in Chapter 2. Fracture mechanics has been used heavily in the aerospace, nuclear, and ship industries, with a recent extension to the ground vehicle industry.

This chapter provides an introduction to the important aspects of linear elastic fracture mechanics (LEFM) and shows how they are used to describe and predict fatigue crack growth life and fracture. It does not contain the mathematics used to develop the theory, but it does provide the background fracture mechanics concepts and numerical tools needed for fatigue design involving crack growth and fracture. Several excellent textbooks on fracture mechanics can be referred to for greater rigor and detail [1–4]. Fracture mechanics uses the stress intensity factor K, the energy release rate G, the crack opening displacement COD, and the J-integral J, along with critical or limiting values of these quantities. This chapter focuses predominantly on the use of the stress intensity factor K and its applications to fatigue crack growth. Concepts related to crack tip plasticity and LEFM limitations are also discussed.

6.1 LEFM CONCEPTS

Fracture mechanics is used to evaluate the strength of a structure or component in the presence of a crack or flaw. Its application to fatigue involves the crack growth process, covering the range from a detectable crack or flaw to final fracture. This process corresponds to the fatigue crack growth regions observed in Figs. 3.1 to 3.4. One of the common methods used to analyze this process is LEFM. This method is used to determine crack growth in materials under the basic assumption that material conditions are predominantly linear elastic during the fatigue process. For crack growth or fracture conditions that violate this basic assumption, elastic-plastic fracture mechanics approaches are used to describe the fatigue and fracture process. Before analyzing crack growth using LEFM concepts, it is necessary to define the basic crack surface displacement modes by which a crack can extend.

6.1.1 Loading Modes

Figure 6.1a shows three modes by which a crack can extend. Mode I is the opening mode, which is the most common, particularly in fatigue, and has received the greatest amount of investigation. Mode II is the in-plane shearing or sliding mode, and mode III is the tearing or antiplane shear mode. Mode I type crack extension is the most common, primarily because cracks tend to grow on the plane of maximum tensile stress. This is the typical mode of crack extension for uniaxial loaded components. Mixed-mode crack extension, i.e., more than one mode present, is associated more closely with the extension of microscopic fatigue cracks, as these cracks tend to grow on planes of maximum shear. An example of mixed-mode I and II crack extension is shown in Fig. 6.1b. This involves the axial loading of a crack inclined with respect to the x-axis. When $\beta = 90°$, K_I is maximum and K_{II} is 0. Even for conditions

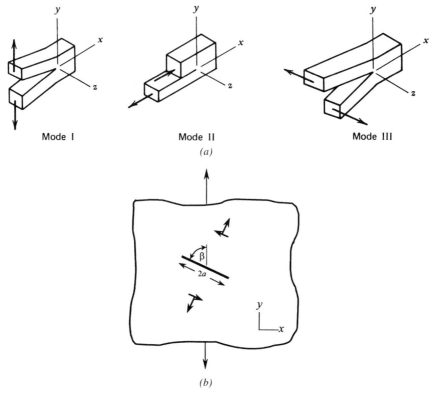

Figure 6.1   Modes of crack extension. (*a*) Three basic modes of loading. (*b*) Mixed-mode I and II loading due to a crack on an inclined plane.

in which β is large but less than 90°, the mode I contribution tends to dominate the crack-tip stress field [5]. Mode III is associated with a pure shear condition, typical of a round notched bar loaded in torsion. Combinations of these crack extension modes can also occur, particularly modes I and III, as shown in Fig. 3.5. Only mode I crack extension will be covered because of its most common occurrence and the fact that the other mode cracks (II and III) in combination with mode I cracks often turn into mode I cracks. K used without a mode subscript I, II, or III normally refers to mode I.

6.1.2   Stress Intensity Factor, K

The groundwork for the development of the stress intensity factor K was laid some 80 years ago by Griffith [6]. He showed that the product of the far field stress, the square root of the crack length, and material properties governed crack extension for brittle materials such as glass. This product was shown to be related to the energy release rate, G, which represents the elastic energy per unit crack surface area required for crack extension. Irwin [7] later made significant advances by applying Griffith's theory to metals with small plastic deformation at the crack tip and used the stress intensity factor K to quantify the crack tip driving force. Using Griffith's energy approach, Irwin showed that

$$G = \frac{K^2}{E} \quad \text{for plane stress and} \quad G = \frac{K^2}{E}(1 - \nu^2) \quad \text{for plane strain} \quad (6.1)$$

From these pioneering works came the widely accepted honor of being called the "early father" and the "modern father" of fracture mechanics for Griffith and Irwin, respectively.

Consider a through-thickness sharp crack in a linear elastic isotropic body subjected to mode I loading. Such a crack is shown schematically in Fig. 6.2. An arbitrary stress element in the vicinity of the crack tip with coordinates r and θ relative to the crack tip and crack plane is also shown. Using the mathematical theory of linear elasticity and the Westergaard stress function in complex form, the stress field at any point near the crack tip can be described

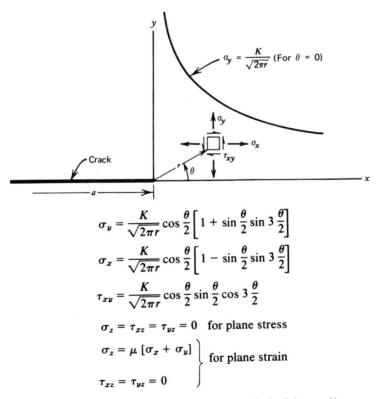

$$\sigma_y = \frac{K}{\sqrt{2\pi r}} \cos \frac{\theta}{2} \left[1 + \sin \frac{\theta}{2} \sin 3 \frac{\theta}{2} \right]$$

$$\sigma_x = \frac{K}{\sqrt{2\pi r}} \cos \frac{\theta}{2} \left[1 - \sin \frac{\theta}{2} \sin 3 \frac{\theta}{2} \right]$$

$$\tau_{xy} = \frac{K}{\sqrt{2\pi r}} \cos \frac{\theta}{2} \sin \frac{\theta}{2} \cos 3 \frac{\theta}{2}$$

$$\sigma_z = \tau_{xz} = \tau_{yz} = 0 \quad \text{for plane stress}$$

$$\left. \begin{array}{l} \sigma_z = \mu \left[\sigma_x + \sigma_y \right] \\[6pt] \tau_{xz} = \tau_{yz} = 0 \end{array} \right\} \quad \text{for plane strain}$$

Figure 6.2   Elastic stresses near the crack tip ($r/a \ll 1$).

[8]. These stresses are given in Fig. 6.2. Higher order terms exist, but these are negligible in the vicinity of the crack tip. It should be noted that by definition the normal and shear stresses involving the z direction (perpendicular to the x–y plane) are zero for plane stress, while the normal and shear strains (and shear stresses) involving the z direction are zero for plane strain.

Figure 6.2 shows that elastic normal and elastic shear stresses in the vicinity of the crack tip are dependent on r, θ, and K only. The magnitudes of these stresses at a given point are thus dependent entirely on K. For this reason, K is called a "stress field parameter" or "stress intensity factor." K is not to be confused with the elastic stress concentration factor K_t, which is the ratio of the maximum stress at a notch to the average or nominal stress at the notch. The value of the stress intensity factor, K, depends on the loading, crack shape, mode of crack displacement, and, finally, the component, specimen, or structure configuration.

The elastic stress distribution in the y direction for $\theta = 0$ is shown in Fig. 6.2. As r approaches zero, the stress at the crack tip approaches infinity, and thus a stress singularity exists at $r = 0$. Since infinite stresses cannot exist in a physical body, the elastic solution must be modified to account for some crack tip plasticity. If, however, the plastic zone radius, r_y, at the crack tip is small relative to local geometry, little or no modification to the stress intensity factor, K, is needed. Thus, an important restriction to the use of linear elastic fracture mechanics is that the plastic zone size at the crack tip must be small relative to the geometrical dimensions of the specimen or part. Crack tip plasticity and LEFM limitations are discussed in Section 6.2. The stress intensity factor, K, is the fundamental parameter of LEFM, and the use and development of various K solutions will now be presented.

6.1.3 *K* Expressions for Common Cracked Members

Values of K for various loadings and configurations can be calculated using the theory of elasticity involving both analytical and computational calculations along with experimental methods. When the crack is small compared to other dimensions of the body or component, the crack is viewed as being contained within an infinite body. Thus, the most common reference value of K is for a through-thickness, two-dimensional center crack of length $2a$ in an infinite sheet subjected to a uniform tensile stress S. For the infinite sheet, K is

$$K = S\sqrt{\pi a} \approx 1.77 S\sqrt{a} \qquad (6.2a)$$

Units of K are MPa$\sqrt{\text{m}}$ and ksi$\sqrt{\text{in.}}$, where 1 MPa$\sqrt{\text{m}} = 0.91$ ksi$\sqrt{\text{in.}}$ (1 ksi$\sqrt{\text{in.}} \approx 1.1$ MPa$\sqrt{\text{m}}$).

The stress intensity factor for other crack geometries, configurations, and loadings are usually modifications of Eq. 6.2a, such that

$$K = S\sqrt{\pi a}\ \alpha \quad \text{or} \quad S\sqrt{\pi a}\ f\left(\frac{a}{w}\right) \quad \text{or} \quad S\sqrt{a}\ Y \qquad (6.2b)$$

where α, $f(a/w)$, and Y are dimensionless parameters, w is a width dimension, and S is the nominal stress assuming that the crack did not exist. For central cracks, the crack length is taken as $2a$ (or $2c$), and for edge cracks the crack length used is just a (or c). Opening mode I stress intensity factor expressions in the form of dimensionless curves for several common configurations of thickness B are given in Fig. 6.3 [1,9]. In each case it is evident that K depends on the crack length to width ratio, a/w. Note that in Fig. 6.3a, as $a/w \rightarrow 0$, K approaches an infinite body solution. These tabulations were obtained from analytical or numerical solutions, often in the form of polynomials. The actual mathematical expression is of greater importance in fatigue crack growth since numerical integration is usually required. The mathematical expressions for K with configurations of Fig. 6.3 are given in Table 6.1. The term $\sqrt{\sec(\pi a/w)}$ shown for a center-cracked plate in tension in Table 6.1 is often used in many finite-width stress intensity factor solutions. For the single or double edge crack in a semi-infinite plate ($a/w \rightarrow 0$)

$$K = 1.12\ S\sqrt{\pi a} = 1.99S\sqrt{a} \approx 2S\sqrt{a} \qquad (6.3)$$

where 1.12 is the free edge correction. Additional stress intensity factor expressions for all three modes, K_I, K_{II}, and K_{III}, can be found in [1–5,8–16]. Superposition of K expressions can also be used for each separate mode. This is discussed in Section 6.1.4. It should be noted that t and B are often used interchangeably for thickness in fracture mechanics, and this is done in this book.

Elliptical-shaped cracks approximate many cracks found in real engineering components and structures and have received widespread analytical, computational, and experimental analysis. Common circular and elliptical embedded and surface cracks are shown in Fig. 6.4. The general reference specimen is the embedded elliptical crack in an infinite body subjected to uniform tension S perpendicular to the crack plane at infinity (Fig. 6.4c). For this embedded configuration

$$K = \frac{S\sqrt{\pi a}}{\Phi}\left[\sin^2\beta + \left(\frac{a}{c}\right)^2\cos^2\beta\right]^{1/4} \qquad (6.4a)$$

where

$$\Phi = \int_0^{\pi/2}\left[1 - \left(1 - \frac{a^2}{c^2}\right)\sin^2\phi\right]^{1/2}d\phi \qquad (6.4b)$$

Here β is the angle shown in Fig. 6.4g, $2a$ is the minor axis, and $2c$ is the major axis. The term Φ is the complete elliptical integral of the second kind and depends on the crack aspect ratio a/c. Values of Φ are given in Fig. 6.4h, where it is seen that Φ varies from 1.0 to 1.571 for a/c ranging from 0 (very

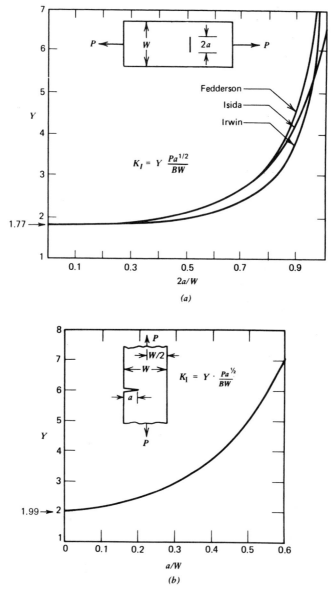

Figure 6.3 K_I for common configurations [1,9] (reprinted by permission of Noordhoff International Publishing Co. and the American Society for Testing and Materials). (*a*) Center-cracked plate in tension [1]. (*b*) Single-edge crack in tension [9]. (*c*) Double-edge crack in tension [9]. (*d*) Single-edge crack in bending [9].

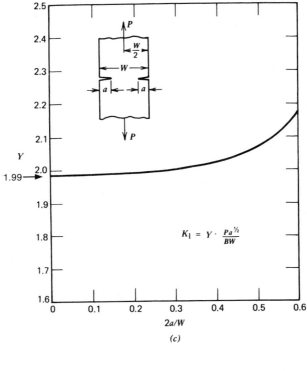

$$K_1 = Y \cdot \frac{P a^{1/2}}{BW}$$

$2a/W$

(c)

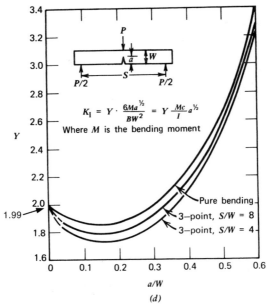

$$K_1 = Y \cdot \frac{6M a^{1/2}}{BW^2} = Y \frac{Mc}{I} a^{1/2}$$

Where M is the bending moment

Pure bending

3—point, $S/W = 8$

3—point, $S/W = 4$

a/W

(d)

Figure 6.3 (*Continued*).

TABLE 6.1 K Expressions for Fig. 6.3

a. Center-cracked plate in tension (for $0 < 2a/w < 0.95$)

Fedderson: $K_I = S\sqrt{\pi a} \left[\sec\left(\dfrac{\pi a}{w}\right) \right]^{1/2}$

Irwin: $K_I = S\sqrt{\pi a} \left[\dfrac{w}{\pi a} \tan\left(\dfrac{\pi a}{w}\right) \right]^{1/2}$

where $S = P/Bw$

b. Single-edge crack in tension (for $0 < a/w < 0.95$)

$K_I = S\sqrt{a} \left[1.99 - 0.41\left(\dfrac{a}{w}\right) + 18.7\left(\dfrac{a}{w}\right)^2 - 38.48\left(\dfrac{a}{w}\right)^3 + 53.85\left(\dfrac{a}{w}\right)^4 \right]$

where $1.12\sqrt{\pi} = 1.99$ $S = P/Bw$

c. Double-edge crack in tension (for $0 < 2a/w < 0.95$)

$K_I = S\sqrt{a} \left[1.98 + 0.36\left(\dfrac{2a}{w}\right) - 2.12\left(\dfrac{2a}{w}\right)^2 + 3.42\left(\dfrac{2a}{w}\right)^3 \right]$

where $S = P/Bw$

d. Single-edge crack in pure bending of a beam (for $0 < a/w < 1$)

$K_I = S\sqrt{a} \left[1.99 - 2.47\left(\dfrac{a}{w}\right) + 12.97\left(\dfrac{a}{w}\right)^2 - 23.17\left(\dfrac{a}{w}\right)^3 + 24.8\left(\dfrac{a}{w}\right)^4 \right]$

where $S = Mc/I = 6M/Bw^2$

---

shallow ellipse) to 1.0 (circle). K varies along the elliptical crack tip according to the trigonometric expression involving the angle β (Eq. 6.4a). The maximum value of K for the embedded elliptical crack exists at the minor axis, and the minimum value is at the major axis. Since fatigue crack growth depends principally on K, the embedded elliptical crack subjected to uniform tension tends to grow to a circle with a uniform K at all points on the crack tip perimeter. For the circular embedded crack in an infinite solid, Eq. 6.4a with $a/c = 1$ and $\Phi = \pi/2$ results in

$$K = S\sqrt{\pi a}\left(\dfrac{2}{\pi}\right) = 2S\sqrt{\dfrac{a}{\pi}} \approx 1.13S\sqrt{a} \qquad (6.5)$$

Surface elliptical cracks tend to grow to other elliptical shapes because of the free surface effect. The surface semicircular or semielliptical crack in a finite thickness solid (Fig. 6.4d) and the quarter-elliptical corner crack (Figs. 6.4e and 6.4f) are very common in fatigue and are more complex than the

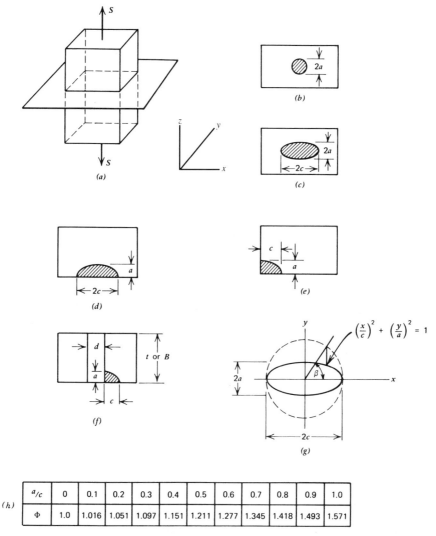

Figure 6.4 Elliptical and circular cracks. Cracks are shown in the $x-y$ plane. (*a*) Tensile loading and crack plane. (*b*) Embedded circular crack. (*c*) Embedded elliptical crack. (*d*) Surface semielliptical crack. (*e*) Quarter-elliptical corner crack. (*f*) Quarter-elliptical corner crack emanating from a hole. (*g*) Elliptical crack parameters. (*h*) values of Φ.

embedded crack in an infinite solid. K at the surface intersection and around the crack tip perimeter has been estimated using numerical methods [13–16] and three-dimensional photoelasticity [17]. References 1 to 5 and 10 to 12 also contain K estimations for these three-dimensional cracks. Expressions developed by Newman and Raju [15] and those summarized in [10–12,16] are perhaps the most widely used for cracks in three-dimensional finite bodies. In addition, extensive solutions exist from computer codes developed over the past several years. A general expression for the mode I semielliptical surface crack in a finite-thickness plate (Fig. 6.4d) is

$$ K = \frac{S\sqrt{\pi a}}{\Phi} M_f M_b \left[\sin^2 \beta + \left(\frac{a}{c}\right)^2 \cos^2 \beta \right]^{1/4} \tag{6.6} $$

where M_f is a front face correction factor and M_b is a back face correction factor. M_f and M_b are functions of β. For this crack in a plate of thickness t, K at the deepest point is approximated as

$$ K \approx \frac{1.12 S\sqrt{\pi a}}{\Phi} \sqrt{\sec(\pi a/2t)} \tag{6.7} $$

where $M_f \approx 1.12$ is analogous to the free edge correction of the single-edge crack and $\sqrt{\sec(\pi a/2t)}$ is the finite-thickness correction factor. As $a/t \to 0$, the finite-thickness correction factor approaches 1. For the quarter-circular corner crack ($a/c = 1$) in an infinite solid in Fig. 6.4e, with two free edges, K is approximated as

$$ K \approx \frac{(1.12)^2 S\sqrt{\pi a}}{\Phi} = (1.12)^2 \, 2S\sqrt{\frac{a}{\pi}} \approx 1.41 S\sqrt{a} \tag{6.8} $$

These simple approximations agree quite well with more complex calculations made using finite-element or other computational analyses that are now common methods of determining K.

6.1.4 Superposition for Combined Mode I Loading

The principle of superposition can be used to determine stress intensity factor solutions when a member is subjected to combined loading conditions. By resolving the loading into components, the resultant stress intensity factor solution can be determined by first finding the stress intensity factor solution caused by each load component acting separately on the member. The resultant stress intensity factor solution can then be determined by algebraically adding the stress intensity factor solution contributions caused by each of the load components. For example, for an eccentrically loaded member that experiences both an axial component and a bending component, the resultant stress intensity solution takes on the form

$$K_I = K_{I(\text{axial})} + K_{I(\text{bending})} \geq 0 \qquad (6.9)$$

It should be noted that $K_I \neq (S_{\text{axial}} + S_{\text{bending}})\sqrt{\pi a}\ \alpha$ as the geometry factor(s) Y, $f(a/w)$, or α, for axial and bending are different. Also, K values of different modes, i.e., I, II, and III, cannot be added together. Mixed-mode crack growth is discussed in Section 10.5.

6.2 CRACK TIP PLASTIC ZONE

Whether fracture occurs in a ductile or brittle manner, or a fatigue crack grows under cyclic loading, the local plasticity at the crack tip controls both fracture and crack growth. It is possible to calculate a plastic zone size at the crack tip as a function of the stress intensity factor and yield strength by using the stress field equations in Fig. 6.2 and the von Mises or maximum shear stress yield criteria. The resultant monotonic plastic zone shape for mode I using the von Mises criterion is shown schematically in Fig. 6.5. For plane stress conditions, a much larger plastic zone exists compared to that of plane

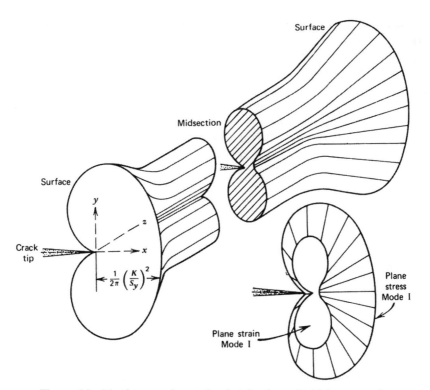

Figure 6.5 Plastic zone size at the tip of a through-thickness crack.

strain conditions, as indicated in the lower right corner. This is due to σ_z having a different value for plane stress than for plane strain. The normal stress component, σ_z, for plane strain conditions is tensile and therefore restricts plastic flow, resulting in a smaller plastic zone compared to plane stress, where $\sigma_z = 0$. Now let us assume that a through-thickness crack exists in a thick plate. The plate's free surfaces have $\sigma_z = \tau_{xz} = \tau_{yz} = 0$ and therefore the free surfaces must be in a plane stress condition. However, the interior region of the plate near the crack tip is closer to plane strain conditions as a result of elastic constraint away from the crack. Thus, the size of the plastic zone along the crack tip varies similarly to that shown schematically in Fig. 6.5. The actual stress–strain distribution within the plastic zone is difficult to obtain; however, this is not significant or necessary for the use of LEFM in design. The schematic in Fig. 6.5 is just one of several models used to describe the plastic zone shape. Other models assume the plastic zone shape to be circular, a strip extending beyond the crack tip, or a butterfly shape. Most plastic zone models were developed using classical yield criteria to determine boundaries where the material begins to yield. Two such models for approximating the plastic zone size are given below.

By substituting the yield strength, S_y, for σ_y, a distance (r) ahead of the crack tip $(\theta = 0)$ can be determined from the equation for σ_y in Fig. 6.2 that identifies the plastic zone boundary. This results in the plastic zone size expression shown in Fig. 6.5. Because of plastic relaxation and redistribution of the stress field in the plastic zone, the actual plane stress plastic zone size, as estimated by Irwin [7], is approximately twice the value shown in Fig. 6.5. Irwin argued that the plasticity at the crack tip causes the crack to behave as if it were longer than its true physical size, and that the stress distribution could not simply terminate above the yield strength, S_y. Thus, the stress distribution for σ_y shown in Fig. 6.2 must shift to the right to accommodate the plastic deformation and satisfy equilibrium conditions. Thus, under monotonic loading, the plane stress plastic zone size, $2r_y$, at the crack tip, in the plane of the crack is

$$2r_y = 2\left[\frac{1}{2\pi}\left(\frac{K}{S_y}\right)^2\right] = \frac{1}{\pi}\left(\frac{K}{S_y}\right)^2 \tag{6.10}$$

The plane strain plastic zone size in the plane of the crack is usually taken as one-third of the plane stress value; thus

$$2r_y \cong \frac{1}{3\pi}\left(\frac{K}{S_y}\right)^2 \tag{6.11}$$

The criteria for deciding between plane stress and plane strain are discussed in Section 6.3. The value r_y is often called the "plastic zone radius."

Another model describing the crack tip plastic zone that has been widely used was formulated by Dugdale [18] for conditions of plane stress. Dugdale modeled the plastic regions extending beyond the physical crack length, $2c$, as narrow strips having a distance R, as shown in Fig. 6.6. Based on his model, an internal stress, σ_y, acting over the distance R recloses the crack to a length of $2c$, where this internal stress equals the material yield strength, S_y. Dugdale showed the plastic zone distance, R, to be

$$R \cong \frac{\pi^2 c}{8}\left(\frac{S}{S_y}\right)^2 \tag{6.12}$$

Rearranging Eq. 6.12 in terms of K and S_y gives the equation

$$R \cong \frac{\pi}{8}\left(\frac{K}{S_y}\right)^2 \tag{6.13}$$

Note that agreement between the plane stress plastic zone size in the plane of the crack from Eq. 6.10 and that from Eq. 6.13 is within 20 percent.

If the plastic zone radius, r_y, at the crack tip is small relative to the local geometry, little or no modification of the stress intensity factor, K, is needed. An approximate suggested restriction for the use of K under monotonic loading without significant violation of LEFM principles is that $r_y \leq a/8$. This general restriction can often be relaxed to approximately $r_y \leq a/4$ under

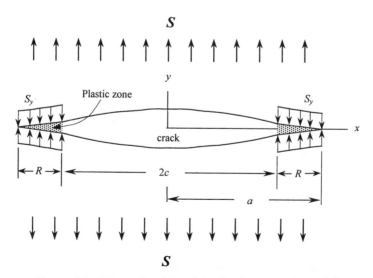

Figure 6.6 Schematic of Dugdale plastic zone strip model.

cyclic loading conditions in which the cyclic plastic zone is smaller than the monotonic plastic zone, as discussed in Section 6.6. Other suggested restrictions include $r_y \leq t/8$ and $(w-a)/8$ where t is the thickness and $(w-a)$ is the uncracked ligament along the plane of the crack. Thus, an important restriction to the use of LEFM is that the plastic zone size at the crack tip must be small relative to the crack length as well as the geometrical dimensions of the specimen or part. A definite limiting condition for LEFM is that net nominal stress in the crack plane must be less than the yield strength. In actual usage, the net nominal stress in the crack plane should often be less than 80 percent of the yield strength [19]. For example, when the nominal stress is equal to 80 percent of the yield strength, a plasticity correction of nearly 20 percent is required for the stress intensity factor, K. In this case, elastic-plastic fracture mechanics may be needed. Plasticity corrections of LEFM as well as elastic-plastic fracture mechanics are discussed in Section 6.9. The plastic zone for cyclic loading is different from that of monotonic loading and is discussed in Section 6.6.

6.3   FRACTURE TOUGHNESS—K_c, K_{Ic}

Critical values of K refer to the condition in which a crack extends in a rapid (unstable) manner without an increase in load or applied energy. Critical values of K are denoted with the subscript c as follows:

$$K_c = S_c \sqrt{\pi a_c}\, f\left(\frac{a_c}{w}\right) \tag{6.14}$$

where S_c is the applied nominal stress at crack instability and a_c is the crack length at instability. K_c is called "fracture toughness" and depends on the material, strain rate, environment (i.e., temperature), thickness, and, to a lesser extent, crack length. Equation 6.14 provides a quantitative design parameter to prevent brittle-type fracture involving applied stress, material selection, and crack size. Fracture toughness, K_c, can be obtained for materials that exhibit limited plasticity from tests using specimens with fatigue cracks with known K expressions such as those in Fig. 6.3 and Table 6.1. Thus, K_c represents the critical value of the stress intensity factor K for a given load, crack length, and geometry required to cause fracture. K_c also typically represents the stress intensity factor at the last cycle of a fatigue fracture. Thus, it can be used to obtain critical crack sizes for fracture under cyclic loading. This is discussed further in Section 6.4.

The general relationship between fracture toughness, K_c, and thickness is shown in Fig. 6.7. The appearance of the fracture surface accompanying the different thicknesses is also shown schematically for single-edge notch specimens. The beach markings at the initial crack tip represent fatigue precracking

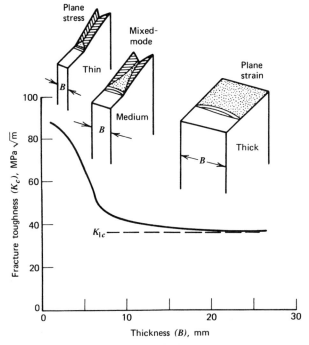

Figure 6.7   Effect of specimen thickness on fracture toughness.

at a low cyclic stress intensity factor range to ensure a sharp crack tip. The fracture toughness values would be higher for dull, or notch type, crack fronts. It is seen that thin parts have a high K_c value accompanied by appreciable "shear lips" or slant fracture. Remember from Chapter 3 that this type of fracture is a "high" energy fracture. As the thickness is increased, the percentage of shear lips or slant fracture decreases, as does K_c. This type of fracture appearance is called "mixed-mode," implying both slant and flat fractures. For thick parts, essentially the entire fracture surface is flat and K_c approaches an asymptotic minimum value. Any further increase in thickness does not decrease the fracture toughness, nor does it alter the fracture surface appearance. The minimum value of fracture toughness is called the "plane strain fracture toughness," K_{Ic}. The subscript I refers to the fact that these fractures occur almost entirely by the mode I crack opening. The term "plane strain" is incorporated here since flat fractures best approach a true plane strain constraint throughout most of the crack tip region. For thin sections where appreciable shear lips occur, the crack tip region most closely experiences a plane stress situation. Thus, plastic zone sizes at fracture are much larger in thin parts than in thick parts. Plane strain fracture toughness K_{Ic} is considered a true material property because it is independent of thickness. However, in

TABLE 6.2   Approximate Thickness Required for Valid K_{Ic} Tests

Steel S_y, MPa (ksi)	Aluminum S_y, MPa (ksi)	Thickness mm (in.)
690 (100)	275 (40)	>76 (3)
1030 (150)	345 (50)	76 (3)
1380 (200)	448 (65)	45 (1 3/4)
1720 (250)	550 (80)	19 (3/4)
2070 (300)	620 (90)	6 (1/4)

order for a plane strain fracture toughness value to be considered valid, it is required that [20]

$$a \text{ and } t \geq 2.5 \left(\frac{K_{Ic}}{S_y} \right)^2 \tag{6.15}$$

Approximate thickness values required for steels and aluminums to obtain valid K_{Ic} values are given in Table 6.2. These thickness values were determined on the basis of a calculated ratio of yield strength to Young's modulus [20]. They are only approximate and are used as guidelines in choosing a first approximate thickness for K_{Ic} testing. A compact tension (C(T)) specimen, similar to that shown in Fig. 4.5i, with a thickness of 76 mm (3 in.) would require width and height dimensions approaching those of this book! Low-strength, ductile materials are subject to plane strain fracture at room temperature only if they are very thick. Therefore, most K_{Ic} data have been obtained for medium- and higher-strength materials or for lower-strength materials at low temperatures.

A general trend of K_{Ic} at room temperature, as a function of yield strength for three major base alloys—aluminum, titanium, and steel—is shown in Fig. 6.8 [21]. As can be seen, a wide range of K_{Ic} values can be obtained for a given base alloy. However, a higher yield or ultimate strength generally produces a decrease in K_{Ic} for all materials, and thus a greater susceptibility to catastrophic fracture. This is an important conclusion that too many engineers overlook. Figure 6.8 does not provide K_{Ic} data for all levels of yield strength because of the large thickness required for low-strength materials. The use of LEFM is not suitable for these low-strength materials under monotonic loading because of the extensive plastic zones occurring at the crack tip, as discussed in Section 6.9. Exceptions occur, however, at very low temperatures, in the presence of corrosive environments, and under fatigue conditions, where only small-scale yielding may occur near the crack tip.

Appreciable K_{Ic} data for specific materials can be found in reference 22, and some representative K_{Ic} values are given in Table A.3. Much variability, however, occurs for a given yield strength, depending on the type and quality of the material. This is best illustrated for steels in Fig. 6.8, where vacuum-

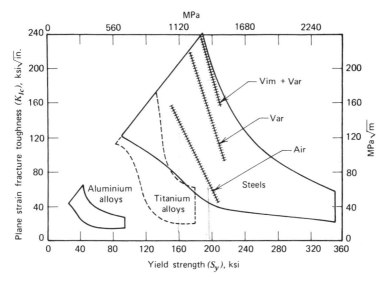

Figure 6.8 Locus of plane strain fracture toughness versus yield strength [21].

induction melting (VIM) plus vacuum-arc melting (VAR) of the base material produce higher K_{Ic} values for a given yield strength than just single vacuum-arc remelting or air melting of steels (AIR). Thus, low-impurity materials provide better fracture toughness. Fracture toughness K_{Ic} or K_c can be very sensitive to metallurgical conditions such as grain orientation, chemistry, and microstructure. Fracture toughness can be enhanced by systematically controlling or varying these variables. The path of the crack during fracture is usually controlled by the microstructure, the dominant constituents being second phases, precipitates, grain size, and texturing or fibering due to mechanical working. Fracture toughness can also be influenced by the effects of crack tip geometry such as blunting, irregular crack front, a change in the mode of crack extension, and branching. The mechanisms of fracture, such as cleavage or microvoid coalescence, can also dictate fracture toughness. For more detailed descriptions of the various metal-toughening mechanisms, the reader is referred to references 1, 2, 4, and 5.

Fracture toughness K_{Ic} or K_c of metals is also dependent on temperature, strain rate, and corrosive environment. Figure 6.9 shows typical temperature results for a low-alloy nuclear pressure vessel steel [23]. As the temperature decreases, K_c usually decreases and yield strength increases. Thus, even though unnotched or uncracked tensile strength increases with decreasing temperature, the flaw or crack resistance can be drastically reduced. An increased strain rate tends to cause changes in K_c similar to those resulting from a decrease in temperature. That is, higher strain rates often produce lower fracture toughness and hence greater crack sensitivity. Corrosive environmental influence on short-term fracture toughness may show small or large changes.

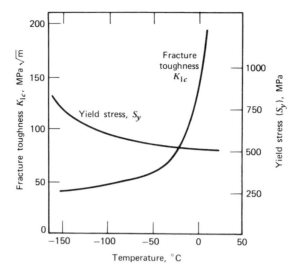

Figure 6.9 Variation of K_{Ic} with temperature for low alloy nuclear pressure vessel steel A533B [23] (reprinted with permission of E. T. Wessel).

However, corrosive environments can cause appreciable decreases in crack tolerance under long-term stress corrosion cracking and fatigue crack growth conditions. Environmental considerations are discussed in Chapter 11.

A general schematic drawing showing how changes in fracture toughness influence the relationship between allowable nominal stress and allowable crack size is presented in Fig. 6.10, which is a plot of

$$S = \frac{K_{Ic}}{\sqrt{\pi a}} \tag{6.16}$$

This equation is obtained from equating the stress intensity factor K for a through-thickness center crack in a wide plate to the fracture toughness K_{Ic}. The allowable stress in the presence of a given crack size is directly proportional to the fracture toughness, while the allowable crack size for a given stress is proportional to the square of the fracture toughness. Thus, increasing K_{Ic} has a much larger influence on allowable crack size than on allowable stress. For monotonic loading of components containing cracks, higher fracture toughness results in larger allowable crack sizes or larger allowable stresses at fracture, as shown in Fig. 6.10. For a K_{Ic} value corresponding to the lower K_{Ic} curve, the dashed region below the lower K_{Ic} curve in Fig. 6.10 identifies the "no fracture" region for a combined stress and crack length, while the region above this curve identifies the "fracture" region for monotonic loading conditions. From Eq. 6.16, a large stress is necessary to cause fracture in the presence of a very small crack. However, remember that a limitation of LEFM

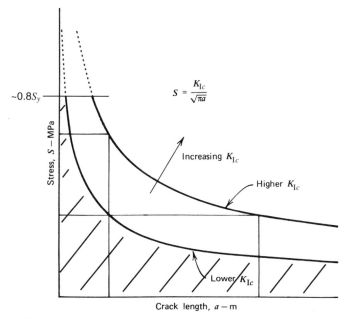

Figure 6.10 Influence of fracture toughness on allowable stress or crack size.

is that the allowable stress should not exceed approximately 80 percent of the yield strength. Thus, $0.8S_y$, as identified in Fig. 6.10, is the upper bound limit of allowable stress using this LEFM approach.

As an example, consider a through-thickness center crack in a wide plate. Let the material under consideration be 4340 tempered steel such that $K_{Ic} = 60$ MPa$\sqrt{m}$ (54.6 ksi$\sqrt{in}$.). Assume that the allowable design stress is 475 MPa (69 ksi). Using Eqs. 6.2a or 6.16 for a through-thickness center crack in a wide plate, it follows that the material could tolerate a crack size, $2a$, of 10.2 mm (0.40 in.). However, if the allowable design stress were reduced to 320 MPa (46 ksi), the allowable crack size, $2a$, would be approximately 22.5 mm (0.89 in.). In order to attain the 22.5 mm (0.89 in.) crack dimension with $S = 475$ MPa (69 ksi), K_{Ic} would have to be 89 MPa$\sqrt{m}$ (81 ksi$\sqrt{in}$.). Thus, the material exhibiting higher fracture toughness has the highest residual strength for a given crack or flaw size or can tolerate larger cracks for a given stress.

Development of fracture toughness testing began over 40 years ago to characterize the resistance to unstable crack growth. Numerous specimen configurations and test procedures were considered. The ASTM E-24 Fracture Committee was responsible for much of the decision making regarding development of standardized fracture toughness test methods. For conditions in which the plane strain fracture toughness K_{Ic} is desired, ASTM Standard E399 contains a detailed description of the specimen geometry, experimental procedure, and data collection and analysis techniques used to determine valid K_{Ic} values [20].

When a specimen or component has a thickness less than that required for plane strain conditions, it will experience either mixed-mode or plane stress conditions, depending on its thickness. Under plane strain conditions, once a critical stress, S_c, is reached, unstable crack growth occurs and fracture is imminent. Under plane stress conditions in which the plastic zone size is greater than that for plane strain conditions, the crack may first extend by slow, stable crack growth prior to unstable fracture. Plane stress fracture toughness, K_c, for thin sheets in general is more difficult to quantify because its value is a function of crack size, specimen thickness, and specimen geometry. ASTM Standard E561 [24] provides a recommended practice for plane stress fracture toughness testing.

Since plane strain fracture toughness is considered a material property, its value may be available for a given material. In a design situation, however, the stress state may be plane stress in which K_c for the particular thickness is required but is often not available. Therefore in design, K_{Ic} is often used rather than K_c because of availability and because K_{Ic} is a more conservative value. However, there are several conditions in which the use of K_{Ic} rather than K_c would involve poor design practice due to inefficiency and cost. One example is the skin material on an aircraft. Therefore, use of K_{Ic} or K_c is dependent on the application and the safety critical aspects of the component or structure.

6.4   FATIGUE CRACK GROWTH, *da/dN–ΔK*

The strength of a component or structure can be significantly reduced by the presence of a crack or discontinuity. However, in most engineering cases, the initial crack or discontinuity size is not critical to cause catastrophic failure. More commonly, subcritical crack growth occurs from the existing crack or discontinuity until a critical crack size is reached, causing fracture. The most common type of subcritical crack growth is due to fatigue in the presence of an existing discontinuity or crack.

Figure 6.11 shows schematically three crack length versus applied cycle curves for three identical test specimens subjected to different repeated stress levels with $S_1 > S_2 > S_3$. All specimens contained the same initial crack length, a_o, and in each test the minimum stress was zero. We see that with higher stresses the crack growth rates that are represented by the slopes of the curves are higher at a given crack length, and the fatigue crack growth life (total number of applied cycles, N_f) is shorter. The crack lengths at fracture are shorter at the higher stress levels. For the given initial crack size, the life to fracture depends on the magnitude of the applied stress and the fracture resistance of the material. We must ask ourselves, how can fatigue crack growth data, such as those in Fig. 6.11, be used in fatigue design? The format

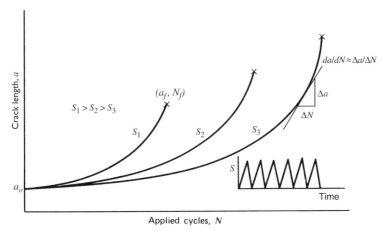

Figure 6.11   Fatigue crack length versus applied cycles. Fracture is indicated by the x.

of Fig. 6.11 is not applicable to fatigue design except under exactly the same conditions used to obtain the data. Applying LEFM concepts to the data in Fig. 6.11 will reduce the data to a format useful in fatigue design. This involves obtaining the fatigue crack growth rate, da/dN, versus the applied stress intensity factor range, ΔK, as shown schematically in Fig. 6.12. The fatigue crack growth rate, da/dN, is simply the slope of the a versus N curve at a given crack length or given number of cycles, as identified by da/dN ($\Delta a/\Delta N$) in Fig. 6.11. Fatigue crack growth rates, da/dN, can be obtained at consecutive positions along these curves as shown in Fig. 6.13 using graphical or numerical methods [25]. The corresponding applied stress intensity factor range, ΔK, is calculated knowing the crack length, a, the applied stress range, ΔS, and the stress intensity factor solution, K, for the part in question. ΔK is defined as

$$\Delta K_I = \Delta K = K_{max} - K_{min} = S_{max} \sqrt{\pi a}\, \alpha - S_{min} \sqrt{\pi a}\, \alpha$$
$$= (S_{max} - S_{min})\sqrt{\pi a}\, \alpha = \Delta S \sqrt{\pi a}\, \alpha \qquad (6.17)$$

Since the stress intensity factor $K = S\sqrt{\pi a}\, \alpha$ is undefined in compression, K_{min} is taken as zero if S_{min} is compressive. The correlation for constant amplitude loading is usually a log-log plot of the fatigue crack growth rate, da/dN, in m/cycle (in./cycle), versus the opening mode stress intensity factor range ΔK_I (or ΔK), in MPa$\sqrt{m}$ (ksi$\sqrt{in}$.). As ΔK primarily depends on ΔS, a, and geometry (for example, α), many models of the form

$$\frac{da}{dN} = f(\Delta S, a, \alpha) = f(\Delta K) \qquad (6.18)$$

have been proposed and developed. Once a versus N data are reduced to da/dN versus ΔK data, the sigmoidal curve generated is essentially independent of initial crack length unless the initial crack is small, as discussed in Section 6.8. Thus, the three *a* versus *N* curves shown in Fig. 6.11, once reduced to *da/dN* versus ΔK data, would fall on the same sigmoidal curve.

The elastic stress intensity factor is often applicable to fatigue crack growth even in low-strength, high-ductility material behavior because the *K* values necessary to cause fatigue crack growth are usually quite low. Thus, plastic zone sizes at the crack tip under fatigue loading are often small enough for even these materials to allow an LEFM approach. At very high crack growth rates, some difficulties can occur as a result of large plastic zone sizes, but this is often not a problem because very little fatigue life may be involved.

6.4.1   Sigmoidal *da/dN–ΔK* Curve

Many fatigue crack growth data have been obtained under constant load amplitude test conditions using sharp cracked specimens similar to those in Figs. 4.5*g* through 4.5*j*. Mode I fatigue crack growth has received the greatest attention because this is the predominant mode of macroscopic fatigue crack growth. K_{II} and K_{III} usually have only second order effects on both crack direction and crack growth rates. The typical log-log plot of *da/dN* versus ΔK shown schematically in Fig. 6.12 has a sigmoidal shape that can be divided

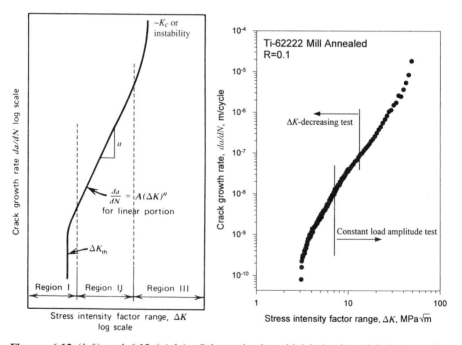

Figures 6.12 (*left*) and **6.13** (*right*)   Schematic sigmoidal behavior of fatigue crack growth rate versus ΔK. Fatigue crack growth rate data for Ti-62222 titanium alloy (courtesy of R. R. Stephens).

into three major regions. Region I is the near threshold region and indicates a threshold value, ΔK_{th}, below which there is no observable crack growth. This threshold occurs at crack growth rates on the order of 1×10^{-10} m/cycle ($\sim 4 \times 10^{-9}$ in./cycle) or less, as defined in ASTM Standard E647 [25]. Below ΔK_{th}, fatigue cracks are characterized as nonpropagating cracks. Microstructure, mean stress, frequency, and environment mainly control region I crack growth. Region II shows essentially a linear relationship between log *da/dN* and log ΔK, which corresponds to the formula

$$\frac{da}{dN} = A(\Delta K)^n \qquad (6.19)$$

first suggested by Paris, Gomez, and Anderson [26]. Here n is the slope of the line and A is the coefficient found by extending the straight line to $\Delta K = 1$ MPa$\sqrt{m}$ (or 1 ksi$\sqrt{in}$.). Region II (Paris region) fatigue crack growth corresponds to stable macroscopic crack growth that is typically controlled by the environment. Microstructure and mean stress have less influence on fatigue crack growth behavior in region II than in region I. In region III the fatigue crack growth rates are very high as they approach instability, and little fatigue crack growth life is involved. This region is controlled primarily by fracture toughness K_c or K_{Ic}, which in turn depends on the microstructure, mean stress, and environment.

For a given material and environment, the fatigue crack growth behavior shown in Fig. 6.12 is essentially the same for different specimens or components because the stress intensity factor range is the principal controlling factor in fatigue crack growth. This allows fatigue crack growth rate versus ΔK data obtained under constant amplitude conditions with simple specimen configurations to be used in design situations. If one knows the stress intensity factor expression, K, for a given component and loading, the fatigue crack growth life of the component can be obtained by integrating the sigmoidal curve between the limits of initial crack size and final crack size. This is illustrated by an example problem in Section 6.4.4. Equation 6.19 and other empirical relationships describing the fatigue crack growth behavior are often used in the integration procedure.

In many cases, integration of Eq. 6.19 by extrapolating to both regions I and III may be satisfactory, as it often gives conservative fatigue crack growth life values. Extrapolation of Eq. 6.19 to region I can be either conservative or nonconservative, while extrapolation to region III is nonconservative due to the nature of the sigmoidal *da/dN* versus ΔK curve. However, the fatigue crack growth life spent in region III is small compared to that in regions I and II; thus, integration of the Paris equation over the entire crack growth regime may be reasonable. Other factors may, however, need to be taken into consideration regarding this approach, such as mean stress effects and small crack growth behavior. These factors will be discussed in later sections.

The greatest usage of Fig. 6.12 and/or Fig. 6.13 has traditionally been in damage-tolerant design for many aerospace applications, as well as nuclear energy systems along with fractographic failure analysis. Fatigue crack growth rate behavior has also become important in material selection and comparative prototype designs in many other applications. However, as with all constant amplitude material fatigue properties, these data alone do not provide sufficient information on fatigue crack growth interaction or load sequence effects, which are discussed in Chapter 9.

6.4.2   Constant Amplitude Fatigue Crack Growth Test Methods

During the 1960s and early 1970s, fatigue crack growth testing involved essentially only regions II and III of the sigmoidal curve. However, since the middle to late 1970s and more recently with sophisticated computer software and control, region I crack growth has received significantly more attention. Standard test methods have been developed for performing constant amplitude fatigue crack growth tests, most notably ASTM Standard E647 [25]. Common test specimens are shown in Figs. 4.5*g* through 4.5*j*, although other types may be used. A typical region II and III fatigue crack growth experiment is performed under constant amplitude cyclic loading with R = constant. Generally, a fatigue precrack is formed at a sharp machined notch at a relatively low ΔK level to provide a sharp crack tip. Once the fatigue precrack has formed, extension of the growing crack is documented in terms of crack length and number of cycles until failure occurs. Crack growth is usually measured with optical, compliance, ultrasonic, eddy current, electrical potential, or acoustic emission techniques. The a versus N data have the appearance of one of the curves in Fig. 6.11. These data are then reduced to generate da/dN versus ΔK data. Data reduction techniques, as recommended in ASTM Standard E647 [25], include a secant (point to point) and incremental polynomial (successive points) methods. This procedure generates what is often referred to as "ΔK-increasing (constant load amplitude) data" and corresponds to regions II and III fatigue crack growth, where crack growth rates above 10^{-8} m/cycle (4×10^{-7} in./cycle) are typically generated.

Fatigue crack growth rates that correspond to near threshold conditions, $\sim 10^{-10}$ m/cycle, are very slow. For example, about 14 hours of continuous testing at a loading frequency of 50 Hz are required to grow a crack 0.25 mm (0.01 in.) at this crack growth rate. Experimentally, it is very difficult to begin a constant amplitude fatigue crack growth test corresponding to region I crack growth rates. To generate region I fatigue crack growth data, one must systematically shed the applied load as the crack grows, referred to as "ΔK-decreasing." Incremental load decreases cannot be too large, typically no greater than 10 percent, as the crack may inadvertently arrest due to crack tip/plasticity interaction. This can lead to fictitiously high threshold values. Details of the recommended load shedding procedure for threshold tests are summarized in ASTM Standard E647. The full range of the sigmoidal curve is often generated by first performing a constant load amplitude (ΔK-increasing) fatigue crack growth test that generates data typically above 10^{-8} m/cycle using one specimen, followed by a ΔK-decreasing test that generates lower crack growth rate data using a second specimen. A typical result is shown in Fig. 6.13 for a medium-strength titanium alloy. As a recommended testing procedure, the constant load amplitude and ΔK-decreasing tests are often started at specific ΔK levels such that there is some data overlap to confirm consistency in the experimental procedure, as identified in Fig. 6.13. Depending on the material and the specimen geometry, the full range of the sigmoidal curve can often be obtained using a single specimen by first load shedding to obtain region I fatigue crack growth rates. This is followed by constant amplitude loading that eventually leads to regions II and III fatigue crack growth rates. This procedure also provides an overlap in data, particularly in regions I and II.

6.4.3   *da/dN–ΔK* for *R* = 0

Conventional *S–N* or *ε–N* fatigue behavior discussed in Chapters 4 and 5 is usually referenced to the fully reversed stress or strain conditions ($R = -1$). Fatigue crack growth data, however, are usually referenced to the pulsating tension condition with $R = 0$ or approximately zero. This is based on the concept that during compression loading the crack is closed, and hence no stress intensity factor, *K,* can exist. The compression loads should thus have little or no influence on constant amplitude fatigue crack growth behavior. In general, this is fairly realistic, but under variable amplitude loading conditions, compression cycles can be important to fatigue crack growth, as seen in Chapter 9.

Many equations depicting fatigue crack growth rates above threshold levels have been formulated. More than 25 years ago, Hoeppner and Krupp [27] listed 33 empirical equations, and many more have been proposed since. However, the Paris equation (Eq. 6.19) is the most popular equation for *R* = 0 loading. Substantial constant amplitude fatigue crack growth data can be found in the *Damage Tolerant Design Handbook* [22]. Barsom [28] has evaluated Eq. 6.19 for a wide variety of steels varying in yield strength from 250 to 2070 MPa (36 to 300 ksi). He shows that the fatigue crack growth rate scatter band for a given ΔK, with many ferritic-pearlitic steels, varies by a factor of about 2. Partial results from Barsom [28] for ferritic-pearlitic steels are shown in Fig. 6.14*a*. He found a factor of 5 scatter band width for martensitic steels, shown in Fig. 6.14*b*. He suggested that conservative values of the upper boundaries of these scatter bands could be used in design situations if actual data could not be obtained.

Figure 6.14 indicates that in general the ferritic-pearlitic steels have better region II fatigue crack growth properties than the martensitic steels. The narrowness of the region II scatter bands for the different classifications of steels suggests that choosing a slightly different steel may not produce large

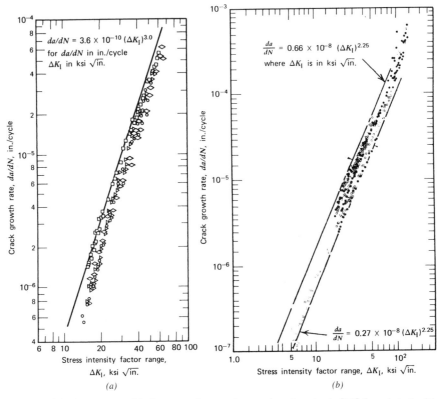

Figure 6.14 Summary of fatigue crack growth rate data for steels [28] (reprinted with permission of the American Society of Mechanical Engineers). (*a*) Ferritic-pearlitic steels. (*b*) Martensitic steels.

changes in constant amplitude region II fatigue crack growth life. However, this does not take into consideration behavior near or at threshold levels, which does show greater variations as a function of microstructure. An approximate schematic sigmoidal scatter band for steels with Barsom's scatter bands superimposed is shown in Fig. 6.15, which also includes austenitic stainless steels from Barsom [28]. It is clear that Barsom's results do not take into consideration complete crack growth rate behavior. Also, despite the numerous tests and materials used to obtain the results presented in Fig. 6.14, there are many exceptions. While values of the exponent *n* typically range between 2 and 5 for metals used in damage-tolerant applications, they have reached as high as 11, as found for A356-T6 cast aluminum alloy [29]. For different aluminum alloys, the width of the scatter band for a given ΔK in region II corresponds to a factor of about 10. In addition, similar analyses have been made for many other types of metal, including titanium and magnesium alloys [30–32]. Typical values of *n* and *A* for these metals can be found in Table 6.3. In general, the crack growth rate exponent *n* is higher for metals that behave in a more brittle

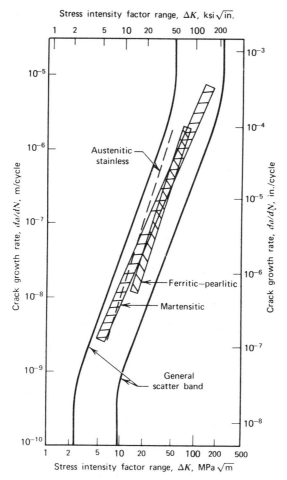

Figure 6.15 Superposition of Barsom's scatter bands on the general fatigue crack growth scatter band for steels.

manner. It should be noted that the results from Barsom's study presented in Table 6.3 are conservative values for n and A, while others are typical average values for fatigue crack growth behavior. The values listed for metals other than steels are provided for general comparison between metal types and should not be used for other metals of the same type, i.e., other aluminum, titanium, or magnesium alloys.

Frequency, wave shape, and thickness effects on constant amplitude fatigue crack growth rates are secondary compared to environmental effects such as corrosion and temperature. They can have influence, but this often is less than that due to different environmental effects. The largest influence comes from variable amplitude loading in various environments. The influence of thickness

TABLE 6.3 **Approximate Region II Fatigue Crack Growth Rate Properties for the Paris Equation (Eq. 6.19) for Various Metals**

Material	Slope, n	Intercept, A (m/cycle)	Intercept, A (in./cycle)
Ferritic-pearlitic steels*	3.0	6.9×10^{-12}	3.6×10^{-10}
Martensitic steels*	2.25	1.35×10^{-10}	6.6×10^{-9}
Austenitic stainless steels*	3.25	5.6×10^{-12}	3.0×10^{-10}
7075-T6 wrought aluminum [22]	3.7	2.7×10^{-11}	1.5×10^{-9}
A356-T6 cast aluminum [29]	11.2	1.5×10^{-20}	7.8×10^{-19}
Ti-6-4 mill annealed titanium [30]	3.2	1.0×10^{-11}	5.2×10^{-10}
Ti-62222 mill annealed titanium [31]	3.2	2.3×10^{-11}	1.2×10^{-9}
AZ91E-T6 cast magnesium [32]	3.9	1.8×10^{-10}	9.4×10^{-9}

* Conservative values suggested by Barsom [28].

can be greatest in region III because of the inverse relationship between fracture toughness and thickness, as shown in Fig. 6.7, which affects the allowable crack size at fracture. High fracture toughness is desirable because of the longer crack size at fracture, which allows easier and less frequent inspection and therefore safer components or structures.

Values of the threshold stress intensity factor range, ΔK_{th}, are given in Table A.4 for selected engineering alloys. These values are usually less than 10 MPa$\sqrt{}$m ($\sim$9 ksi$\sqrt{}$in.) for steels and less than 4 MPa$\sqrt{}$m ($\sim$3.5 ksi$\sqrt{}$in.) for aluminum alloys. Values of ΔK_{th} are substantially less than the K_{Ic} values given in Fig. 6.8 and Table A.3. In fact, ΔK_{th} can be as low as just several percent of K_{Ic}. The threshold stress intensity factor, ΔK_{th}, has often been considered analogous to the unnotched fatigue limit, S_f, since an applied stress intensity factor range below ΔK_{th} does not cause fatigue crack growth. Use of ΔK_{th} in design may be appropriate for conditions involving high-frequency or high-cycle applications such as turbine blades, or for metals with very high da/dN versus ΔK slopes such as cast aluminum, particularly in situations where small cracks or metallurgical discontinuities exist. Figure 6.16 shows the use of ΔK_{th} as a design parameter for no crack growth using a single-edge crack in an infinitely wide plate subjected to $R = 0$ loading. In Fig. 6.16, any combination of ΔS and crack length, a, that falls below the curve [assume that $\Delta K_{th} = 5.5$ MPa$\sqrt{}$m (5 ksi$\sqrt{}$in.)] does not cause fatigue crack growth. This criterion comes from setting ΔK in Eq. 6.17 equal to ΔK_{th} and using the proper stress intensity factor K from Fig. 6.3b. It should be noted from Fig. 6.16 that for crack lengths greater than about 2.5 mm (0.1 in.), ΔS_{th} is about 55 MPa (8 ksi) or less. Thus, to prevent cracks from propagating, the tensile stress must be kept very small. It should also be noted from Fig. 6.16 that for crack lengths smaller than about 0.1–1 mm (0.004–0.04 in.), the tensile stress rises rapidly, approaching gross yielding ($\approx 0.8S_y$). In addition, cracks approaching this size, $a \leq 1$ mm (0.04 in.), may fall into the category of

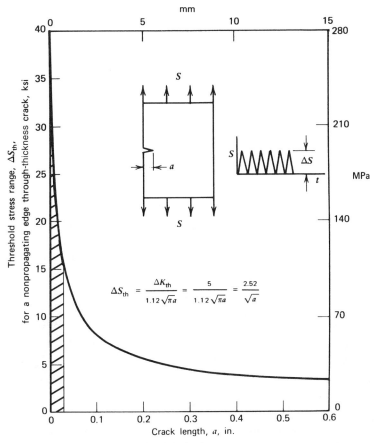

Figure 6.16 Threshold stress range for nonpropagating edge through-thickness cracks in wide plates, $R = 0$, $\Delta K_{th} = 5.5$ MPa$\sqrt{m}$ (5 ksi$\sqrt{in}$). The hashed region, where $a \leq 1$ mm ($a \leq 0.04$ in.), indicates LEFM breakdown due to small crack size effects.

small fatigue crack growth behavior which may violate LEFM assumptions. Behavior of small cracks is discussed in Section 6.8.

6.4.4 Crack Growth Life Integration Example with No Mean Stress Effects

Let us assume that a very wide SAE 1020 cold-rolled thin plate is subjected to constant amplitude uniaxial cyclic loads that produce nominal stresses varying from $S_{max} = 200$ MPa (29 ksi) to $S_{min} = -50$ MPa (-7.3 ksi). The monotonic properties for this steel are $S_y = 630$ MPa (91 ksi), $S_u = 670$ MPa (97 ksi), $E = 207$ GPa (30×10^6 psi), and $K_c = 104$ MPa$\sqrt{m}$ (95 ksi$\sqrt{in}$.). What fatigue life would be attained if an initial through-thickness edge crack existed and was 1 mm (0.04 in.) in length?

Before we can solve this problem, several questions must be answered: Is LEFM applicable? What is the applicable stress intensity factor expression for this component and loading? What crack growth rate equation should be used? How do we integrate this equation? What value of ΔK will cause fracture? Does corrosion or temperature play an important role?

In solving this problem, let us assume that a corrosive environment is not involved and that room temperature prevails. Limitations on the use of LEFM for this example include factors such as stress level and plastic zone size. The applied maximum stress level is approximately 30 percent of the yield strength, below the general restriction of $0.8S_y$. The initial monotonic plastic zone size can be estimated using Eq. 6.10. First, the initial value of K_{max} is calculated, with $S_{max} = 200$ MPa and $a_i = 1$ mm. This gives

$$K_{max} = 1.12S\sqrt{\pi a} = (1.12)(200)\sqrt{\pi(0.001)} = 12.6 \text{ MPa}\sqrt{m}$$

Since the problem states that the plate is thin, using this K_{max} value in Eq. 6.10 for plane stress gives

$$2r_y = \frac{1}{\pi}\left(\frac{K}{S_y}\right)^2 = \frac{1}{\pi}\left(\frac{12.6}{630}\right)^2 = 0.000127 \text{ m} = 0.127 \text{ mm}$$

where the plastic zone radius, r_y, is 0.0635 mm. This value is less than the general restriction in comparison to the crack length, as presented in Section 6.2 ($r_y \leq a/8$); thus, application of LEFM is reasonable.

The Paris crack growth rate equation (Eq. 6.19) is often a reasonable expression for region II, and even region III, crack growth behavior. Integration of the Paris equation involves numerical methods unless α from Eq. 6.17 is independent of crack length a. Since K for an infinite plate with a single-edge crack has α equal to a constant of 1.12 from Table 6.1 or Fig. 6.3b, it would be desirable to assume the infinite plate K solution. This is a very reasonable assumption as long as the crack length does not exceed about 10 percent of the plate width (see Fig. 6.3b, where Y is nearly constant for $a/w < 0.1$). Direct integration is preferable for the illustration, and hence the infinite plate is assumed. Since specific crack growth rate data are not given for the SAE 1020 cold-rolled steel, a reasonable first approximation would be to use the conservative material constants from Table 6.3 for ferritic-pearlitic steels suggested by Barsom [28]. Although this equation was developed for $R = 0$, the small compressive stress, -50 MPa, will not have much effect on crack growth behavior and can be neglected; thus, $\Delta S = 200 - 0 = 200$ MPa is very reasonable. The initial value of ΔK is equal to K_{max} in this case since $\Delta K = K_{max} - K_{min}$ and K_{min} is effectively 0. Thus, $\Delta K = 12.6$ MPa$\sqrt{m} - 0 = 12.6$ MPa$\sqrt{m}$. This initial ΔK is above threshold levels based on data in Table A.4. Given this initial value of ΔK, it is important to recognize what the starting crack growth rate is and what crack growth rate equation should be used. Use of Eq. 6.19 will provide an estimate of the fatigue crack growth rate for this material with $\Delta S = 200$ MPa at $\Delta K = 12.6$ MPa$\sqrt{\text{m}}$. Thus, substitution of the appropriate properties from Table 6.3 and $\Delta K = 12.6$ MPa$\sqrt{\text{m}}$ in Eq. 6.19 gives

$$\frac{da}{dN} = A(\Delta K)^n = 6.9 \times 10^{-12}(12.6)^3 = 1.4 \times 10^{-8} \text{ m/cycle}$$

This initial fatigue crack growth rate falls within the limits of the Paris region of crack growth rates, as seen in Fig. 6.15. Therefore, it would seem quite reasonable that the Paris equation is applicable for subsequent fatigue crack growth life calculations. It should be noted that if the initial ΔK and da/dN values fall below the power law region, as defined by Eq. 6.19, the designer could still use Eq. 6.19, or use another empirical relationship that better describes the overall behavior of the material. Based on the sigmoidal shape of the fatigue crack growth curve, extrapolation of Eq. 6.19 to the near-threshold region results in conservative fatigue crack growth life calculations for long crack growth behavior. Very little life is involved in region III; thus, extrapolation to this region will have little effect on the total life. The final crack length a_f (a_c is also commonly used) can be obtained from setting $K_{\max}$ at fracture equal to K_c. Thus, the following equations apply:

$$\Delta K = \Delta S\sqrt{\pi a}\,\alpha \tag{6.20a}$$

$$\frac{da}{dN} = A(\Delta K)^n = A(\Delta S\sqrt{\pi a}\,\alpha)^n = A(\Delta S)^n\,(\pi a)^{n/2}\,\alpha^n \tag{6.20b}$$

and

$$a_f = \frac{1}{\pi}\left(\frac{K_c}{S_{\max}\,\alpha}\right)^2 \tag{6.20c}$$

Integrating Eq. 6.20b,

$$N_f = \int_0^{N_f} dN = \int_{a_i}^{a_f} \frac{da}{A(\Delta S)^n\,(\pi a)^{n/2}\,\alpha^n}$$

$$= \frac{1}{A(\Delta S)^n\,(\pi)^{n/2}\,\alpha^n}\int_{a_i}^{a_f}\frac{da}{a^{n/2}} \tag{6.20d}$$

If n ≠ 2

$$\int_{a_i}^{a_f}\frac{da}{a^{n/2}} = \frac{a^{-(n/2)+1}}{-n/2+1}\bigg]_{a_i}^{a_f} = \frac{a_f^{(-n/2)+1} - a_i^{(-n/2)+1}}{-n/2+1} \tag{6.20e}$$

Therefore

$$N_f = \frac{a_f^{(-n/2)+1} - a_i^{(-n/2)+1}}{(-n/2 + 1)A(\Delta S)^n (\pi)^{n/2} \alpha^n} \tag{6.20f}$$

Equation 6.20f in another form is given by

$$N_f = \frac{2}{(n - 2)A(\Delta S)^n (\pi)^{n/2}\alpha^n} \left[\frac{1}{a_i^{(n-2)/2}} - \frac{1}{a_f^{(n-2)/2}} \right] \tag{6.20g}$$

Equations 6.20f or 6.20g are the general integration of the Paris equation when α is independent of crack length a and when n is not equal to 2. **These two equations are not correct if α changes significantly with a between the limits a_i and a_f.** Also note that for fatigue crack growth life between crack lengths a_1 and a_2, N_f, a_i and a_f in Eqs. 6.20d to 6.20g can be replaced by $N_{1-2} = \Delta N$, a_1 and a_2, respectively. For cases where α is a function of a, integration is usually necessary, using either standard numerical techniques or computer programs.

Using Eq. 6.20c to obtain the critical crack length at fracture:

$$a_f = \frac{1}{\pi} \left(\frac{K_c}{S_{max} \alpha} \right)^2 = \frac{1}{\pi} \left(\frac{104}{200 \times 1.12} \right)^2 = 0.068 \text{ m} = 68 \text{ mm} = 2.7 \text{ in.}$$

Substitution of the appropriate values into Eq. 6.20g results in

$$N_f = \frac{2}{(3 - 2)(6.9 \times 10^{-12})(200)^3(\pi)^{3/2} (1.12)^3} \left[\frac{1}{(0.001)^{(3-2)/2}} - \frac{1}{(0.068)^{(3-2)/2}} \right]$$

$$= 4631 \left[\frac{1}{0.0316} - \frac{1}{0.2608} \right] = 129\,000 \text{ cycles}$$

Now let us assume that the fracture toughness, K_c, was incorrect by a factor of ± 2, that is, $K_c = 208$ MPa$\sqrt{m}$ (180 ksi$\sqrt{in.}$) or 52 MPa$\sqrt{m}$ (42.5 ksi$\sqrt{in.}$). The final crack length from Eq. 6.20c would result in $a_f = 275$ mm (10.8 in.) and 17 mm (0.68 in.), respectively. The final life, N_f, from Eq. 6.20g would be 138 000 and 111 000 cycles, respectively. Thus, increasing or decreasing the fracture toughness by a factor of 2 caused increases or decreases in the final crack length by a factor of 4, respectively. However, changes in fatigue crack growth life were less than 15 percent, which are very small differences. If the initial crack length, a_i, was 5 mm (0.2 in.), the life would have been only 48 000 cycles instead of 129 000 cycles for the original problem with a_i equal to 1 mm (0.04 in.). This illustrates the importance of minimizing initial flaw or crack lengths to obtain long fatigue life and shows that appreciable changes in fracture toughness will alter final crack lengths but may not have appreciable effects on fatigue life. High fracture toughness in fatigue design, however, is still very desirable because of the randomness of many load histories and the larger crack lengths before fracture, which allow much better inspection success.

6.5 MEAN STRESS EFFECTS

The general influence of mean stress on fatigue crack growth behavior is shown schematically in Fig. 6.17. The stress ratio $R = K_{min}/K_{max} = S_{min}/S_{max}$ is used as the principal parameter. Most mean stress effects on crack growth

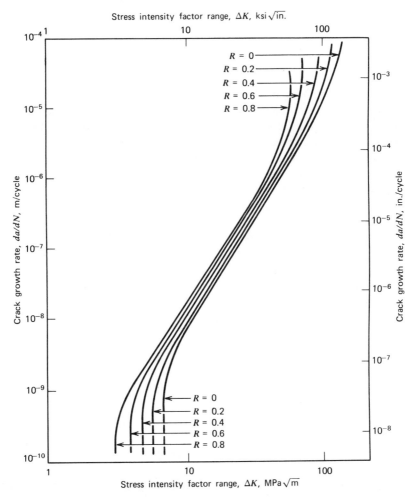

Figure 6.17 Schematic mean stress influence on fatigue crack growth rates.

have been obtained for only positive stress ratios, i.e., $R \geq 0$. Figure 6.17 indicates that increasing the R ratio (which means increasing the mean stress) has a tendency to increase the crack growth rates in all portions of the sigmoidal curve. The increase in region II, however, may be small. In region III, where fracture toughness K_c or K_{Ic} controls, substantial differences in crack growth rates occur for different R ratios. The upper transition regions on the curves are shifted to lower ΔK values as R, and hence K_{max}, increases. It should be recognized that the effect of the R ratio on fatigue crack growth behavior is strongly material dependent. Thus, the differences shown between the R ratio curves in Fig. 6.17 could be less or greater than those shown in one, two, or all three regions of fatigue crack growth curves, depending on the metal.

A commonly used equation depicting mean stress effects in regions II and III is the Forman equation [33]:

$$\frac{da}{dN} = \frac{A'(\Delta K)^{n'}}{(1 - R)K_c - \Delta K} = \frac{A'(\Delta K)^{n'}}{(1 - R)(K_c - K_{max})} \tag{6.21}$$

where A' and n' are empirical material fatigue crack growth rate constants and K_c is the applicable fracture toughness for the material and thickness. The Forman equation is a modification of the Paris equation that incorporates mean stress and region III fatigue crack growth behavior. It should be recognized that A' and n' are not equivalent to the coefficient and slope values A and n in the Paris equation. As K_{max} approaches K_c, the denominator approaches zero; thus, the crack growth rate, da/dN, becomes very large. This describes region III type crack growth behavior. If the material constants A' and n' are known, various stress ratio curves generally collapse into a single curve.

Another common empirical relationship used to describe mean stress effects with $R \geq 0$ is the Walker relationship, which when applied to the stress intensity factor range ΔK takes the form

$$\frac{da}{dN} = \frac{A(\Delta K)^n}{(1 - R)^{n(1-\lambda)}} = A''(\Delta K)^n \tag{6.22a}$$

where A and n are the Paris coefficient and slope for $R = 0$, respectively, and λ is a material constant. Equations 6.19 and 6.22a are similar, with the coefficients of the equations, A and A'', being the only difference for a specific stress ratio as

$$A'' = \frac{A}{(1 - R)^{n(1-\lambda)}} \tag{6.22b}$$

The stress ratio, R, does not affect the slope, n. Similar to Eq. 6.21, use of Eq. 6.22a has the general benefit of collapsing fatigue crack growth data for various stress ratios, R, into a single straight line for region II crack growth. Because the effect of R on fatigue crack growth is material dependent, it is necessary to determine the material constant λ. Values of λ for various metals range from 0.3 to nearly 1, with a typical value around 0.5 [34]. For metals for which data at various stress ratios are not available, a value of $\lambda = 0.5$ is often used. Lower values of λ indicate a stronger influence of R on fatigue crack growth behavior.

The effect of mean stress on ΔK_{th} values can be substantial, as indicated in Fig. 6.17 and Table A.4. The ΔK_{th} values for nine materials with different positive R ratios are included in Table A.4. For R increasing from zero to about 0.8, the threshold ΔK_{th} decreases by a factor of about 1.5 to 2.5. This has the effect of shifting the curve in Fig. 6.16 toward the abscissa by these same factors, which reduces ΔS_{th} for a given crack length by the same factors. Use of a form similar to the Walker equation to describe the effect of the R ratio on the threshold gives

$$\Delta K_{th(R \neq 0)} = \Delta K_{th(R=0)}(1 - R)^{(1 - \gamma)} \tag{6.23}$$

where γ is an empirical constant fitted to test data for positive R values other than zero.

The effect of negative R ratios, which includes compression in the cycle, has received less investigation and is less understood. The results of many negative R ratio tests on wrought and cast steels, cast irons, and aluminum alloys subjected to constant amplitude conditions in regions II and III indicate that fatigue crack growth rates based on ΔK values (which neglect compressive nominal stresses) are similar to $R = 0$ results or are increased by no more than a factor of 2 [35–37]. For region I, near threshold fatigue crack growth rates at $R = -1$ were shown to be higher than the rates obtained by tests carried out at low positive stress ratios, as well as showing slightly lower threshold values [37–39]. This behavior is explained by the presence of smoother crack surfaces due to the compressive loads, which leads to less surface roughness-induced crack closure. The mechanisms associated with this effect are discussed in Section 6.7.

Crack Growth Life Integration Example with Mean Stress Effects   Using the example problem in Section 6.4.4, let us assume for now that all given conditions are the same except that the stress ratio, R, is 0.33 instead of -0.25, where $S_{max} = 300$ MPa and $S_{min} = 100$ MPa. What fatigue life is attained given the same initial crack size?

First, since $R > 0$, we must choose a relationship that accounts for mean stress effects. Use of the Forman equation, Eq. 6.21, requires the need for additional stress ratio data that are not readily available. However, the Walker relationship allows approximations of the material constants based on $R = 0$ fatigue crack growth data. Therefore, we use the Walker relationship. Although λ is not known for this material and stress ratio, we will assume that

$\lambda \approx 0.5$ without additional data, which should provide a reasonable estimate. The material constant A'' for $R = 0.33$ can be estimated from Eq. 6.22b:

$$A'' = \frac{A}{(1 - R)^{n(1-\lambda)}} \approx \frac{6.9 \times 10^{-12}}{(1 - 0.33)^{3(1-0.5)}} = 1.27 \times 10^{-11}$$

Thus, the estimated power law relationship for the $R = 0.33$ condition using Eq. 6.19 is

$$\frac{da}{dN} = 1.27 \times 10^{-11} (\Delta K)^3$$

The final crack length will be shorter than that calculated in Section 6.4.4 because the maximum applied stress is greater. From Eq. 6.20c

$$a_f = \frac{1}{\pi} \left(\frac{K_c}{S_{max} \, \alpha} \right)^2 = \frac{1}{\pi} \left(\frac{104}{300 \times 1.12} \right)^2 = 0.0305 \text{ m} = 30.5 \text{ mm} = 1.2 \text{ in.}$$

Using the direct integration method from Section 6.4.4 gives

$$N_f = \frac{2}{(3 - 2)(1.27 \times 10^{-11})(200)^3(\pi)^{3/2}(1.12)^3} \left[\frac{1}{(0.001)^{(3-2)/2}} - \frac{1}{(0.0305)^{(3-2)/2}} \right]$$

$$= 2516 \left[\frac{1}{0.0316} - \frac{1}{0.1746} \right] = 65 \ 000 \text{ cycles}$$

Therefore, for the same alternating stress, where $\Delta S = (300 - 100) = 200$ MPa, but where $R = 0.33$ rather than -0.25, the fatigue life decreased by almost a factor of 2 (from 129 000 to 65 000 cycles). To validate this analysis, we need to check the initial crack growth rate to see if the Walker equation is applicable. Also, are there LEFM limitations based on the plastic zone size or maximum stress? First, from the estimated power law expression shown above

$$\frac{da}{dN} = 1.27 \times 10^{-11} (\Delta K)^3 = 1.27 \times 10^{-11} (12.6)^3 = 2.5 \times 10^{-8} \text{ m/cycle}$$

This fatigue crack growth rate is greater than that calculated in the example in Section 6.4.4. Therefore, it is reasonable to assume that the Walker relationship can be used to describe the fatigue crack growth behavior for this example. The initial monotonic plastic zone size using Eq. 6.10, where $K_{max} = 1.99 S_{max} \sqrt{a} = 1.99(300)\sqrt{0.001} = 18.9$ MPa$\sqrt{m}$, is

$$2r_y = \frac{1}{\pi} \left(\frac{K}{S_y} \right)^2 = \frac{1}{\pi} \left(\frac{18.9}{630} \right)^2 = 0.000286 \text{ m} = 0.286 \text{ mm}$$

where the plastic zone radius, r_y, is 0.143 mm. This value is roughly 14 percent of the initial crack size, which is close to the general restriction presented in Section 6.2 for monotonic loading ($r_y \le a/8$) but less than the restriction for cyclic loading ($r_y \le a/4$), as is the case here. Also, the maximum stress level is well below $0.8S_y$; thus, the results and use of LEFM appear reasonable.

While the above example problem and that in Section 6.4.4 show a comparison for two conditions based on the same alternating stress, i.e., $\Delta S = 200$ MPa, this may not be realistic in a design situation. For example, when $R = 0.33$, in order to have the same alternating stress as in the example in Section 6.4.4 for $R = -0.25$, $S_{max} = 300$ MPa and $S_{min} = 100$ MPa. This maximum stress level may or may not be feasible, depending on all the design constraints. Thus, in damage-tolerant design, one must consider both the stress ratio and the stress level (mean and range of stresses). As another example, consider the following: What maximum and minimum stress levels based on a stress ratio of 0.33 will give the same life, 129 000 cycles, as that calculated in the example problem of Section 6.4.4 for a stress ratio of -0.25? One must recognize that a_f in Eqs. 6.20f or 6.20g is dependent on the magnitude of S_{max}, which at this point is unknown. However, in Section 6.4.4 it was shown that an increase or decrease in the final crack length by a factor of 4 changed the fatigue life by less than 15 percent. Thus, a first approximation could be to assume a final crack length the same as that calculated in the above example, i.e., $a_i = 30.5$ mm. A trial-and-error approach using Eqs. 6.20c and either 6.20f or 6.20g will provide the desired results. Therefore, to solve this problem, one need only rearrange Eq. 6.20g such that ΔS is on the left-hand side of the equation and recognize that $\Delta S = (S_{max} - S_{min})$ and $S_{min}/S_{max} = 0.33$. This gives

$$(\Delta S)^3 = \frac{2\left[\dfrac{1}{(0.001)^{(3-2)/2}} - \dfrac{1}{(0.0305)^{(3-2)/2}}\right]}{(3-2)(1.27 \times 10^{-11})(129\ 000)(\pi)^{3/2}(1.12)^3}$$

or

$$\Delta S = 159\ \text{MPa}$$

from which $S_{max} = 239$ MPa. Substitution of $S_{max} = 239$ MPa into Eq. 6.20c yields $a_f = 0.048$ m, which when substituted back into Eq. 6.20g to solve for ΔS yields $\Delta S = 162$ MPa where $S_{max} = 243$ MPa. One more iteration shows this value to be correct. Thus, $S_{max} = 243$ MPa and $S_{min} = 81$ MPa. Therefore, for the same fatigue life (129 000 cycles), the maximum stress could be increased from 200 to 243 MPa, an increase of over 20 percent, if the loading conditions were modified such that the stress ratio was increased from -0.25 to 0.33. However, the stress range, ΔS, was reduced from 200 MPa to 162 MPa. Again, this shows the importance of mean stress and stress range effects on fatigue crack growth behavior.

6.6 CYCLIC PLASTIC ZONE SIZE

In Section 6.2, Eqs. 6.10 to 6.13 describe the monotonic plastic zone size developed at the crack tip. This is the plastic zone size developed due to the maximum load applied in the loading cycle. Figures 6.18a and 6.18b show a schematic of this using the Irwin model [7] described in Section 6.2. The maximum load in the loading cycle, point A in Fig. 6.18a, produces the monotonic plastic zone size and stress distribution shown in Fig. 6.18b. However, as the maximum load is reduced during the loading cycle to the minimum load, point B in Fig. 6.18a, the local stress near the crack tip is reduced to a value less than that observed for the monotonic plastic zone size. This unloading causes the development of the cyclic plastic zone size, $2r_y'$, and the corresponding stress distribution, shown in Fig. 6.18c. Thus, Fig. 6.18 illustrates the monotonic (Fig. 6.18b) and cyclic plastic zone (Fig. 6.18c) sizes at the maximum and minimum loads of a single load cycle. For the assumed elastic-perfectly plastic behavior, the maximum stress developed in Fig. 6.18b is S_y. The maximum stress change developed during unloading can be as large as $2S_y$. Thus,

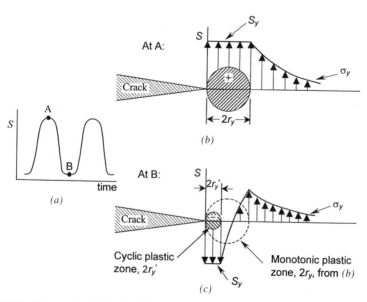

Figure 6.18 Schematic of the plastic zone at the tip of an advancing crack. (a) Loading cycle. (b) Monotonic plastic zone. (c) Cyclic plastic zone.

Fig. 6.18c shows the sum of the inelastic loading stress distribution plus the elastic unloading stress distribution. Note that the stresses within the cyclic plastic zone, $2r_y'$, are compressive, while outside the cyclic plastic zone the compressive stress decreases and then becomes tensile. A key point to recognize is that the sign of the inelastic stress distribution associated with the cyclic plastic zone size is opposite to the sign of the applied stress during loading. Thus, if a region yields in tension during loading, as shown in Fig. 6.18b, after unloading a portion of that region is in compression, as shown in Fig. 6.18c. The size of the cyclic plastic zone where yielding occurs can be approximated by using $2S_y$ for S_y and ΔK for K in Eq. 6.10 for plane stress conditions [40]. This gives

$$2r_y' \approx \frac{1}{\pi}\left(\frac{\Delta K}{2S_y}\right)^2 = \frac{1}{4\pi}\left(\frac{\Delta K}{S_y}\right)^2 \tag{6.24}$$

where $2r_y'$ is the cyclic plastic zone size under plane stress conditions for $R \geq 0$. For $R = 0$, where $K_{max} = \Delta K$, the size of the plastic zone is only one-quarter of the size that existed at the peak tensile load. Equations 6.10 and 6.24 are estimates for plane stress conditions. The cyclic plastic zone for plane strain conditions using analyses similar to those used in Section 6.2 yields a value one-third as large as that of the corresponding cyclic plastic zone for plane stress. Therefore, for plane strain conditions, the cyclic plastic zone is

$$2r_y' \approx \frac{1}{3\pi}\left(\frac{\Delta K}{2S_y}\right)^2 = \frac{1}{12\pi}\left(\frac{\Delta K}{S_y}\right)^2 \tag{6.25}$$

If the material cyclic hardens or softens due to conditions presented in Chapters 3 and 5, S_y should be replaced with the cyclic yield strength, S_y'.

The primary controlling parameter for crack growth is the stress intensity factor, which also dictates the crack tip plastic zone size. Excessive plasticity, where the plastic zone size or applied stress is a large fraction of the crack size or the yield strength, respectively, violates the basic assumptions of LEFM, whether due to monotonic or cyclic loading. However, because the cyclic plastic zone size is usually much smaller than the monotonic plastic zone size, LEFM can often be applied to fatigue crack growth situations with good success, even for materials that exhibit significant plasticity, or for region III crack growth where the extent of plasticity is greatest. However, in this region of crack growth, a small fraction of the total fatigue life is usually involved; thus, accuracy is not as important. Because r_y' is always less than r_y for cyclic loading, limitations of LEFM associated with r_y, as presented in Section 6.2, can often be extended to r_y' for fatigue crack growth. Excessive plasticity is usually of greater significance in fracture toughness applications regarding fracture conditions than in fatigue crack growth. However, validity of fatigue crack growth analysis is extremely important in fatigue design. Excessive plasticity effects during fatigue crack growth are discussed in Section 6.9.

6.7 CRACK CLOSURE

Crack closure was first quantified and its importance demonstrated in 1970 by Elber [41], who showed that a fatigue crack could be closed even under a tensile load. Based on experimental results using thin sheets of 2024-T3 aluminum alloy, Elber argued that a reduction in the crack tip driving force occurred as a result of residual tensile deformation left in the wake of a fatigue crack tip as the crack grew into the crack tip compressive stress region, $2r_y'$. The residual tensile deformation caused the crack faces to close prematurely before the minimum load was reached. Since the time of Elber's work, extensive research on crack closure has been performed and documented. Several symposia and conferences have been organized during the past 25 years dedicated solely to crack closure, culminating in various proceedings [42,43]. Many fatigue crack growth effects have been explained, at least in part, by closure concepts. However, many details of the mechanisms of crack closure are still only partly understood.

Figure 6.19 shows the various definitions of the stress intensity factor range during cyclic loading. Two points are defined on the K versus time plot, one where the crack opens on the loading portion of the cycle, K_{op}, and one where the crack closes on the unloading portion of the cycle, K_{cl}. In some cases, these values may occur at the same K value; in other cases, there may be a slight offset where K_{op} is slightly larger than K_{cl}. In this book, K_{op} will be used in defining the controlling parameter for fatigue crack growth. As can be seen from Fig. 6.19, the opening stress intensity factor, K_{op}, is typically

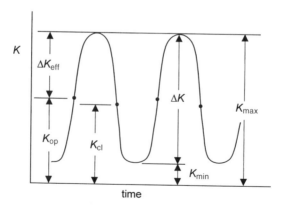

Figure 6.19   Schematic representation of opening and effective stress intensity factors.

greater than the minimum stress intensity factor, K_{min}. While the nominal stress intensity factor range, ΔK, is defined by $K_{max} - K_{min}$, the effective stress intensity factor range, ΔK_{eff}, as defined by Elber, is $K_{max} - K_{op}$. Thus, the effective crack tip driving force, ΔK_{eff}, is less than the nominal crack tip driving force, ΔK, and the damaging portion of the loading cycle is restricted to the opening portion of the stress intensity factor range, ΔK_{eff}. Put another way, the portion of the loading cycle that occurs below K_{op} has little or no influence on fatigue crack growth. One parameter used to define the measurement of closure is the closure ratio, given by

$$U = \frac{\Delta K_{eff}}{\Delta K} = \frac{(1 - K_{op}/K_{max})}{(1 - R)} \tag{6.26}$$

which indicates the portion of the loading cycle where the crack is open (or opening). For loading conditions where there is little or no closure, U approaches 1, while for loading conditions that show extensive crack closure, U is small. Typically at high stress ratios (e.g., for $R > 0.5$), crack growth generally displays limited crack closure, while at low stress ratios ($R < 0.5$), crack growth tends to exhibit higher levels of crack closure.

The effect of crack closure on region I behavior is probably of greatest interest, as it is in this region where much of the crack growth life is expended and where crack closure is most significant. In addition, region I crack growth is influenced by many variables, including stress ratio, environment, and microstructure, which have a direct influence on the effect of crack closure. The practical significance of crack closure is related to the growth, retardation, or arrest of fatigue cracks under in-service load histories. In-service load histories are seldom constant amplitude loading, and in most cases a variable amplitude load history exists. In either case, sufficient knowledge of the crack opening load may be required for adequate life predictions. This is because K_{op} ($\propto S_{op}$) is required to determine ΔK_{eff} and because ΔK_{eff} may be the desired parameter for correlating fatigue crack growth rates under various load histories, including both constant and variable amplitude loading. The effect of crack closure on variable amplitude loading is presented in Chapter 9.

Since the time of Elber's pioneering work, additional crack closure mechanisms have been identified that describe many of the characteristics associated with fatigue crack growth. The most extensively studied types of closure include plasticity-induced, oxide-induced, and roughness-induced closure. Other, less well studied types of crack closure include viscous fluid-induced and phase transformation-induced closure. These various forms of fatigue crack closure have been categorized and reviewed extensively by a number of investigators [5,44–46]. A brief introduction of the three main types will now be presented.

Elber's work fundamentally described what is now referred to as plasticity-induced closure. It is most prevalent in metals at low stress ratios under plane stress conditions, although it can still be significant for plane strain. This form of closure arises from the presence of the compressive plastic zone itself, as shown in Fig. 6.18c. Not only is a compressive plastic zone developed at the tip of a crack with the application of a load cycle, but there is also a "wake" of plastically deformed material along the crack faces as the crack extends forward. If the load cycle has a low stress ratio, the compressive residual stress will bring the crack faces together before the minimum load is reached, creating closure. Based on plasticity-induced closure arguments, ΔK_{eff} has been shown to correlate crack growth behavior reasonably well, as fatigue crack growth curves for various stress ratios collapse into a single narrow scatter band.

Oxide formation in various aggressive environments can affect crack closure levels in a number of ways. The oxide can form as either a general uniform layer on the crack faces or as rough, bulky deposits. At low stress ratio and near-threshold levels there is a greater propensity for repeated crack face contact to occur. This behavior is enhanced by the combination of mode I, II, or III displacement and/or fracture surface roughness. At low stress intensity factor levels there is a continual breaking and reforming of the oxide layer along the crack faces. At high stress ratios, where crack closure is less prevalent, oxide debris and buildup are not as significant.

Crack deflection due to microstructural contributions can have a significant effect on threshold behavior. This form of crack closure is generally referred to as roughness-induced closure. An increase in surface roughness due to changes in slip character and grain size promotes crack deflection. In the presence of mixed-mode loading, premature crack face contact will occur, leading to higher crack opening levels. Numerous studies have shown the dependence of threshold behavior on microstructure for various stress ratios [47–49]. Materials with very small grain size showed no crack closure even at low stress ratios. This led to low threshold values and no stress ratio dependence. For coarser-grained materials, more serrated and faceted morphology was observed that led to higher opening loads. In regions II and III, where crack opening displacements are higher, roughness-induced closure has much less effect.

It should be recognized that the various forms of closure can operate synergistically. For example, oxide-induced closure requires a fretting action. Therefore, closure must already exist in another form, such as plasticity-induced or roughness-induced closure.

The methods used to measure crack closure range from simple and inexpensive to complex and expensive. These methods include direct observation, indirect observation, and compliance measurement. Direct observations may include techniques that use various microscopes, interferometry, lasers, or other detection devices. Indirect observations such as striation counting and spacing, crack growth rate changes, and high stress ratio tests have also been used to evaluate crack closure. Compliance measurements that indicate a variation in the load-displacement response involve techniques that use clip gages, strain gages, electrical potential, ultrasound, and eddy current. Compliance methods are generally the simplest and most often used. The closure phenomenon is quite complex, and much of the controversy that has developed regarding crack closure may be related to the methods used for its experimental measurement. Round robin studies sponsored by ASTM [50,51] showed variations in crack opening loads in excess of 20 percent from one test facility to the next even under controlled test conditions. However, even though quantitative measurements and predictions of crack closure have produced mixed results, qualitative indications that help improve the understanding of the effects of variables such as microstructure, environment, and load history can be quite significant.

6.8 SMALL FATIGUE CRACKS AND LEFM LIMITATIONS

In many design situations, LEFM analyses allow a direct comparison of fatigue crack growth behavior between engineering components or structures and laboratory specimens using the stress intensity factor range, ΔK. This holds true for conditions in which LEFM similitude is maintained, primarily based on small-scale crack tip yielding. This is controlled primarily by the plastic zone size to crack length ratio and the magnitude of the operating stress. Loss of similitude with ΔK can occur when operating stress levels are too high, resulting in excessive plasticity, or when cracks are small in comparison to either the plastic zone or microstructural dimensions. Thus, LEFM assumptions are violated for these conditions. This section focuses on small cracks. Section 6.9 will evaluate excessive plasticity effects and the use of elastic-plastic fracture mechanics.

Figure 6.20 provides a schematic comparison between stress range, ΔS, and crack length, a, on a log-log scale, as first presented by Kitagawa and Takahashi [52]. The sloped straight line labeled A-A, which defines the long crack threshold, $\Delta K_{th} = \Delta S_{th}\sqrt{\pi a_{th}}\ \alpha$, represents the stress–crack length combination below which a crack should not grow, based on LEFM. At high stress range levels, extending beyond that shown in Fig. 6.20, LEFM assumptions are violated due to excessive plasticity. The horizontal line labeled B-B identifies the fatigue limit or fatigue strength range, below which fatigue life is very long, as discussed in Chapter 4. Experimental work has been shown to deviate from these two lines between certain small crack lengths, labeled a_1 and a_2 in Fig. 6.20, and to follow the "solid curve" which merges with the threshold and fatigue limit lines. For the solid curve segment labeled C-C, a_1 defines the smallest crack length capable of lowering the fatigue limit and a_2 defines the crack size at which small crack effects on LEFM analysis end. Thus, the experimentally determined threshold stress range, ΔS_{th}, shown for an arbitrary crack length (a_{th}) in Fig. 6.20, is less than the stress range predicted from either the fatigue limit (labeled point 1) or the long crack threshold (labeled point 2). This holds true up to a crack length of a_2, as shown in Fig. 6.20. This behavior leads to cracks growing below the long crack threshold and the

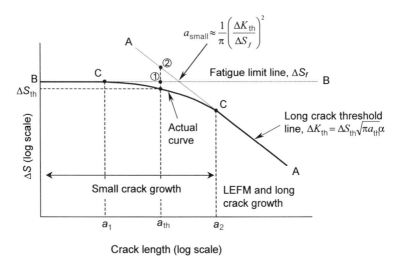

Figure 6.20   Schematic of the Kitagawa-Takahashi diagram showing the relationship between stress range and crack length.

fatigue limit and can lead to nonconservative life predictions based on LEFM and/or fatigue limit analysis. One simple method used to approximate the crack length for which small crack growth behavior could pose a concern for a material consists of setting the equations for the threshold line and the fatigue limit line equal to each other. Combining these equations and solving for the crack length gives

$$a_{\text{small}} \approx \frac{1}{\pi}\left(\frac{\Delta K_{\text{th}}}{\Delta S_f}\right)^2 \tag{6.27}$$

where a_{small} approximates the crack length below which small crack growth behavior can be expected and ΔS_f is the fatigue limit range, i.e., twice the fatigue limit, S_f.

Pearson [53] first reported the anomalous behavior between small and long cracks in an aluminum alloy. Many sources on the subject of small cracks exist due to the large amount of research performed in this area over the last 25 years [54–57]. With the extensive attention small cracks have received, various definitions of small cracks have been proposed. Based on the general classification of small fatigue cracks as defined in ASTM Standard E647 [25], small cracks are defined as follows:

Microstructurally small if their length is comparable to the scale of some micro-structural dimension, for example the grain size.

Physically small if their length is typically between 0.1 and 1 or 2 mm (0.004 and 0.04 or 0.08 in.).

Mechanically small if their length is small compared to the scale of local plasticity, for example a crack growing from a notch exhibiting local plasticity.

Mechanically small cracks, typically those growing from notches, are discussed in Section 7.4.

Before the specific characteristics of these small cracks are presented, let us look at the practical significance of small crack behavior based on its influence on fatigue crack growth life. Several studies have shown that the fatigue crack growth rates of small cracks can be significantly higher than the corresponding growth rates of long cracks for the same nominal stress intensity factor range ΔK. Also, small cracks grow at values of ΔK that are below the long crack threshold stress intensity factor range, ΔK_{th}. This in effect shifts the region I "small" crack growth curve to the left of the sigmoidal fatigue crack growth curve presented in Section 6.4.1. This general behavior is shown schematically in Fig. 6.21 for both microstructurally and physically small cracks. In general, most of the fatigue life of a crack occurs when the crack is small. Current design philosophy using LEFM analysis for long cracks, presented in Section 6.4.3, can usually provide accurate estimates of fatigue lives for cracks that are typically greater than about 1 mm (~0.04 in.) in length. However, as discussed in Chapter 3, cracks nucleate and grow at dimensions much smaller than this size. Thus, life predictions based on sigmoidal *da/dN* versus ΔK, as presented in Section 6.4.3, when extended to small crack behavior may lead to nonconservative estimates. This is of paramount importance for safety critical components, such as gas turbine engine discs

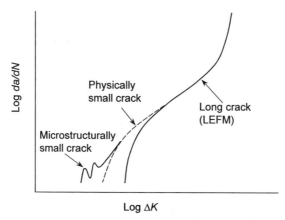

Figure 6.21   Schematic of typical crack growth rate behavior of small and long cracks.

and blades, where fatigue life is typically dominated by microstructurally and/ or physically small crack growth and critical crack sizes can be very small in the presence of high stress levels.

Microstructurally small cracks are characterized as such because they can be influenced significantly by the microstructure. As the schematic in Fig. 6.21 shows, fatigue crack growth of microstructurally small cracks is usually discontinuous in nature, with accelerations and decelerations in their growth rate. This discontinuous growth is associated with the interaction between the crack tip and microstructural barriers of the material such as grain boundaries and sometimes second phase particles. As a small crack grows within a grain, the crack growth rate is typically higher than that observed for the general "long crack" da/dN versus ΔK curve. As the crack grows and approaches a grain boundary, the crack growth can retard because of slip incompatibility with adjacent grains. This causes the crack growth rate to decrease. Further cycling provides enough damage near the crack tip for the crack to grow eventually into the adjacent grain or grains; thus, the crack growth rate increases once again. Once the small crack grows to a size that is several times larger than the grain size, typically around 10 grain sizes [58], microstructural barriers do not provide sufficient constraint, as they did when the crack was smaller. Thus, the microstructurally small crack curve generally merges with the physically small and then long crack curve, as depicted in Fig. 6.21. The growth of microstructurally small cracks typically violates the basis of continuum mechanics and LEFM due to lack of homogeneity and isotropy at the small microstructural scale. The use of probabilistic and statistical methods, and/or damage mechanics, appears to be a potential solution for better characterizing the fatigue crack growth of microstructurally small cracks.

Physically small cracks, typically less than about 1 or 2 mm (~0.04 or 0.08 in.), often do not violate LEFM limitations based on stress level or near-tip plastic zone size, yet still propagate at faster rates than long cracks subjected to the same nominal crack tip driving force. The general fatigue crack growth behavior of physically small cracks is shown in the schematic of Fig. 6.21 and in Fig. 6.22 [59]. Their growth behavior is similar to the general long crack region I fatigue crack growth behavior, except that the growth rates are higher and ΔK_{th} is lower. As discussed in the previous section, crack closure can contribute significantly to the effective ΔK level. The anomalous behavior of physically small cracks appears to be associated with the lack of premature contact between the crack faces behind the crack tip as it advances, the opposite of that typically observed at low stress ratios for long crack behavior. Thus, the crack closure effects presented in Section 6.7 are less pronounced for physically small cracks that are not long enough for the crack faces to come in contact prior to the application of the minimum load. The growth behavior of physically small cracks is similar to that of long cracks at higher stress ratios. This can be understood by studying Fig. 6.17 regarding mean stress effects. Small crack growth at low applied stress ratios in region I in Fig. 6.21 closely resembles the high stress ratio curve in Fig. 6.17. Also, in regions II and III, the crack growth rate is increasing rapidly. Thus, a physically small crack operating in regions II or III becomes a long crack very quickly.

A simple approximation for this type of small crack growth behavior is extrapolation of the Paris equation (Eq. 6.19) to region I growth [60]. Although not an accurate fit to most long crack growth data in region I, extrapolation of the Paris equation may provide better life predictions by taking into account much of the nonconservative behavior associated with small crack growth. However, one must be careful when performing this operation, as extrapolation could result in either conservative or nonconservative behavior, depending on the small crack behavior observed. A schematic depicting this extrapolation is shown in Fig. 6.23. Examination of the data in Fig. 6.22 reveals that the physically small crack data fall below extrapolation of the Paris line to region I. Thus, for this material, conservative life predictions would prevail using this simple approximation. Use of Eqs. 6.19, 6.21, or 6.22 to make life predictions for combined small and long crack growth behavior is given as a homework problem.

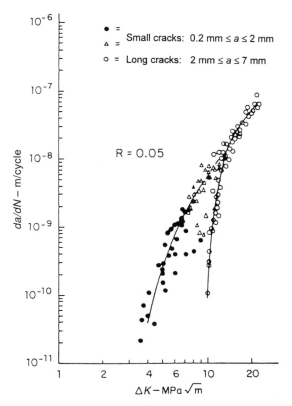

Figure 6.22 Fatigue crack growth behavior for small and long cracks in SAE 1020 hot-rolled steel [59] (Copyright ASTM. Reprinted with permission).

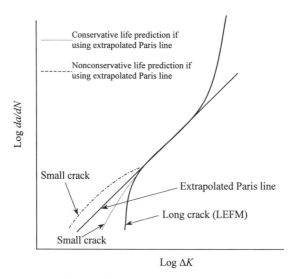

Figure 6.23 Schematic of life prediction assessment using the extrapolated power law expression from long crack growth behavior.

6.9 PLASTICITY EXTENSION OF LEFM AND ELASTIC-PLASTIC FRACTURE MECHANICS

LEFM was originally developed to describe fracture behavior and later was extended to fatigue crack growth behavior, mostly for elastic conditions. It has been shown that LEFM concepts can be used where limited plasticity exists, but only in the crack tip region. To account for this effect, the models presented in Section 6.2 assume that the crack length is enlarged slightly from a to $a + r_y$. The stress intensity factor equation is then modified to

$$K_{\text{eff}} = S\sqrt{a + r_y}\, f\!\left(\frac{a + r_y}{w}\right) \qquad (6.28)$$

which is simply a modification of Eq. 6.2b taking into account a plasticity correction. It should be noted that K_{eff} defined in Eq. 6.28 for small-scale plasticity correction is not the same as K_{eff} defined in Section 6.8 for crack closure. Since calculation of the plastic zone size, r_y, requires the value of K_{eff}, a trial-and-error procedure is often necessary to obtain K_{eff} unless $f[(a + r_y)/w]$ does not change significantly for the new crack length $(a + r_y)$. The plastic zone correction is typically small when $S \ll S_y$ but increases appreciably when the applied stress, S, exceeds approximately $0.8S_y$. Therefore, conditions that exhibit appreciable plasticity, i.e., when a plasticity correction of about 20 percent or more is calculated, violate the basic assumptions of LEFM. This is typical of metals with low strength and high ductility at fracture and may also occur with fatigue crack growth. Thus the evolution of elastic-plastic fracture mechanics (EPFM).

While the stress intensity factor, K, is the unique characterizing parameter of the near tip stress field in LEFM, a corresponding parameter used in EPFM is the J-integral, as formulated by Rice [61], Eshelby [62], and Hutchinson [63]. The crack opening displacement (COD) approach is another technique used to characterize fracture and fatigue crack growth for conditions with significant plasticity. This approach was first introduced by Wells [64], prior to the introduction of the J-integral. However, it has not received the same attention as the J-integral and will not be discussed here. Consideration of elastic-plastic fracture and nonlinear behavior is prominent in the nuclear power industry, specifically for pressure vessels and piping. Metals used to fabricate these structures are usually steels that have ductile behavior and high toughness where fracture or fatigue crack growth is accompanied by extensive plastic deformation. Specific concerns include thermal shock during a loss of coolant accident, radiation damage, stress corrosion cracking, and crack growth in welds and bi-material joints. Similar to the extensive use of K in the aerospace industry, because of the catastrophic consequences of failure in the nuclear power industry, a great deal of research has focused on the development of nonlinear (elastic-plastic) fracture mechanics methods to assess fracture risks in nuclear power plant structures and components. Elastic-plastic fracture concepts also are used in elevated temperature applications, critical structures in the defense industry, and offshore petrochemical rigs, structures, and components, as well as, more recently, in safety critical components and structures in other industries. While this chapter is primarily intended to present the fundamentals of LEFM as applied to fatigue, a brief discussion of EPFM concepts will be provided regarding the use of the J-integral and extension of these concepts to elastic-plastic fatigue crack growth. More details on nonlinear behavior and EPFM are found in references such as 4, 65, and 66.

The J-integral concept is based on an energy balance approach, much the same as K, as presented in Section 6.1.2. Rice [61] showed that for nonlinear elastic behavior, a line integral related to energy could be used to describe a cracked body as follows:

$$J = \int_{\Gamma}(Wdy - T_i \frac{\partial u_i}{\partial x}ds) \tag{6.29}$$

where W is the strain energy density, T_i is the stress vector, u_i is the displacement vector, and ds is an increment along a contour surrounding a crack tip. Γ is a contour beginning on the lower crack surface and ending on the upper crack surface, traveling counterclockwise. A schematic of the line contour surrounding a crack tip is shown in Fig. 6.24. Mathematics show that the J-integral is path independent, i.e., results using the contour Γ_{I} or Γ_{II} are

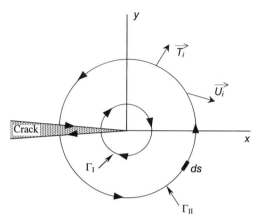

Figure 6.24   Line contours surrounding a crack tip for *J*-integral formulation.

equal. This allows a contour to be chosen that is remote to the crack tip and contains only elastic displacements and loads. Therefore, by selectively choosing a contour for which elastic loads and displacements are known, the elastic-plastic energy release rate, or *J*, can be obtained. For the elastic-plastic case

$$J = J_e + J_p \tag{6.30}$$

where J_e and J_p are the elastic and plastic components of the *J*-integral, respectively. For the linear elastic case, $J = J_e = G$. Furthermore, from this equivalence it can be shown that

$$J = G = \frac{K^2}{E} \qquad \text{for plane stress and}$$

$$J = G = \frac{K^2}{E}(1 - \nu^2) \quad \text{for plane strain} \tag{6.31}$$

for linear elastic conditions. Thus, compatibility exists between the *J*-integral and LEFM concepts. Also, a critical value of *J*, defined as J_c, predicts the onset of crack extension, analogous to K_c in LEFM. ASTM Standard E1820 [67] was developed to provide procedures and guidelines for the determination of fracture toughness of metallic materials using *K*, COD, and *J* parameters. The standard includes methods for obtaining J_c and J_{Ic} for opening mode (mode I) loading.

The mathematics used to obtain expressions for the *J*-integral for use in fracture and fatigue crack growth are relatively complex and are beyond the scope of this book. Experimental methods to determine *J*-integral solutions, originally developed by Begley and Landes [68], employed the compliance method utilizing the load–displacement curve. In general, finite element techniques are now employed to aid the development of expressions for J. Various references provide J-integral solutions for a number of geometries and crack lengths [4,65,69,70].

The definition of J was formulated for nonlinear elastic materials. Nonlinear elastic behavior is similar to elastic-plastic behavior when the applied load is only in one direction. Thus, one can model elastic-plastic behavior as nonlinear elastic behavior only if no unloading occurs. In principle, this would restrict the use of J up to the beginning of crack extension but not for crack growth or reversal of the load. However, the use of J, particularly ΔJ, has been extended to describe fatigue crack growth under large-scale plasticity. Replacing W, T_i, and u_i in Eq. 6.29 with ΔW, ΔT_i, and Δu_i yields ΔJ where Eqs. 6.30 and 6.31 can then be written in terms of ranges. Thus,

$$\Delta J = \int_\Gamma \left(\Delta W dy - \Delta T_i \frac{\partial \Delta u_i}{\partial x} ds \right) \tag{6.32}$$

and

$$\Delta J = \Delta J_e + \Delta J_p \tag{6.33}$$

For plane stress and linear-elastic conditions, ΔJ reduces to the following equation:

$$\Delta J = \Delta J_e = \frac{\Delta K^2}{E} \tag{6.34}$$

Early experimental work using ΔJ to characterize fatigue crack growth was performed by Dowling and Begley [71] and Brose and Dowling [72] for a wide range of elastic to nearly fully plastic conditions. Fatigue crack growth data were shown to correlate well with ΔJ for this wide range of conditions. Subsequent studies also showed that fatigue crack growth data could be characterized using ΔJ [73,74]. ΔJ has been used successfully to characterize elastic-plastic fatigue crack growth with significant plasticity, and similar to LEFM, its use has been extended to include additional applications. This has led to some scrutiny. This scrutiny is based on differences observed between the loading and unloading portions of the fatigue cycle. The stress–strain relationships for loading and unloading involving plastic deformation are different, i.e., they are path dependent. This violates the principles used in the development of J. Metals in which cyclic saturation (cyclic stability) is reached rapidly have been modeled accurately using ΔJ [65,66]. However, many metals do not achieve cyclic stability, even at failure; thus, ΔJ may not uniquely correlate the fatigue crack growth rates. The presence of crack closure under large-scale plasticity has also raised questions regarding the use of ΔJ. After nearly 25 years, many significant advances have been made regarding the use of J and ΔJ, yet there are still limitations regarding the use of ΔJ to correlate fatigue crack growth adequately under significant plasticity.

6.10 SUMMARY

LEFM and its use to describe fatigue crack growth have received considerable attention over the past 40 years. Advanced testing facilities have allowed a broad range of fatigue crack growth mechanisms to be studied and understood more thoroughly. The development of LEFM by Griffith and Irwin was instrumental in providing a quantitative understanding of the reduced strength of components in the presence of flaws or cracks. The stress intensity factor, which is the fundamental parameter of LEFM, is a function of applied stress, crack length, and geometry, and for various crack geometries and configurations takes the form of one of the following:

$$K = S\sqrt{\pi a}\,\alpha \quad \text{or} \quad S\sqrt{\pi a}\,f\!\left(\frac{a}{w}\right) \quad \text{or} \quad S\sqrt{a}\,Y \qquad (6.2b)$$

Determination of fracture toughness, K_c or K_{Ic}, of metals has provided engineers with a quantitative design criterion to prevent failure of cracked components. Limitations on the use of LEFM due to plasticity have been established that restrict the net nominal stress to less than approximately 80 percent of the yield strength, and the plastic zone to be small relative to the crack length and local geometry. Understanding the limitations of LEFM such as plastic zone size and crack length effects has aided the development of other methods, such as EPFM, to predict fatigue crack growth behavior. LEFM applications to fatigue crack growth and damage-tolerant design have allowed life estimations or predictions to be made. In order to perform these calculations, information such as the stress intensity factor, fracture toughness, maximum and minimum stresses, and initial crack size is needed. The crack growth life is more sensitive to initial crack size than fracture toughness, which illustrates the importance of minimizing initial crack or discontinuity size. Empirical relationships describing fatigue crack growth developed by Paris, Forman, Walker, and others, along with computer solutions have been instrumental in allowing fatigue life calculations to be made. Region II fatigue crack growth essentially shows a linear relationship between log da/dN and log ΔK which corresponds to the equation

$$\frac{da}{dN} = A(\Delta K)^n = A(\Delta S)^n\,(\pi a)^{n/2}\,\alpha^n \qquad (6.20b)$$

Integration of Eq. 6.20b results in the general fatigue life equation:

$$N_f = \int_0^{N_f} dN = \frac{1}{A(\Delta S)^n \, (\pi)^{n/2}} \int_{a_i}^{a_f} \frac{da}{\alpha^n a^{n/2}}$$

Stress ratio, frequency, microstructure, material, and the environment affect fatigue crack growth rates and can cause significant shifts in the fatigue crack growth curves. The threshold stress intensity factor range, ΔK_{th}, defines a lower limit of crack growth rate, typically around 10^{-10} m/cycle, below which cracks are assumed to be nonpropagating.

Concepts such as the J-integral and EPFM, crack closure, and small crack growth have provided a great deal of stimulus to the engineering community for more than 25 years. Accurate fatigue life predictions for critical engineering components are essential, and integration of these concepts into life prediction models has had some success. However, the complex issues associated with each and the loss of similitude have raised significant questions and concerns. Despite these, LEFM is still applicable to many fatigue situations. The importance of testing and inspection, discussed in Chapter 2, is reemphasized here.

6.11 DOS AND DON'TS IN DESIGN

1. Do recognize that the presence of cracks or crack-like manufacturing and metallurgical discontinuities can significantly reduce the strength of a component or structure and that LEFM can aid both qualitatively and quantitatively in estimating static strength, as well as fatigue crack growth life and final fracture. The stress intensity factor K describes the stress field at the tip of a fatigue crack.

2. Do consider that fracture toughness depends much more on metallurgical discontinuities and impurities than does ultimate or yield strength. Low-impurity alloys have better fracture toughness.

3. Don't expect that doubling the thickness or ultimate strength of a component will double the fracture load. Cracks can exist, and fracture toughness may drop appreciably with both thickness and ultimate strength increases.

4. Do recognize the importance of distinguishing between plane stress and plane strain in fracture mechanics analysis, as fracture toughness, crack tip plasticity, and LEFM limitations can be significantly different in the two conditions.

5. Do determine reasonable estimates of stress intensity factors K for components containing cracks by modeling these cracks with simplified K solutions available in handbooks or computer software through computational, analytical, or experimental methods.

6. Don't neglect the importance of nondestructive flaw or crack inspection for both initial and periodic inspection periods.

7. Do note that most fatigue crack growth usually occurs in mode I even under mixed-mode conditions, and hence that the opening mode stress intensity factor range, ΔK_I, is often the predominant controlling factor in fatigue crack growth.

8. Do investigate the possibility of using LEFM principles in fatigue crack growth life predictions even in low-strength materials; crack tip plasticity can be small even in low-strength materials under fatigue conditions. If plasticity is large, EPFM principles may be required.

9. Don't neglect the importance of mean stress effects in fatigue crack growth applications. Various relationships exist to account for these effects in predicting fatigue life.

10. Do understand the importance of crack closure, especially in region I, and its effect on fatigue life.

11. Do recognize, using LEFM principles, that small cracks less than approximately 1 or 2 mm (0.04 or 0.08 in.) can grow faster than long cracks at the same nominal crack tip driving force, ΔK. Investigate whether small crack behavior is significant or can be neglected.

12. Do consider the possibility of inspection before fracture. High fracture toughness materials may not provide appreciable increases in fatigue crack growth life, but they do permit longer cracks before fracture, which makes inspection and detection of cracks more reliable.

REFERENCES

1. D. Broek, *Elementary Engineering Fracture Mechanics,* 4th ed., Kluwer Academic Publications, Dordrecht, the Netherlands, 1986.

2. R. W. Hertzberg, *Deformation and Fracture Mechanics of Engineering Materials,* 4th ed., John Wiley and Sons, New York, 1996.

3. H. L. Ewalds and R. J. H. Wanhill, *Fracture Mechanics,* Edward Arnold Publishers, Baltimore, 1984.

4. T. L. Anderson, *Fracture Mechanics: Fundamentals and Applications,* 2nd ed., CRC Press, Boca Raton, FL, 1995.

5. S. Suresh, *Fatigue of Materials,* 2nd ed., Cambridge University Press, Cambridge, 1998.

6. A. A. Griffith, "The Phenomena of Rupture and Flow in Solids," *Philos. Trans., R. Soc. Lond.,* Ser. A., Vol. 221, 1920, p. 163.

7. G. R. Irwin, "Analysis of Stresses and Strains Near the End of a Crack Traversing a Plate," *J. Appl. Mech.,* Vol. 24, 1957, p. 361.

8. V. Weiss and S. Yukawa, "Critical Appraisal of Fracture Mechanics," *Fracture Toughness Testing and Its Applications,* ASTM STP 381, ASTM, West Conshohocken, PA, 1965, p. 1.

9. W. F. Brown, Jr., and J. E. Srawley, "Plane Strain Crack Toughness Testing of High Strength Metallic Materials," ASTM STP 410, ASTM, West Conshohocken, PA, 1966, p. 1.

10. H. Tada, P. C. Paris, and G. R. Irwin, *The Stress Analysis of Cracks Handbook*, 2nd ed., Paris Productions, St. Louis, MO, 1985.

11. G. C. Sih, *Handbook of Stress Intensity Factors*, Institute of Fracture and Solid Mechanics, Lehigh University, Bethleham, PA, 1973.

12. D. P. Rooke and D. J. Cartwright, *Compendium of Stress Intensity Factors*, Her Majesty's Stationery Office, London, 1976.

13. F. W. Smith and D. R. Sorenson, "The Semi-Elliptical Surface Crack: A Solution by the Alternating Method," *Int. J. Fract.*, Vol. 12, No. 1, 1976, p. 47.

14. T. A. Cruse and G. J. Meyers, "Three-Dimensional Fracture Mechanics Analysis," *J. Struct. Div. ASCE*, Vol. 103, No. ST2, 1977, p. 309.

15. J. C. Newman, Jr., and I. S. Raju, "Stress-Intensity Factor Equations for Cracks in Three-Dimensional Finite Bodies," *Fracture Mechanics, Vol I: Theory and Analysis*, ASTM STP 791, J. C. Lewis and G. Sines, eds., ASTM, West Conshohocken, PA, 1983, p. 238.

16. Y. Murakami, ed. in chief, *Stress Intensity Factors Handbook*, Vol. 2, Pergamon Press, Oxford, 1987.

17. C. W. Smith, "Fracture Mechanics," *Experimental Techniques in Fracture Mechanics*, Vol. 2, A. S. Kobayashi, ed., Iowa State University Press, Ames, 1975, p. 3.

18. D. S. Dugdale, "Yielding of Steel Sheets Containing Slits," *J. Mech. Phys. Solids*, Vol. 8, 1960, p. 100.

19. *Fracture Toughness Testing and Its Application*, ASTM STP 381, ASTM, West Conshohocken, PA, 1965.

20. *Standard Test Method for Plane-Strain Fracture Toughness of Metallic Materials*, ASTM E399, Vol. 03.01, ASTM, West Conshohocken, PA, 2000, p. 431.

21. W. S. Pellini, "Criteria for Fracture Control Plans," NRL Report 7406, May 1972.

22. *Damage Tolerant Design Handbook, A Compilation of Fracture and Crack Growth Data for High Strength Alloys*, CINDAS/Purdue University, Lafeyette, IN, 1994.

23. E. T. Wessel, "Variation of K_c with Temperature for Low Alloy Nuclear Pressure Vessel Steel A533B," *Practical Fracture Mechanics for Structural Steel*, Chapman and Hall, London, 1969.

24. *Standard Practice for R-Curve Determination*, ASTM E561, Vol. 03.01, ASTM, West Conshohocken, PA, 2000, p. 522.

25. *Standard Test Method for Measurement of Fatigue Crack Growth Rates*, ASTM E647, Vol. 03.01, ASTM, West Conshohocken, PA, 2000, p. 591.

26. P. C. Paris, M. P. Gomez, and W. E. Anderson, "A Rational Analytical Theory of Fatigue," *Trend Eng.*, Vol. 13, No. 9, 1961, p. 9.

27. D. W. Hoeppner and W. E. Krupp, "Prediction of Component Life by Application of Fatigue Crack Growth Knowledge," *Eng. Fract. Mech.*, Vol. 6, 1974, p. 47.

28. J. M. Barsom, "Fatigue-Crack Propagation in Steels of Various Yield Strengths," *Trans. ASME, J. Eng. Ind.*, Ser. B, No. 4, 1971, p. 1190.

29. R. I. Stephens, ed., *Fatigue and Fracture Toughness of A356-T6 Cast Aluminum Alloy*, SP-760, Society of Automotive Engineers, Warrendale, PA, 1988.

30. J. Feiger, *"Small Fatigue Crack Growth from Notches in Ti-6Al-4V Under Constant and Variable Amplitude Loading,"* M. S. thesis, University of Idaho, 1999.

31. H. O. Liknes and R. R. Stephens, "Effect of Geometry and Load History on Fatigue Crack Growth in Ti-62222," *Fatigue Crack Growth Thresholds, Endurance Limits, and Design,* ASTM STP 1372, J. C. Newman and R. S. Piascik, eds., ASTM, West Conshohocken, PA, 1999, p. 175.

32. D. L. Goodenberger and R. I. Stephens, "Fatigue of AZ91E-T6 Cast Magnesium Alloy," *J. Eng. Mater. Tech.,* Vol. 115, 1993, p. 391.

33. R. G. Forman, V. E. Kearney, and R. M. Engle, "Numerical Analysis of Crack Propagation in Cyclic-Loaded Structures," *Trans. ASME, J. Basic Eng.,* Vol. 89, No. 3, 1967, p. 459.

34. N. E. Dowling, *Mechanical Behavior of Materials: Engineering Methods for Deformation, Fracture, and Fatigue,* 2nd ed., Prentice-Hall, Upper Saddle River, NJ, 1998.

35. R. I. Stephens, E. C. Sheets, and G. O. Njus, "Fatigue Crack Growth and Life Predictions in Man-Ten Steel Subjected to Single and Intermittent Tensile Overloads," *Cyclic Stress-Strain and Plastic Deformation Aspects of Fatigue Crack Growth,* ASTM STP 637, ASTM, West Conshohocken, PA, 1977, p. 176.

36. R. I. Stephens, P. H. Benner, G. Mauritzson, and G. W. Tindall, "Constant and Variable Amplitude Fatigue Behavior of Eight Steels," *J. Test. Eval.,* Vol. 7, No. 2, 1979, p. 68.

37. J. M. Barsom and S. T. Rolfe, *Fracture and Fatigue Control in Structures,* 3rd ed., ASTM, West Conshohocken, PA, 1999.

38. P. Au, T. H. Topper, and M. L. El Haddad, *Behavior of Short Cracks in Airframe Components,* AGARD Conference Proceedings No. 328, AGARD, France, 1983, p. 11.

39. A. F. Blom, "Near-Threshold Fatigue Crack Growth and Crack Closure in 17-4 PH Steel and 2024-T3 Aluminum Alloy," *Fatigue Crack Growth Threshold Concepts,* D. L. Davidson and S. Suresh, eds., Metallurgical Society of AIME, Warrendale, PA, 1984, p. 263.

40. J. R. Rice, "Mechanics of Crack Tip Deformation and Extension by Fatigue, *Fatigue Crack Propagation,* ASTM STP 415, ASTM, West Conshohocken, PA, 1967, p. 249.

41. W. Elber, "Fatigue Crack Closure Under Cyclic Tension," *Eng. Fract. Mech.,* Vol. 2, 1970, p. 37.

42. J. C. Newman and W. Elber, eds., *Mechanics of Fatigue Crack Closure,* ASTM STP 982, ASTM, West Conshohocken, PA, 1988.

43. J. C. Newman and R. C. McClung, eds., *Advances in Fatigue Crack Closure Measurement and Analysis,* Vol. 2, ASTM STP 1343, ASTM, West Conshohocken, PA, 1999.

44. D. L. Davidson and S. Suresh, eds., *Fatigue Crack Growth Threshold Concepts,* TMS-AIME, New York, 1984.

45. S. Suresh, G. F. Zaminski, and R. O. Ritchie, "Oxide-induced Crack Closure: An Explanation for Near-Threshold Corrosion Fatigue Crack Growth Behavior," *Metallurg. Trans.,* Vol. 12A, 1981, p. 1435.

46. R. O. Ritchie, S. Suresh, and C. M. Moss, "Near-Threshold Fatigue Crack Growth in 2 1/4 Cr-1 Mo Pressure Vessel Steel in Air and Hydrogen, *J. Eng. Mater. Tech.,* Vol. 102, 1980, p. 293.

47. K. Minakawa and A. J. McEvily, "On Crack Closure in the Near-threshold Region," *Scripta Metallurg.*, Vol. 15, 1981, p. 633.

48. N. Walker and C. J. Beevers, "A Fatigue Crack Closure Mechanism in Titanium," *Fatigue Eng. Mater. Struct.*, Vol. 1, 1979, p. 135.

49. G. T. Gray, J. C. Williams, and A. W. Thompson, "Roughness-Induced Crack Closure: An Explanation for Microstructurally Sensitive Fatigue Crack Growth, *Metallurg. Trans.*, Vol. 14A, 1983, p. 421.

50. E. P. Phillips, "Results of the Round Robin on Opening Load Measurements Conducted by ASTM Task Group E24.04.04 on Crack Growth Measurement and Analysis," NASA Technical Memorandum 1016001, NASA, 1989.

51. E. P. Phillips, "Results of the Second Round Robin on Opening-Load Measurement Conducted by ASTM Task Group E24.04.04 on Crack Growth Measurement and Analysis," NASA Technical Memorandum 109032, NASA, 1993.

52. H. Kitagawa and S. Takahashi, "Applicability of Fracture Mechanics to Very Small Cracks or the Cracks in the Early Stage," *Proceedings of the Second International Conference on Mechanical Behavior of Materials,* American Society for Metals, Metals Park, OH, 1976, p. 627.

53. S. Pearson, "Initiation of Fatigue Cracks in Commercial Aluminum Alloys and the Subsequent Propagation of Very Short Cracks," *Eng. Fract. Mech.*, Vol. 7, 1975, p. 235.

54. S. Suresh and R. O. Ritchie, "Propagation of Short Fatigue Cracks," *Int. Metals Rev.*, Vol. 29, 1983, p. 445.

55. K. J. Miller and E. R. de les Rios, eds., *The Behavior of Short Fatigue Cracks,* Mechanical Engineering Publication, London, 1986.

56. R. O. Ritchie and J. Lankford, eds., *Small Fatigue Cracks,* The Metallurgical Society, Warrendale, PA, 1986.

57. J. Larson and J. E. Allison, eds., *Small Crack Test Methods,* ASTM STP 1149, ASTM, West Conshohocken, PA, 1992.

58. P. E. Irving and C. J. Beevers, "Microstructural Influences on Fatigue Crack Growth in Ti–6Al–4V," *Mater. Sci. Eng.*, Vol. 14, 1974, p. 229.

59. R. I. Stephens, H. W. Lee, R. Bu, and G. K. Werner, "The Behavior of Short and Long Fatigue Cracks at Threshold and Near-Threshold Levels," *Fracture Mechanics: Eighteenth Symposium,* D. T. Read and R. P. Reed, eds., ASTM STP 945, ASTM, West conshohocken, PA, 1988, p. 881.

60. C. R. Owen, R. J. Bucci, and R. J. Kegaritse, "An Aluminum Quality Breakthrough for Aircraft Structural Reliability," *J. Aircraft,* Vol. 26, 1989, p. 178.

61. J. R. Rice, "A Path-Independent Integral and the Approximate Analysis of Strain Concentration by Notches and Cracks," *J. Appl. Mech., Trans. ASME,* Vol. 35, 1968, p. 379.

62. J. D. Eshelby, "The Continuum Theory of Lattice Defects," *Solid State Physics,* Vol. 3, Academic Press, New York, 1956, p. 79.

63. J. W. Hutchinson, "Singular Behavior at the End of a Tensile Crack in a Hardening Material," *J. Mech. Phys. Solids,* Vol. 16, 1968, p. 13.

64. A. A. Wells, "Unstable Crack Propagation in Metals: Cleavage and Fast Fracture," *Proceedings of the Crack Propagation Symposium,* Vol. 1, paper 84, Cranfield, England, 1961, p. 210.

65. A. Saxena, *Nonlinear Fracture Mechanics for Engineers,* CRC Press, Boca Raton, FL, 1998.

66. M. F. Kanninen and C. H. Popelar, *Advanced Fracture Mechanics,* Oxford University Press, New York, 1985.

67. *Standard Test Method for Measurement of Fracture,* ASTM E1820, Vol. 03.01, ASTM, West Conshohocken, PA, 2000, p. 1000.

68. J. A. Begley and J. D. Landes, "The J-Integral as a Fracture Criterion," *Fracture Toughness, Part II,* ASTM STP 514, ASTM, West Conshohocken, PA, 1972, p. 1.

69. V. Kumar, M. D. German, and C. F. Shih, "An Engineering Approach for Elastic-Plastic Fracture Analysis," EPRI Report NP-1931, Electric Power Research Institute, Palo Alto, CA, July 1981.

70. A. Zahoor, *Ductile Fracture Handbook,* Vol. III, EPRI Report NP-6301-D, Electric Power Research Institute, Palo Alto, CA, 1989.

71. N. E. Dowling and J. A. Begley, "Fatigue Crack Growth During Gross Plasticity and the J-Integral," *Mechanics of Crack Growth,* ASTM STP 590, ASTM, West Conshohocken, PA, 1976, p. 82.

72. W. R. Brose and N. E. Dowling, "Fatigue Crack Growth Under High Stress Intensities in 304 Stainless Steel," *Elastic-Plastic Fracture,* ASTM STP 668, ASTM, West Conshohocken, PA, 1979, p. 720.

73. D. A. Jablonski and B. H. Lee, "Computer Controlled Fatigue Crack Growth Rate Testing Using the J-Integral," *Fatigue Life: Analysis and Prediction, Proceedings of the ASM International Conference on Fatigue, Corrosion Cracking, Fracture Mechanics, and Failure Analysis,* Materials Park, OH, 1985.

74. J. R. Rice, "Elastic-Plastic Crack Growth," *Mechanics of Solids,* H. G. Hopkins and M. J. Sewell, eds., Pergamon Press, Oxford, 1982, p. 539.

PROBLEMS

1. Using the principle of superposition, determine K_I for the member shown, which has an offset force P equal to 100 kN. The crack is a through-thickness crack. Repeat the calculations for $a = 30$ mm.

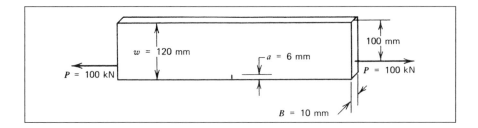

2. Superimpose a plot of $K_I/(S\sqrt{\pi a})$ versus angle β (for β between 0° and 90°) for an elliptical embedded crack with $a/c = 1$ (circle), 0.5, and 0.1

(long, shallow crack), assuming that S and a are constant. Comment on the influence of the crack shape on K_I for a given stress and crack length a.

3. Compare the approximate K_I value for a quarter-circular corner crack with the results of problem 2.

4. For the A533B steel of Fig. 6.9, what force will cause fracture at $0°$ and $-50°C$ for a circular corner crack of radius $a = 10$ mm in a thick, axially loaded 50×100 mm cross section bar? What plastic zone sizes exist in the two fracture conditions? Is the LEFM model justifiable in solving this problem? What force would cause general yielding if the crack were not present? Discuss the significance of the above questions and answers.

5. A gas turbine component is made of recrystallized, annealed Ti–6Al–4V with $K_{Ic} = 85$ MPa$\sqrt{m}$ and $S_y = 815$ MPa. A surface semicircular crack $(a/c = 1)$ similar to that in Fig. 6.4d is found during a routine maintenance inspection. If the component thickness is 25 mm, comment on the stress state (i.e., plane stress or plane strain). If a stress is applied normal to the crack plane like that in Fig. 6.4a, what maximum stress is required to cause fracture if $a = 8$ mm and $K_c = 105$ MPa$\sqrt{m}$? If the thickness were doubled, what maximum stress would cause fracture? Comment on the conditions required for fracture at each thickness and whether LEFM is valid for each case.

6. An 18 Ni (200) maraging steel thick rod fractured under an axial load. It was suggested to increase the strength and use an 18 Ni (300) maraging steel. Using Table A.3, comment on the above suggestion for (a) no cracks present, and (b) a crack present. For a given crack size, what is the ratio of the load-carrying capacity of the two steels? How would you overcome the original fracture problem, assuming that the crack was present at fracture?

7. Assume the fracture toughness data of Fig. 6.7 and assume that a uniaxially loaded part is 2.5 mm thick. It is desired to increase the fracture load capacity by a factor of 3. If the thickness were tripled and no cracks existed, this could be a satisfactory solution. However, if a through crack existed in both parts, would tripling the thickness give the desired tripling in fracture load? What load change at fracture would be achieved?

8. Determine an approximate allowable centrally located through-thickness crack size, $2a$, in a very wide, thick plate of 2024-T3 and 7075-T6 aluminum alloys subjected to a nominal stress of 270 MPa. Repeat the calculation for nominal stresses equal to $S_y/2$. What is the significance of the above results?

9. For the region II fatigue crack growth rate properties provided in Table 6.3, plot the corresponding equations on a log da/dN versus log ΔK plot for 10^{-8} m/cycle $\leq da/dN \leq 10^{-5}$ m/cycle. Comment on the differences of these results between materials.

10. Fractographic electron microscopy often reveals indications of fatigue crack growth striations that can be used to estimate prevailing stress intensity factor levels. Figure 3.15a shows one such electron fractograph. The component was a uniaxially loaded, single-edge cracked plate of width 20 mm, thickness 5 mm, and crack length 6 mm. For $R = 0$ constant amplitude loading, what approximate axial force was applied to the plate? Assume that the Paris equation constants are $A = 10^{-12}$ and $n = 3.5$, where da/dN is in m/cycle and ΔK is in MPa$\sqrt{m}$.

11. The following crack growth data were obtained with SAE 0030 cast steel at $-34°C$ using compact tension specimens with $H/W = 0.49$, $W = 64.8$ mm, thickness $B = 8.2$ mm, $P_{max} = 6.14$ kN, and $P_{min} = 0.089$ kN. K_I for this specimen is given by

$$K_I = \frac{P\sqrt{a}}{Bw}\left[30.96 - 195.8\left(\frac{a}{w}\right) + 730.6\left(\frac{a}{w}\right)^2 - 1186.3\left(\frac{a}{w}\right)^3 + 754.6\left(\frac{a}{w}\right)^4\right]$$

Plot a log-log curve of da/dN versus ΔK. Reduce the data using a numerical scheme. See ASTM Standard E647. Obtain the Paris equation coefficient A and exponent n.

Cycles	Crack Length (mm)	Cycles	Crack Length (mm)
0	22.0	751 200	27.7
37 100	22.1	801 600	28.9
68 300	22.3	863 400	29.8
111 200	22.5	908 800	30.9
173 500	22.7	942 900	32.0
238 000	23.0	978 000	33.5
286 400	23.3	1 006 700	35.1
371 000	23.8	1 028 900	36.6
446 200	24.1	1 059 700	39.2
501 200	24.9	1 069 400	42.0
560 900	25.8	1 073 200	44.1
691 800	27.0	1 074 900	45.0

12. A very wide 25 mm thick plate of 7075-T6 aluminum contains a single-edge crack of length $a = 1$ mm. The plate is subjected to an alternating stress with $S_{max} = 175$ MPa and $S_{min} = 0$. Assuming that the crack growth expressions in Table 6.3 are applicable, determine

(a) whether plane stress or plane strain conditions exist.

(b) the critical crack length at fracture.

(c) the number of cycles needed to cause fracture.

(*d*) If the plate thickness were reduced to 5 mm and the applied stress levels remained the same, would you expect the total life to fracture to change significantly? Comment on your rationale.

13. A uniaxially loaded very wide sheet of low carbon steel with $S_y = 350$ MPa, $S_u = 550$ MPa, $K_c = 110$ MPa$\sqrt{m}$, and $\Delta K_{th}(R = 0) = 5$ MPa$\sqrt{m}$ is subjected to the following load spectrum:

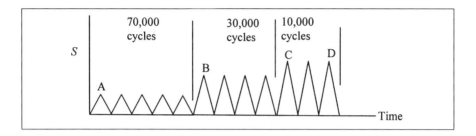

The three maximum stress levels, S_{max}, are 150, 200, and 250 MPa, and $R = 0$. The initial center crack length $2a = 5.0$ mm. Assume that Barsom's crack growth expressions are applicable. Determine:

(*a*) K_I at time A and compare with ΔK_{th} and K_c.

(*b*) What value of S_{max} will not cause crack growth at time A.

(*c*) The crack length at times B, C, and D.

(*d*) K_I at B, C, and D and compare with K_c.

(*e*) The plane stress plastic zone size, $2r_y$, at A, B, C, and D.

(*f*) The total number of cycles required to cause fracture if $S_{max} = 250$ MPa is continued beyond D until fracture.

(*g*) Comment on the applicability of LEFM in (*a*) through (*f*).

14. Conversion of the exponential Paris equation for fatigue crack growth rules from SI units to American/British units or vice versa has too frequently been done incorrectly. Verify that the conversions in Table 6.3 have been done correctly.

15. For Problem 1, use a computer numerical integration program and determine the approximate number of cycles to grow the crack to fracture if the load cycles between 0 and 100 kN. Some material properties may have to be estimated or found elsewhere. The following plate materials are used:

(*a*) Mild steel.

(*b*) D6AC.

(*c*) 2024-T3 aluminum.

(*d*) Ti–6Al–4V.

16. Consider 7075-T6 aluminum with fatigue crack growth properties as given in Table 6.3. Assuming that $\lambda = 0.5$ in the Walker equation,

   (a) estimate da/dN versus ΔK equations in the form of Eq. 6.22a for $R = 0.4$ and 0.8.

   (b) plot the equations from (a) on a log-log scale along with the $R = 0$ equation. Keep in mind the proper region II limits of the equation.

   (c) For a given ΔK, by what factor does the crack growth rate, da/dN, increase if R is increased from 0 to 0.4? If it is increased from 0 to 0.8?

17. Repeat Problem 13 but assume that S_{min} is 100 MPa during the entire load spectrum. Let $\lambda = 0.5$ and $\gamma = 0.3$ for use in Eqs. 6.22a and 6.23, if applicable. What is the crack length at times B, C, and D in comparison to those in Problem 13c?

18. A medium-strength steel displays the following region II Paris relationship for long crack behavior at $R = 0$:

$$\frac{da}{dN} = 2.4 \times 10^{-11} (\Delta K)^{2.75}$$

where da/dN is in m/cycle and ΔK is in MPa$\sqrt{m}$. Data on physically small cracks were generated for the same material at $R = 0$, and fitting a power law expression to the data yielded the following relationship:

$$\frac{da}{dN} = 1.8 \times 10^{-10} (\Delta K)^{1.75}$$

   (a) Plot the equation for these two relationships, the long crack equation between 10^{-8} m/cycle $< da/dN < 10^{-5}$ m/cycle and the small crack equation between 1 MPa$\sqrt{m}$ and where it merges with the long crack Paris equation. If $\Delta K_{th} = 5$ MPa$\sqrt{m}$ for the long crack data, also sketch the approximate sigmoidal portion of the long crack growth curve in region I. Complete the approximate sigmoidal long crack growth curve if $K_c = 95$ MPa$\sqrt{m}$.

   (b) Based on your plot, will extrapolation of the Paris equation to region I predict conservative or nonconservative fatigue life if a physically small crack exists in a component made from this material?

19. A uniaxially loaded very wide sheet of the material from Problem 18 is subjected to constant amplitude loading at $R = 0$ with $S_{max} = 110$ MPa. Let $K_c = 95$ MPa$\sqrt{m}$ and $S_y = 440$ MPa. Laboratory experiments have shown that for this material, physically small crack growth occurs up to a length of 1 mm. If the wide sheet contains an initial edge crack with $a_i = 0.3$ mm,

(a) Calculate the fatigue life of the sheet using only the long crack growth behavior.

(b) Calculate the fatigue life, taking into consideration the small crack growth behavior.

(c) Comment on your results and the use of this life prediction technique.

20. A very wide, 60 mm thick plate of mill annealed Ti–6Al–4V with a through-thickness central crack of length 15 mm is subjected to a nominal stress of 700 MPa normal to the crack plane. Determine:

(a) the applied stress intensity level, K.

(b) the effective stress intensity level, K_{eff}, based on plasticity correction.

(c) If the plate were cyclically loaded between $S_{min} = 0$ and $S_{max} = 725$ MPa, determine the total number of cycles needed to cause fracture.

(d) Comment on your results and on the applicability of LEFM.

CHAPTER 7

NOTCHES AND THEIR EFFECTS

Notch effects have been a key problem in the study of fatigue for more than 125 years, since Wöhler showed that adding material to a railway axle might make it weaker in fatigue. He stated that the radius at the shoulder between a smaller and a larger diameter is of prime importance to the fatigue life of axles and that fatigue cracks will start at the transition from a smaller to a larger section.

Notches cannot be avoided in many structures and machines. A bolt has notches in the thread roots and at the transition between the head and the shank. Rivet holes in sheets, welds on plates, and keyways on shafts are all notches. Although notches can be very dangerous, they can often be rendered harmless by suitable treatment.

To understand the effects of notches and the means to overcome them, one must consider five parameters in addition to the behavior of smooth specimens of the same material:

Concentrations of stress and of strain
Stress gradients
Mean stress effects and residual stresses
Local yielding
Nucleation and growth of cracks

In some cases, one of the five parameters by itself may explain the difference in behavior between a smooth part and a notched part that has an equal cross section at the root of the notch. Even when several parameters are involved, it is often possible to use "variable constants" or "notch factors" that correlate the test results. We intend to avoid the use of adjustable factors and must therefore consider the effects of all five parameters. In the end, this will be less difficult than the adjustment of a single notch factor.

7.1 CONCENTRATIONS AND GRADIENTS OF STRESS AND STRAIN

Notches concentrate stresses and strains. The degree of concentration is a factor in the fatigue strength of notched parts. It is measured by the elastic stress concentration factor, K_t, defined as the ratio of the maximum stress, σ, or strain, ε, at the notch to the nominal stress, S, or strain, e.

$$K_t = \frac{\sigma}{S} = \frac{\varepsilon}{e} \quad \text{as long as} \quad \frac{\sigma}{\varepsilon} = \text{constant} = E \tag{7.1}$$

where σ and ε represent local stress and strain at the notch and S and e represent nominal stress and strain, respectively. Let us consider a sheet with a central circular hole. K_t depends on the ratio of hole diameter to sheet width. Figure 7.1 shows K_t plotted versus the ratio of hole diameter to sheet width. Two curves are shown. In the upper curve the nominal stress is defined

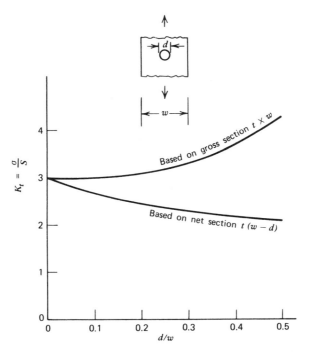

Figure 7.1 Elastic stress concentration factor for a central circular hole in a sheet.

as load divided by total or gross area ($w \times t$). In the lower curve the nominal stress is defined as load divided by net area, i.e., the area remaining after the hole has been cut out. In this book we use the net area to define the nominal stress when using stress concentration factors unless otherwise stated. However, in calculating the stress intensity factor from the nominal stress, we use the gross area as if the crack did not exist, as discussed in Chapter 6.

Figure 7.2 shows the stresses near a circular hole in the center of a wide sheet in tension. This problem was investigated and solved as early as 1898 by Kirsch [1]. The following equations for the axial stress, σ_y, and the transverse stress, σ_x, on a transverse line through the center of the hole for plane stress are taken from Grover [2].

$$\frac{\sigma_y}{S} = 1 + 0.5 \left(\frac{r}{x}\right)^2 + 1.5 \left(\frac{r}{x}\right)^4 \tag{7.2}$$

$$\frac{\sigma_x}{S} = 1.5 \left(\frac{r}{x}\right)^2 - 1.5 \left(\frac{r}{x}\right)^4 \tag{7.3}$$

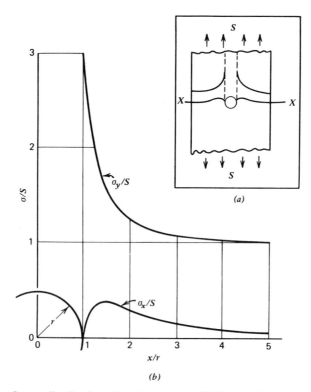

Figure 7.2 Stress distribution along the section X-X near a hole in a wide sheet.

where S = nominal stress = load/area

σ_y = axial stress

σ_x = transverse stress

x = distance from center of hole

r = radius of hole

Values of σ_y/S and σ_x/S are plotted versus x/r in Fig. 7.2. We see that σ_y/S decreases quite rapidly as the distance from the edge of the hole is increased. K_t at the edge of the hole is 3, while at a distance of $0.25r$ the value of σ_y/S is only about 2. At a distance of $2r$ it is only 1.07. In other words, the stress at the edge of the hole is three times the nominal stress, but at a distance from the hole edge equal to the diameter, it is only about 7 percent higher than the nominal stress. Also, the state of stress at the notch is uniaxial and away from the notch it is biaxial for this plane stress condition.

The slope of the σ_y versus x curve at the edge of the hole is another measure of the rapid decrease in stress as we move away from the edge of the hole. The rapid decrease in stress with increasing distance from the notch and the existence of biaxial or triaxial states of stress at a small distance from the notch are typical of stress concentrations. They explain why we cannot expect to predict the behavior of notched parts with great accuracy by applying stress concentration factors to the fatigue strength values obtained from smooth parts. Numerical values of stress gradients or simple design formulas for stress distribution near notches are not often readily available in the literature. For deep, narrow notches with semicircular ends, a formula analogous to LEFM formulas has been given for the stress distribution [3]. It is

$$\sigma_y = \sigma_{max} \left(\frac{0.5r}{0.5r + d} \right)^{1/2} \tag{7.4}$$

where d is the distance from the edge of the notch of radius r.

The stress concentration produced by a given notch is not a unique number, as it depends on the mode of loading and on the type of stress that is used to calculate K_t. For instance, for a circular hole in a wide sheet, we have the following stress concentration factors:

In tension	3
In biaxial tension	2
In shear	4 based on maximum tension
	2 based on maximum shear

Elastic stress concentration factors are obtained from the theory of elasticity, from numerical solutions, or from experimental measurements. The most common and most flexible numerical method is the finite element method.

When using this method, a model with relatively fine mesh in the areas of steep stress gradients is required to ensure computational accuracy. Experimental measurement techniques that are widely used include brittle coatings, photoelasticity, thermoelasticity, and strain gages. In the brittle coating technique, a brittle coating is sprayed on the surface and allowed to dry. Crack patterns developed by the loading and their relation to a calibration coating indicate regions and magnitudes of stress concentrations. For the photoelasticity technique, a specimen with geometry identical to that of the actual notched part is made of a certain transparent material. Changes in the optical properties of the transparent material subjected to loading, measured by a polariscope, indicate stress distributions and magnitudes. In the thermoelasticity technique, stress distribution is obtained by monitoring small temperature changes of the specimen or component subjected to cyclic loading. The most common experimental measurement technique used with actual parts involves an electrical resistance strain gage bonded to the surface in the region of interest. An applied load causes dimensional changes in the gage resulting in changes in electrical resistance, which in turn indicates the existing strain at the gage location.

It is important to know where the most dangerous stress concentrations exist in a part. Charts of stress concentration factors are available in the literature [4]. Examples of such charts for stepped shafts in tension, bending, and torsion are shown in Fig. 7.3, and for a plate with opposite U-shaped notches in tension and bending are shown in Fig. 7.4. It is important, however, to remember that elastic stress concentration factors for homogeneous isotropic materials depend only on geometry (independent of material) and mode of loading, and that they apply only when the notch is in the linear elastic deformation condition.

For qualitative estimates, some engineers like to use an imperfect analogy between stresses or strains and liquid flow. Restrictions or enlargements in a pipe produce local increases in flow velocity somewhat similar to the local increases in stresses produced by changes in cross section. The designer will try to "streamline" the contours of parts, as indicated in Fig. 7.5. Consider, for instance, an elliptic hole in a wide sheet. Placed lengthwise with the forces or flow, it produces less stress concentration and less flow interference than when it is placed crosswise. The formula for K_t produced by an elliptic hole with principal axes $2a$ and $2b$ is

$$K_t = 1 + 2\frac{b}{a} = 1 + 2\sqrt{\frac{b}{r}} \tag{7.5}$$

where b is the axis transverse to the tension force and r is the radius of curvature at the endpoint of b. With an ellipse 30 mm long and 10 mm wide the stress concentration is

$K_t = 7$ if placed crosswise
$K_t = 1.67$ if placed lengthwise

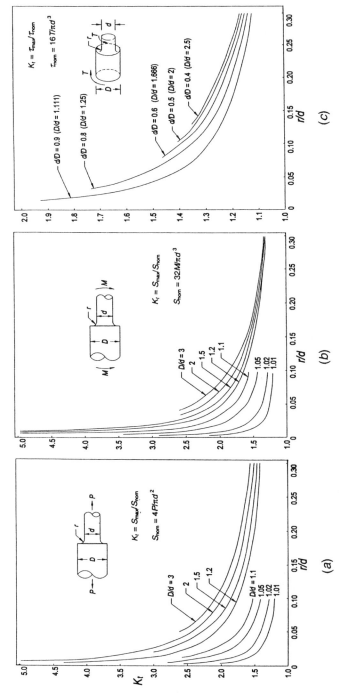

Figure 7.3 Stress concentration factors for a stepped shaft [4]. (*a*) Tension. (*b*) Bending. (*c*) Torsion (reprinted by permission of John Wiley and Sons, Inc.).

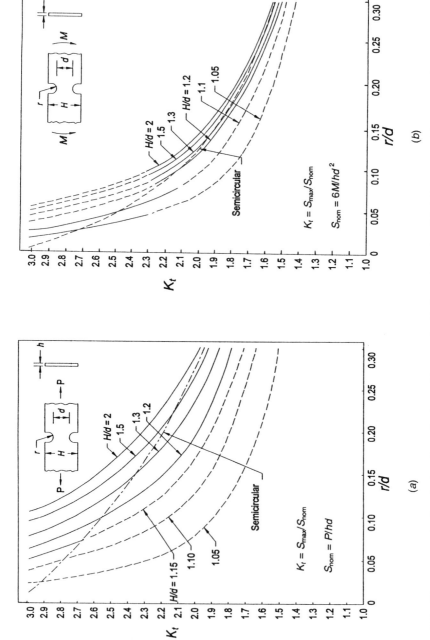

Figure 7.4 Stress concentration factors for a flat bar with opposite U-shaped notches [4]. (*a*) Tension. (*b*) Bending (reprinted by permission of John Wiley and Sons, Inc.).

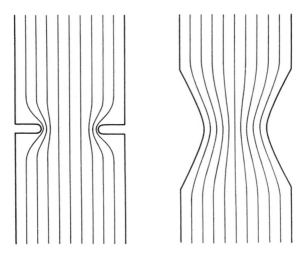

Figure 7.5 The crowding and bending of flow lines near obstructions helps to visualize the concentration of stresses and strains near notches. The large section and the small section are the same in both cases, but the transitions are different.

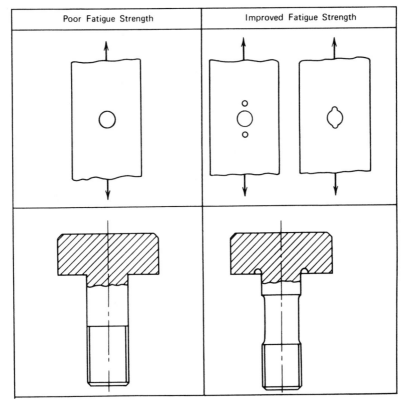

Figure 7.6 Mitigation of notch effects by design [6] (reprinted with permission of Melbourne University Press).

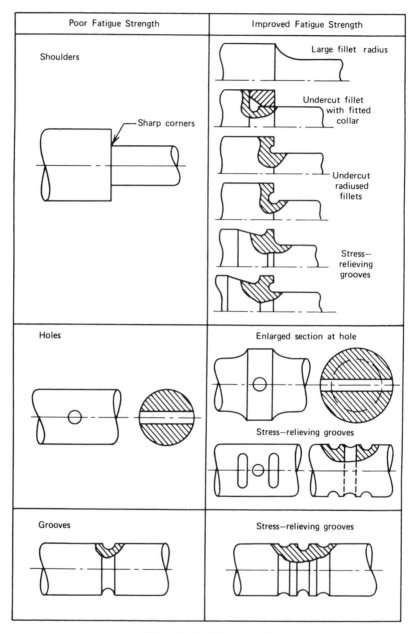

Poor Fatigue Strength	Improved Fatigue Strength
Shoulders	Large fillet radius
	Undercut fillet with fitted collar
Sharp corners	Undercut radiused fillets
	Stress—relieving grooves
Holes	Enlarged section at hole
	Stress—relieving grooves
Grooves	Stress—relieving grooves

Figure 7.6 (*Continued*)

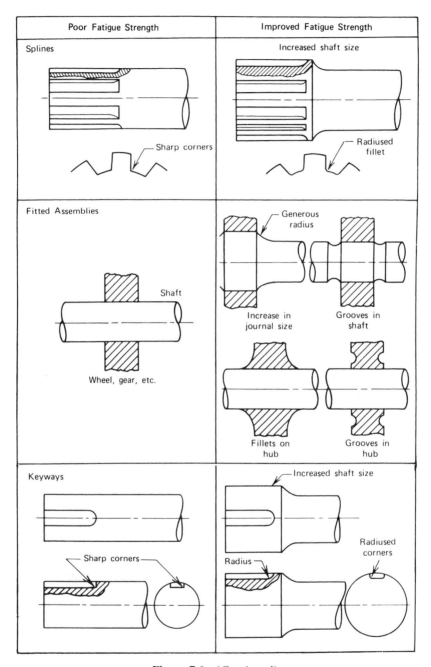

Poor Fatigue Strength	Improved Fatigue Strength
Splines	Increased shaft size
Sharp corners	Radiused fillet
Fitted Assemblies	Generous radius
Shaft	Increase in journal size — Grooves in shaft
Wheel, gear, etc.	Fillets on hub — Grooves in hub
Keyways	Increased shaft size
Sharp corners	Radius — Radiused corners

Figure 7.6 (*Continued*)

(From Eq. 7.5, the radii of curvature are 1.7 and 45 mm.) This example illustrates a way of mitigating stress concentrations. If a hole is needed, the stress concentration can be reduced by elongating it to an ellipse in the direction of loading or by adding smaller relief holes above and below it aligned with the direction of loading [5]. Additional examples of mitigating stress concentrations are given in Fig. 7.6 [6].

7.2   *S–N* APPROACH FOR NOTCHED MEMBERS

The effect of the notch in the stress–life approach is taken into account by modifying the unnotched *S–N* curve through the use of the fatigue notch factor, K_f. This factor is defined and described in the next section.

7.2.1   Notch Sensitivity and the Fatigue Notch Factor, K_f

When the theory of elasticity was used to compute stress concentrations, there was hope that the fatigue strength of a notched component could be predicted as the strength of a smooth component divided by a factor computed from the theory. The facts, however, are different. Notched fatigue strength depends not only on the stress concentration factor, but also on other factors such as the notch radius, the material strength, and the mean and alternating stress levels. The ratio of smooth to net notched fatigue strengths, based on the ratio of alternating stresses, is called K_f.

$$K_f = \frac{\text{Smooth fatigue strength}}{\text{Notched fatigue strength}} \qquad (7.6)$$

Therefore, the fatigue notch factor, K_f, is not necessarily equal to the elastic stress concentration factor. Within reasonable limits, we can predict K_f by the methods explained in the following sections.

As a base for estimating the effect of other parameters, we estimate the fatigue notch factor for zero mean stress and long life (10^6–10^8 cycles). The difference between K_f and K_t is believed to be related to the stress gradient and localized plastic deformation at the notch root. The reasoning for stress gradient influence is that the notch stress controlling the fatigue life may not be the maximum stress on the surface of the notch root, but rather an average stress acting over a finite volume of the material at the notch root. This average stress is lower than the maximum surface stress, calculated from K_t. Also, when small cracks nucleate at the notch root, they grow into regions of lower stress due to the stress gradient. The localized plastic deformation and notch blunting effect due to yielding at the notch root reduce the notch root stress amplitude, particularly at short lives. Therefore, fatigue behavior at the notch root is complex and is not a simple function of the notch geometry, characterized by the stress concentration factor, K_t. Values of K_f for $R = -1$ generally range between 1 and K_t (where K_t is defined based on the net area), depending on the notch sensitivity of the material, q, which is defined by

$$q = \frac{K_f - 1}{K_t - 1} \tag{7.7}$$

A value of $q = 0$ (or $K_f = 1$) indicates no notch sensitivity, whereas a value of $q = 1$ (or $K_f = K_t$) indicates full notch sensitivity. The fatigue notch factor can then be described in terms of the material notch sensitivity as

$$K_f = 1 + q(K_t - 1) \tag{7.8}$$

Neuber [7] developed the following approximate formula for the notch factor for $R = -1$ loading:

$$q = \frac{1}{1 + \sqrt{\rho/r}} \quad \text{or} \quad K_f = 1 + \frac{K_t - 1}{1 + \sqrt{\rho/r}} \tag{7.9}$$

where r is the radius at the notch root. The characteristic length, ρ, depends on the material. Values of ρ for steel alloys are shown in Fig. 7.7 [8], and a few values of ρ for aluminum alloys are as follows [2]:

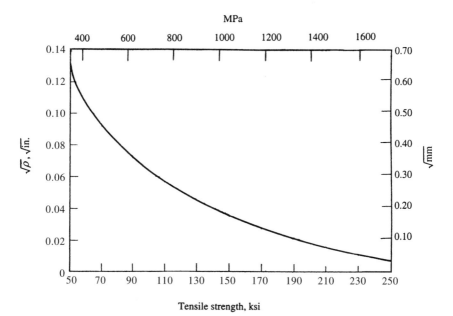

Figure 7.7 Neuber's material characteristic length $\sqrt{\rho}$ versus tensile strength for steel alloys [8].

S_u, MPa (ksi)	150 (22)	300 (43)	600 (87)
ρ, mm (in.)	2 (0.08)	0.6 (0.025)	0.4 (0.015)

Peterson [9] has observed that good approximations for $R = -1$ loading can also be obtained by using the somewhat similar formula

$$q = \frac{1}{1 + a/r} \quad \text{or} \quad K_f = 1 + \frac{K_t - 1}{1 + a/r} \tag{7.10}$$

where a is another material characteristic length. Values for a are given by Peterson [9]. An empirical relationship between S_u and a for steels is given as [10]

$$a = 0.0254 \left(\frac{2070}{S_u}\right)^{1.8} \quad \text{with } S_u \text{ in MPa and } a \text{ in mm} \tag{7.11a}$$

or

$$a = 0.001 \left(\frac{300}{S_u}\right)^{1.8} \quad \text{with } S_u \text{ in ksi and } a \text{ in in.} \tag{7.11b}$$

For aluminum alloys, a is estimated as 0.635 mm (0.025 in.) [9].

It should be kept in mind that the formulas used to estimate K_f, such as those of Neuber and Peterson, are empirical. These formulas and the associated material characteristic lengths express the fact that for large notches with large radii we must expect K_f to be almost equal to K_t, but for small sharp notches we may find that $K_f \ll K_t$ (little notch effect) for metals with ductile behavior, although K_f remains large for high-strength metals. In general, hard metals are usually more notch sensitive than soft metals.

Looking at a numerical example, we calculate K_f (according to Peterson's formula) for a 25 mm hole and for a 1 mm hole in mild steel with $S_u = 350$ MPa and in hard steel with $S_u = 1750$ MPa. In all cases $K_t = 2.5$. We find the following values for K_f from Eq. 7.11:

	$S_u = 350$ MPa	$S_u = 1750$ MPa
25 mm diameter hole	$K_f = 2.4$	$K_f = 2.5 = K_t$
1 mm diameter hole	$K_f = 1.7$	$K_f = 2.4$

Thus, for the 1 mm diameter hole in the mild steel, K_f is two-thirds of K_t, whereas for the 25 mm diameter hole in the hard steel $K_f = K_t$.

7.2.2   Effects of Stress Level on Notch Factor

For a fatigue life of 10^6 to 10^8 cycles with $R = -1$ we can estimate the notched fatigue strength as S_f/K_f. The nominal net stress calculated for these conditions is less than the yield strength. We can estimate the notch factor, K_f, as shown in Section 7.2.1.

In monotonic testing, notches may increase or decrease the nominal strength. A sharp groove in a notched tensile bar with ductile behavior produces a higher nominal ultimate tensile strength than a uniform bar of the same minimum diameter. The smooth bar necks down and finally fails with a much reduced area. The notched bar cannot neck down because of the unyielded metal above and below the notch. Its ultimate tensile strength is greater. For a material behaving in a brittle manner, necking does not occur in the smooth bar and greater notch sensitivity exists. Therefore, the ultimate tensile strength of the notched bar made of such a material is generally lower than that of the smooth bar. Data on the behavior of notched parts and smooth parts at different constant amplitude stress levels generally show that the notched fatigue strength is a greater fraction of the smooth fatigue strength as the nominal stresses get higher. Where data are available, the designer will of course use them, but in the absence of data the behavior of notched parts must be estimated.

We have seen in the previous section how one can estimate K_f, the fatigue notch factor, for fully reversed stresses at long life. In the absence of other data, one can estimate the monotonic tensile strength of the notched part for a metal behaving in a ductile manner to be equal to the strength of the smooth part in monotonic testing. This corresponds, strictly speaking, to one-quarter of a fully reversed cycle but it may be assumed to be 1 cycle. If a monotonic test can be made, one would, of course, use its result rather than the estimate.

Two endpoints of the notched *S–N* curve are thus estimated, one at 1 cycle and the other at 10^6 to 10^8 cycles. A straight line between these points, Basquin's equation, in a log *S*–log *N* plot is a reasonable approximation unless other data are available. If the *S–N* line for smooth parts shows a pronounced curvature, we would expect the curvature to appear also on the line for the notched part. An *S–N* curve for fully reversed stresses for notched parts in the usual laboratory environment can thus be estimated from the following data:

   The ultimate tensile strength, S_u, or true fracture strength, σ_f, of the material
   The long life, fully reversed fatigue strength, S_f, for smooth specimens of comparable size
   The material characteristic length, a or ρ
   The elastic stress concentration factor, K_t, and notch root radius, r.

For example, for an 80 mm wide sheet of 1020 hot-rolled steel with a 10 mm central hole, we would construct the *S–N* curve in the following way. Use Fatigue strength at approximately 10^6 cycles and ultimate tensile strength from Table A.1. They are 241 and 448 MPa (35 and 65 ksi), respectively. Since fatigue strengths listed in Table A.1 are for bending loads, they should be reduced by 10 to 25 percent for axial loading as indicated in Section 4.4.2. Here we use 15 percent reduction resulting in an unnotched fatigue limit of 0.85 (241) = 205 MPa. We therefore define the upper line in Fig. 7.8, from 448 MPa at 1 cycle to 205 MPa at 10^6 cycles.

The elastic stress concentration factor is 2.7 from Fig. 7.1. The material characteristic length, ρ, is 0.24 mm (from Fig. 7.7). The fatigue notch factor then is

$$K_f = 1 + \frac{K_t - 1}{1 + \sqrt{\rho/r}} = 1 + \frac{2.7 - 1}{1 + \sqrt{0.24/5}} = 2.4$$

The S–N line for the sheet with the hole then goes from 448 MPa at 1 cycle to 205/2.4 = 85 MPa at 10^6 cycles, as shown in Fig. 7.8. It should be noted that for metals behaving in a brittle manner, the notch effect at short lives is usually more pronounced than that shown in Fig. 7.8. Also, an alternative estimate of the S–N curve for a notched member made of a ductile material assumes equal fatigue strengths of the notched and smooth members at 10^3 reversals. This reflects the notch-strengthening effect under monotonic loading, previously discussed. This alternative estimate is further discussed in Section 7.2.3 and illustrated with an example problem in Section 7.2.4.

7.2.3   Mean Stress Effects and Haigh Diagrams

We know that stress range or strain range is the most important parameter of the fatigue life of smooth specimens. For notched parts this is true only if

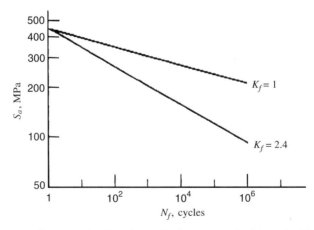

Figure 7.8 S–N diagrams for 1020 hot-rolled sheet steel with and without a notch.

the local maximum stress is tensile and above a small threshold. If there is no tensile stress, there will be no fatigue failure. Can this be true for notched parts but false for smooth parts? No, but cycling between a maximum and a minimum compressive stress is a fairly common regime for notched parts; for smooth parts, such a regime would usually produce buckling or yielding if the alternating stress is larger than the fully reversed fatigue strength.

Figure 7.9 shows the fatigue strengths of smooth and notched specimens of aluminum alloy 7075-T6 at 10^4 and at 10^7 cycles plotted versus the mean stress [2]. The elastic stress concentration factor was 3.4. From Fig. 7.9, values of the fatigue notch factor, K_f, are as follows:

	At 10^4 Cycles	At 10^7 Cycles
At zero mean stress	$K_f = 51/22 = 2.3$	$K_f = 22/10 = 2.2$
At 172 MPa (25 ksi) mean stress	$K_f = 42/13 = 3.2$	$K_f = 17/3 = 5.7$

Obviously, not only is the fatigue notch factor not equal to the elastic stress concentration factor, but it also changes with the mean stress and cycles to failure. Figure 7.10 shows lines for median fatigue life of 10^6 cycles for smooth parts and for notched parts with $K_f = 2.9$ for this material. They are plotted in terms of alternating stress $S_a = (S_{max} - S_{min})/2$ versus mean stress $S_m = (S_{max} + S_{min})/2$. Note the different character of the line for notched parts compared to the line for smooth parts and the great variation in the ratio K_f of their fatigue strengths: K_f is $175/60 = 2.9$ at zero mean stress, increases to a maximum of $150/31 = 4.8$ at 110 MPa mean stress, and decreases slowly at greater tensile mean stress. On the compression side, K_f decreases rapidly to

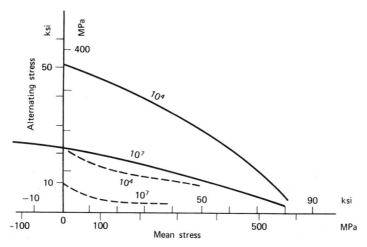

Figure 7.9 Constant life diagram for 7075-T6 wrought aluminum alloy with $S_u = 570$ MPa (82 ksi) [2]. (——) Unnotched, (---) notched, $K_t = 3.4$.

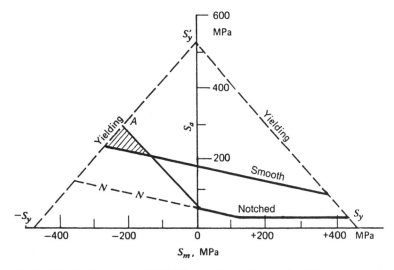

Figure 7.10 Haigh diagram for 7075-T6 aluminum alloy at 1 million cycles, with and without a notch.

1 at 140 MPa compressive mean stress and then becomes less than 1 at greater compressive mean stress, which means that a part with a groove may be stronger than a smooth part with the same minimum cross section (cross-hatched region). Can this be true? Yes, and we explain why in the next chapter.

The equality of notched and smooth fatigue strengths in the presence of compressive mean stress conforms with general experience. The cross hatched area above the dashed line N–N corresponds to nonpropagating cracks, which have often been observed in sharply notched parts.

The diagram in Fig. 7.10 is typical. Similar diagrams can easily be constructed for other materials and other values of K_f. The important points to remember are as follows:

1. Mean stress has more effect in notched parts than in smooth specimens.
2. Tensile mean stress can increase the fatigue notch factor, K_f, above the stress concentration factor, K_t, and can be fatal in fatigue loading.
3. Compressive mean stress can significantly reduce and even eliminate the effects of stress concentrations and save parts.

In the next chapter, we see that mean stresses inherent in the unloaded part due to residual stresses are often much greater than mean stresses caused by external loads.

The S–N approach to the combined effects of the mean stress and the notch is based on the use of available or estimated Haigh diagrams (constant life diagrams), such as that shown in Fig. 7.10. These diagrams show the combinations of nominal alternating stress, net regional mean stress (net regional mean stress equals nominal mean stress plus or minus residual stress near the surface), and notch factor that corresponds to the fatigue limit or to a life of about 10^6 to 10^8 cycles. From such diagrams, one point for an S–N curve is obtained. A second point for the S–N curve is obtained from knowledge or an estimate of the stress corresponding to a very short life, usually either 1 or 1000 cycles. These points are joined by a straight line on log S–log N coordinates. When test data are available, one will, of course, use them. When such data are not available, one must rely on estimates.

To estimate a Haigh diagram, the following data must be either available or estimated:

The monotonic yield strength, S_y
The cyclic yield strength, S'_y
The unnotched, fully reversed fatigue limit, S_f, or the fatigue strength at
    about 10^6 to 10^8 cycles
The true fracture strength, σ_f
The critical alternating tensile stress, S_{cat}
The fully reversed, long life fatigue notch factor, K_f

The critical alternating tensile stress, S_{cat}, is the stress below which cracks will not grow. It is closely related to the threshold stress intensity factor range. It was estimated by Fuchs [11] as 70 MPa (10 ksi) for hard steel, 30 MPa (4 ksi) for mild steel, and 20 MPa (3 ksi) for high-strength aluminum. However, if any margin for safety is used, S_{cat} will be practically zero and can be taken as zero in design.

Construction of the Haigh diagram is shown in Fig. 7.11. The abscissa represents mean stress (net regional mean stress). The ordinate represents alternating stress. The points located on these axes are $\pm S_y$, S'_y, S_f, and σ_f. These points define the behavior of smooth parts, similar to Fig. 4.14, based

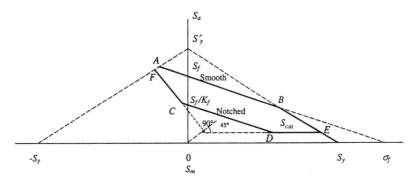

Figure 7.11 Haigh diagram.

on the Morrow equation discussed in Chapter 4. The lines connecting these points are drawn in Fig. 7.11 from $-S_y$ to S'_y to $+S_y$ and from S_f to σ_f, which intersects the first set of lines at A and B. Any combination of mean and alternating stresses outside the triangle from $-S_y$ to S'_y to $+S_y$ corresponds to gross yielding. Any combination above the line AB will produce a median fatigue life of less than 10^6 to 10^8 cycles for smooth parts.

For a notched part, three additional lines are drawn. One is parallel to AB, passing through S_f divided by K_f. This is shown as line CD in Fig. 7.11. It defines the development of cracks by local high stress. The other two lines correspond to S_{cat}. With tensile mean stress of more than S_{cat}, an alternating stress of $S_a = S_{cat}$ is all tensile alternating [$S_{cat} = (S_{max} - S_{min})/2$ for $S_{min} \geq 0$]. With $R \leq 0$, the alternating stress must be $(2S_{cat} - S_m)$ to produce S_a alternating tensile stress. (See Problem 7.11.) The line corresponding to S_{cat} is horizontal in most of the area of tensile mean stress (line DE); the line is inclined at 45° in the area of compressive mean stress or $R \leq 0$ (line CF). Small cracks will not grow as long as the combinations of mean and alternating stresses are below these two lines.

For a part with a notch, the estimated Haigh diagram is shown by lines $FCDE$. The presence of tensile mean stress reduces the amount of alternating stress that can be tolerated. The noteworthy feature is that maximum alternating stress can be tolerated with a sufficient compressive mean stress. If the mean stress, usually present as a residual stress, extends to a reasonable depth, it will arrest cracks. Because of the residual stress concentration in notches and the stress gradient in notches, the nominal stress is an excellent approximation of the relevant regional stress, as shown by Fuchs [12]. The high notched fatigue strength obtainable at point F in Fig. 7.11 also confirms the validity of this approach. The long life estimated by this simplified method thus includes both the prevention of crack formation (by limiting the alternating shear stress) and the prevention of crack growth (by limiting the alternating tensile stress).

A simpler approach for estimating the long life fatigue strength with a tensile or compressive mean stress which does not require construction of the Haigh diagram involves the use of the modified Goodman equation, discussed in Chapter 4 for unnotched members (Eq. 4.5a). For a notched member, the long life smooth fatigue strength is simply divided by the fatigue notch factor, K_f:

$$\frac{S_a}{(S_f/K_f)} + \frac{S_m}{S_u} = 1 \tag{7.12}$$

where S_a and S_m are net section alternating and mean stresses, respectively. The estimates for both smooth and notched parts based on the modified Goodman equation, along with the yield limits, are shown in Fig. 7.12. The estimated long life fatigue strength based on this equation, however, is usually less conservative than the estimate from the Haigh diagram in the tensile mean stress region but more conservative in the compressive mean stress region.

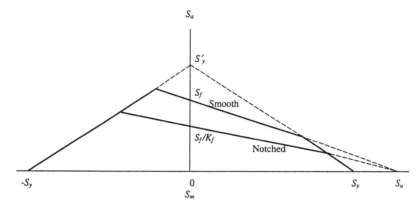

Figure 7.12 Modified Goodman diagram for smooth and notched members.

With diagrams like those shown in Figs. 7.11 and 7.12, we can estimate the long life fatigue strength of parts with notches for any combination of mean stress and alternating stress. A point on the *S–N* curve can thus be estimated for long fatigue life. For an estimate of a short life at a high stress, static fracture is one point that can be used. It may be considered to correspond to 1 cycle, 1 reversal, or 1/4 cycle of fully reversed loading. For our purpose, the difference between 1/4 and 1 cycle can be disregarded. The best estimate is based on a monotonic fracture test. A conservative estimate for metals with ductile behavior equates the nominal fracture stress S in the part to S_u, the ultimate tensile strength, or to σ_f, the true fracture strength. Notched parts may show different nominal stresses at fracture than smooth tensile test bars, and extrapolation of fatigue test data from smooth ductile specimens to life at 1/2 cycle usually shows a higher strength than S_u [13]. The extrapolated strength is σ_f', the fatigue strength coefficient. More accurate predictions can be obtained by assuming that smooth and notched parts of metals with ductile behavior have equal nominal fatigue strengths at 1000 reversals or 500 cycles [14,15]. After a fatigue strength for long life, on the order of millions of cycles, and a fatigue strength for 1 or 1000 cycles have been estimated, one can interpolate by assuming a straight line between these points on an *S–N* diagram on logarithmic scales (Basquin's Eq. 4.7). This methodology is illustrated in the example problem in Section 7.2.4.

Finally, we should consider the uncertainties that can exist in the models discussed. Data from different sources may not agree, and methods proposed by different investigators may not agree. The uncertainties are caused by the inherent scatter in test data, by the great influence of small details on fatigue life, and by our lack of quantitative data about the basic mechanisms that operate in developing small fatigue cracks. These uncertainties should not deter us from making estimates, but they should cause us to allow sufficient margins for safety and to avoid overelaborate computations that might obscure the core of the matter by the complexity of the equations and the number of digits printed out by the computer.

7.2.4 Example of Life Estimation with the *S–N* Approach

A notched part for which fatigue data have been published is shown in Fig. 7.13. It is made from quenched and tempered hot-rolled RQC-100 steel. The elastic stress concentration factor, K_t, is 3. The nominal stress, $S = P/A + Mc/I$, is $11.25P$ (MPa), where P is the load in kN. Relevant data for RQC-100 steel in the long transverse direction are given in Table A.2: $S_u = 931$ MPa, $S_y = 883$ MPa, $\sigma_f = 1330$ MPa, $S_y' = 600$ MPa, $\sigma_f' = 1240$ MPa, and $b = -0.07$. We want to construct the *S–N* lines for (*a*) completely reversed, constant amplitude loading and (*b*) constant amplitude loading with a minimum nominal stress of 50 MPa.

The *S–N* curve for smooth specimens in terms of reversals to failure is given by Basquin's equation (Eq. 5.15), S_a or $S_{Nf} = \sigma_f' (2N_f)^b = 1240 (2N_f)^{-0.07}$. For the notched part with $K_t = 3$ and a notch radius of 4.75 mm (3/16 in.), we calculate q and K_f from Neuber's formula (Eq. 7.9) with $\rho = 0.049$ mm from Fig. 7.7 for $S_u = 931$ MPa, as

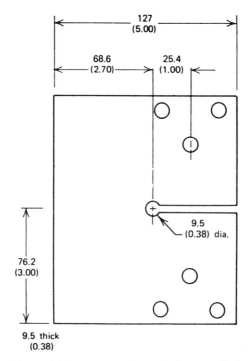

Figure 7.13 Test specimen design, mm (in.).

$$q = \frac{1}{1 + \sqrt{0.049/4.75}} = 0.91 \quad \text{and} \quad K_f = 1 + \frac{3 - 1}{1 + \sqrt{0.049/4.75}} = 2.82$$

A value of $q = 0.91$ indicates high notch sensitivity, which is expected from the relatively high strength of the material and the relatively large notch radius. The fully reversed $(R = -1)$ fatigue strength at 1 million cycles can be estimated as $S_f = 1240 (2 \times 10^6)^{-0.07} = 449$ MPa. The long life, fully reversed $(R = -1)$ fatigue strength of the notched part then is estimated to be $S_f/K_f = 449/2.82 = 159$ MPa.

(a) A straight line from point A, 1240 MPa at 1 reversal (1 cycle = 2 reversals), to point B, 449 MPa at 2 million reversals, is the $R = -1$ S–N curve for smooth specimens shown in Fig. 7.14. The $R = -1$ S–N curve for the notched part is drawn in two alternative ways as straight lines. One assumes equal strength of notched and smooth parts at 1 reversal; the other assumes that smooth and notched parts have equal strength at 1000 reversals. The former assumption would be more correct for materials with brittle behavior, the latter for materials with ductile behavior. The first of these lines goes from point A to point C at 2 million reversals and 159 MPa; the second goes from point C to point D at 1000 reversals on line AB.

This example was chosen because the test results and material data are well documented in the literature. Test data for the notched part are also shown in Fig. 7.14. Crossing of the line for notched parts and that for smooth specimens at 1000 reversals appears to be justified in this instance, as two test points are near the crossing. The discrepancies between S–N model estimates and experimental values are small here. Generally, however, greater differences may be expected with these simplified, but common, models.

(b) The Haigh diagram can be used when the mean stress is different from zero. $S_{cat} = 30$ MPa is assumed for the RQC-100 steel. The diagram is shown in Fig. 7.15. On this diagram, a line of constant minimum stress is straight and inclined at 45° upward to the right $(S_a = S_m - S_{min} = S_m - 50$ MPa). Starting at 50 MPa mean and zero alternating stress, we intersect the constant life line for the notched part at C (105 MPa) and that for the smooth part at D (330 MPa). We construct the S–N line for the smooth part from 330 MPa alternating stress at 2 million reversals to 1240 MPa at 1 reversal. For the notched part we draw the line from 105 MPa at 2×10^6 reversals to 1240 MPa at 1 reversal. These lines are shown in Fig. 7.16.

If the simpler approximation based on the modified Goodman equation is used, the notched fatigue strength at 2 million reversals is calculated as

$$\frac{S_a}{S_f/K_f} + \frac{S_m}{S_u} = \frac{S_a}{449/2.82} + \frac{S_a + 50}{931} = 1 \quad \text{or} \quad S_a = 129 \text{ MPa}$$

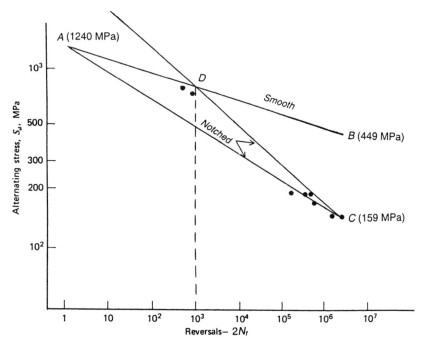

Figure 7.14 Estimated $S-N$ curves and measured fatigue life for fully reversed ($R = -1$) loading.

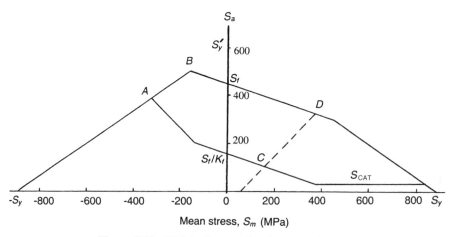

Figure 7.15 Haigh diagram for the example part.

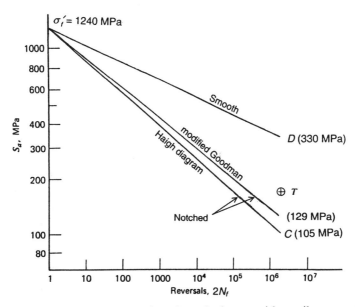

Figure 7.16   *S–N* lines for smooth and notched parts with tensile mean stress.

The *S–N* line for the notched part based on this approximation is also shown in Fig. 7.16. Life estimation of this part when the mean stress is compressive is given as a homework problem in Chapter 8 (Problem 8.8).

The nominal alternating stress at 1 reversal is 1240 MPa for these examples because the mean stress disappears in these parts when the deformation is mainly plastic. When the outer fibers of solid sections loaded in bending or torsion are strained beyond yielding, initial mean and residual stresses can also be reduced to practically zero in the surface region. This is not true of mean stress in axial tests where the whole section must support the mean load. In this case, a more reasonable model may be to apply the mean stress correction to the entire life regime. This can be done by using Eqs. 4.8 and 4.7 in a manner similar to the example problem in Section 4.6 to account for the mean stress effect. A single test point for the loading of the notched part in example (*b*) is available in the literature [16]. It is shown as point *T* in Fig. 7.16. The agreement is typical of that in fatigue life predictions unless they are fitted to specific test data. When specific test data can be used to interpolate a prediction, they are, of course, the best available information.

7.3   NOTCH STRAIN ANALYSIS AND THE STRAIN–LIFE APPROACH

As was discussed in Chapter 5, a common application of the strain–life approach is in fatigue analysis of notched members. This is because the deformation of the material at the notch root is often inelastic involving plastic strain and, therefore, fatigue behavior is best described in terms of strain. Also, notch stresses and strains are explicitly considered in the strain–life approach, whereas the S–N approach is only in terms of nominal stresses. Therefore, the mean stress at the notch root which can have a significant effect on fatigue behavior can be directly accounted for in the strain–life approach. This is particularly important in situations where significant notch plastic deformation and mean stresses are present, such as in variable amplitude loading with overloads and during low cycle fatigue. Variable amplitude loading is discussed in Chapter 9.

Application of the strain–life approach involves two steps. First, it requires determination of local (notch) stresses and strains. Life prediction can then be made using the local stresses and strains, based on the strain–life equation and analysis discussed in Chapter 5. Notch stress/strain analysis is presented and discussed in the following sections and illustrated with an example problem. This analysis is first presented for monotonic loading and then extended to constant amplitude cyclic loading.

7.3.1 Notch Stresses and Strains

Notch root and nominal stresses and strains are represented by (σ, ε) and (S, e), respectively. These notations are defined and shown in Fig. 7.17. As indicated in the figure, definitions of the nominal stress, S, and the nominal strain, e, can be based on either the gross or the net cross-sectional area. Uniformly repeated load cycles impose uniformly repeated strain cycles on the metal at the root of the notch as long as most of the part remains elastic. If the metal at the notch root is strained beyond the yield strength, it may strain harden and cyclic harden or soften, as discussed in Chapters 3 and 5. The material at the notch root is under "strain control."

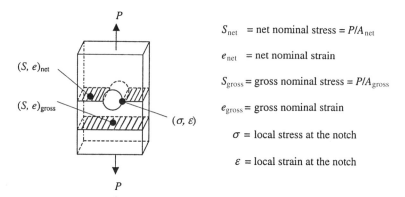

S_{net} = net nominal stress = P/A_{net}

e_{net} = net nominal strain

S_{gross} = gross nominal stress = P/A_{gross}

e_{gross} = gross nominal strain

σ = local stress at the notch

ε = local strain at the notch

Figure 7.17 Definitions of notch and nominal stresses and strains.

As long as stresses and strains at the notch root remain elastic, we have

$$\sigma = K_t S \qquad \varepsilon = K_t e \tag{7.13}$$

As mentioned earlier, K_t can be defined based on either net or gross cross-sectional area. Either definition can be used as long as both K_t and S (or e) are based on gross or net area since $(K_t S)_{net} = (K_t S)_{gross}$. The most commonly used definitions of K_t and nominal stress or strain are based on the net area, however.

The loads on notched parts are often sufficiently high such that the local stress calculated from the nominal stress and the stress concentration factor by the formula $\sigma = K_t S$ is considerably above the yield strength. When the local stress exceeds the yield strength it will be less than $K_t S$, and we can no longer use K_t to relate notch stress to nominal stress. Also, stresses are no longer proportional to strains. We then define strain and stress concentration factors as

$$K_\varepsilon = \frac{\varepsilon}{e} \tag{7.14}$$

$$K_\sigma = \frac{\sigma}{S} \tag{7.15}$$

A schematic showing variations of stress and strain concentration factors with notch stress is presented in Fig. 7.18. During elastic deformation ($\sigma < S_y$), $K_\sigma = K_\varepsilon = K_t$, since stresses and strains are related by the modulus of elasticity, E. As notch stress increases and the deformation becomes inelastic, K_σ reduces and K_ε increases due to increased plastic deformation at the notch. The relation between σ and ε is given by the monotonic stress–strain curve, which is

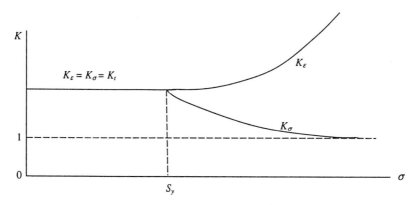

Figure 7.18   Stress and strain concentration factor variations versus notch stress.

often represented by the Ramberg-Osgood equation given in Section 5.1 and repeated here for convenience:

$$\varepsilon = \varepsilon_e + \varepsilon_p = \frac{\sigma}{E} + \left(\frac{\sigma}{K}\right)^{1/n} \tag{7.16}$$

Values for n and K can be taken from Table A.2.

The question to be answered by notch-strain analysis is, given the nominal elastic stress, S, or strain, e, what are the local stress, σ, and the local strain, ε, at the notch root surface and near the surface? Answers to this question can be obtained by experimental methods, finite element methods, analytical models, and, for some cases, by the equations of the theory of plasticity. Experimental methods were discussed in Section 7.1. For finite element methods, in addition to the requirement of small element size in high-stress gradient regions for solution accuracy, a realistic representation of the nonlinear material stress–strain behavior (such as the Ramberg-Osgood equation, if appropriate) is required. For many fatigue problems, however, one may be satisfied with estimating the conditions at the surface of the notch root by considering analytical models. Analytical models require the value of the elastic stress concentration factor, K_t. For complex geometries where K_t may be difficult to define or obtain, analytical models can be used in combination with linear finite element analysis. In this case, linear finite element analysis provides the knowledge of elastic stress and strain at the notch root.

Analytical models commonly used include the linear rule, Neuber's rule, and strain energy density or Glinka's rule. The linear rule and Neuber's rule represent two extreme cases of plane strain and plane stress, respectively. The linear rule is expressed as

$$K_\varepsilon = K_t = \frac{\varepsilon}{e} \quad \text{or} \quad \varepsilon = K_t e \tag{7.17}$$

For nominal elastic behavior, $e = S/E$. The notch strain, ε, can be computed directly and, if desired, the notch stress, σ, can then be obtained from the stress–strain curve, or Eq. 7.16. For cyclic loading, notch and nominal stresses and strains are replaced by their respective ranges. The linear rule or strain concentration invariance agrees with measurements in plane strain situations, such as circumferential grooves in shafts in tension or bending [17].

7.3.2 Neuber's Rule

Neuber derived the following relation for longitudinally grooved shafts in torsion [18]:

$$K_\varepsilon K_\sigma = K_t^2 \tag{7.18}$$

or

$$\varepsilon\sigma = K_t^2 \, eS \qquad (7.19)$$

According to this relation, the geometrical mean of the stress and strain concentration factors under plastic deformation conditions remains constant and equal to the theoretical stress concentration factor, K_t. Neuber's rule has been the most widely used notch stress/strain analysis model and agrees with measurements in plane stress situations, such as thin sheets in tension. Application of this rule requires the solution of two simultaneous equations, Eqs. 7.16 and 7.19. On σ–ε coordinates, Eq. 7.16 describes the stress–strain curve and Eq. 7.19 describes a hyperbola. The intersection of these two curves defines the desired values σ and ε. Figure 7.19 shows the application for monotonic loading using a graphical method, where point A is the solution by Neuber's rule.

For nominal elastic behavior, $e = S/E$ and Neuber's rule reduces to

$$\varepsilon\sigma = \frac{(K_t S)^2}{E} \qquad (7.20)$$

Combining Eqs. 7.20 and 7.16 results in

$$\frac{\sigma^2}{E} + \sigma \left(\frac{\sigma}{K}\right)^{1/n} = \frac{(K_t S)^2}{E} \qquad (7.21)$$

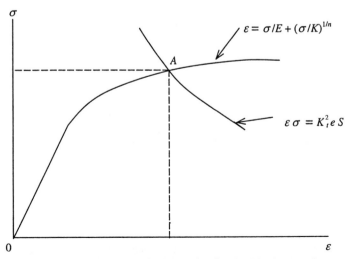

Figure 7.19 Notch strain determination by Neuber's rule.

This equation can be solved for notch stress, σ, by iteration or numerical techniques. If nominal behavior is inelastic, Eq. 7.16 must be applied to relate nominal stress and strain and used with Eq. 7.19 rather than Eq. 7.20 to solve for notch stress and strain. For large-scale nominal plastic deformation (i.e., general notch region yielding), a modification of Neuber's rule is proposed in reference 19. In this case, however, general yielding and buckling should also be considered as potential failure modes.

For cyclic loading, the monotonic stress–strain curve is replaced by the hysteresis curve (represented by Eq. 5.13) and the strains and stresses are replaced by the strain ranges and stress ranges. In addition, Topper et al. [20] have suggested use of the fatigue notch factor, K_f, in place of the theoretical stress concentration factor, K_t, for cyclic loading when using Neuber's rule. Their recommendation is based upon better agreement with experimental fatigue life results. Neuber's rule used for computing $\Delta\varepsilon$ then is

$$\Delta\varepsilon\,\Delta\sigma = K_f^2\,\Delta e\,\sigma S \qquad (7.22)$$

For nominal elastic behavior, $\Delta e = \Delta S/E$ and

$$\Delta\varepsilon\,\Delta\sigma = \frac{(K_f\,\Delta S)^2}{E} \qquad (7.23)$$

Equation 7.21 for monotonic loading can also be extended to cyclic loading by replacing stress and strain with their respective ranges and K_t with K_f as

$$\frac{(\Delta\sigma)^2}{E} + 2\Delta\sigma\left(\frac{\Delta\sigma}{2K'}\right)^{1/n'} = \frac{(K_f\,\Delta S)^2}{E} \qquad (7.24)$$

Application of Neuber's rule is illustrated by Fig. 7.20 for constant stress amplitude cyclic loading. For initial loading from zero to maximum nominal

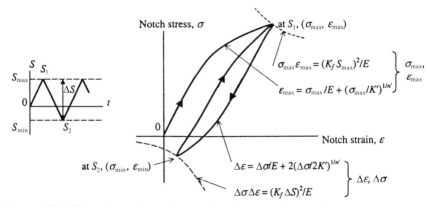

Figure 7.20 Illustration of notch stress/strain determination by Neuber's rule for constant amplitude cyclic loading.

stress level ($S_1 = S_{max}$), the maximum notch strain and stress (ε_{max}, σ_{max}) can be found from intersection of the Neuber hyperbola for maximum notch stress/strain with the cyclic stress–strain curve. Use of the cyclic stress–strain curve assumes that the deformation behavior is cyclically stable. For unloading from S_1 to S_2 (or $\Delta S = S_1 - S_2$), point S_1 is considered the reference point and the notch deformation curve follows the hysteresis loop curve of the material. At point S_2, notch strain and stress ranges ($\Delta \varepsilon$, $\Delta \sigma$) are found by intersection of the Neuber hyperbola in terms of cyclic stresses and strains with the hysteresis loop curve. It is important to note that for unloading, the point of reversal S_1 is used as the origin of the Neuber hyperbola and the hysteresis curve, not $\sigma = 0$. For continued constant amplitude loading, as shown in Fig. 7.20, the notch stress and strain will continue to follow the closed hysteresis loop shown.

Knowing $\Delta \varepsilon$, $\Delta \sigma$, and σ_{max}, we can obtain $\varepsilon_a = \Delta \varepsilon / 2$, and $\sigma_m = (\sigma_{max} + \sigma_{min})/2 = \sigma_{max} - \Delta \sigma / 2$. Notch strain amplitude, ε_a, and notch mean stress, σ_m, are then used for life prediction analysis using Eqs. 5.22–5.24, as discussed in Chapter 5. Also, similar analysis is used for variable amplitude loading, discussed in Chapter 9. Such a point-by-point analysis can be implemented by a computer code. It should also be noted that if the nominal behavior is inelastic, Eq. 7.22 is used and the nominal stress range, ΔS, and the nominal strain range, Δe, are related by the hysteresis curve, Eq. 5.13, rather than by Hooke's law. This, however, is often not the case.

7.3.3 Strain Energy Density or Glinka's Rule

A more recent notch stress/strain analysis rule is strain energy density or Glinka's rule [21]. This rule is based on the assumption that the strain energy density at the notch root is nearly the same for linear elastic notch behavior (W_e) and elastic-plastic notch behavior (W_p), as long as the plastic deformation zone at the notch is surrounded by an elastic stress field. This is shown schematically in Fig. 7.21. For nominal elastic stress, S, the nominal strain energy density, W_S, is given by (with $e = S/E$ and $de = dS/E$)

$$W_S = \int_0^e S \, de = \int_0^S \frac{S}{E} \, dS = \frac{S^2}{2E} \tag{7.25}$$

At the notch root with a stress concentration factor of K_t, strain energy density, assuming linear elastic behavior ($\sigma = K_t S$ and $\varepsilon = \sigma/E$), is

$$W_e = \int_0^{\varepsilon_e} \sigma \, d\varepsilon = \int_0^{\sigma_e} \frac{\sigma}{E} \, d\sigma = \frac{\sigma_e^2}{2E} = \frac{(K_t S)^2}{2E} \tag{7.26}$$

For elastic-plastic behavior at the notch root, the stress–strain relationship can be expressed by Eq. 7.16 and the strain energy density is given by

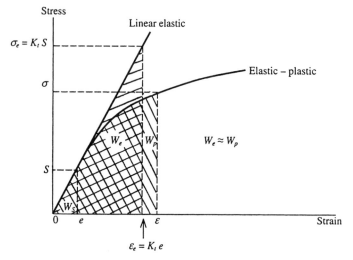

Figure 7.21   Graphical representation of the strain energy density or Glinka's rule.

$$W_p = \int_0^\varepsilon \sigma \, d\varepsilon = \frac{\sigma^2}{2E} + \frac{\sigma}{n+1}\left(\frac{\sigma}{K}\right)^{1/n} \tag{7.27}$$

Setting $W_p = W_e$ results in

$$\frac{\sigma^2}{E} + \frac{2\sigma}{n+1}\left(\frac{\sigma}{K}\right)^{1/n} = \frac{(K_t S)^2}{E} \tag{7.28}$$

Comparing Eqs. 7.21 and 7.28, we can see that the only difference from Neuber's rule is the factor $[2/(n+1)]$. Since $n < 1$, smaller notch stress (and therefore smaller notch strain) is predicted based on this equation than from Neuber's rule, resulting in longer fatigue life compared to that from Neuber's rule.

For a given nominal stress S, notch stress σ can be calculated from Eq. 7.28. For cyclic loading, Eq. 7.28 is written in terms of stress and strain ranges, and material monotonic deformation properties (K and n) are replaced by cyclic deformation properties (K' and n'), analogous to Eq. 7.24, resulting in the following equation:

$$\frac{(\Delta\sigma)^2}{E} + \frac{4\Delta\sigma}{n'+1}\left(\frac{\Delta\sigma}{2K'}\right)^{1/n'} = \frac{(K_t \Delta S)^2}{E} \tag{7.29}$$

This equation relates nominal stress range, ΔS, to notch stress range, $\Delta\sigma$. Notch strain range, $\Delta\varepsilon$, is then found from the hysteresis loop, Eq. 5.13,

$$\Delta\varepsilon = \frac{\Delta\sigma}{E} + 2\left(\frac{\Delta\sigma}{2K'}\right)^{1/n'} \qquad (5.13)$$

and used in the strain–life equations to find the fatigue life, similar to the procedure for Neuber's rule.

7.3.4 Plane Stress versus Plane Strain

The notch strain analysis presented and discussed in the previous sections assumed a uniaxial state of stress at the notch root. This condition exists under a plane state of stress such as that in a thin plate. For plane strain conditions such as those in a thick plate, however, the state of stress is no longer uniaxial, as shown in Fig. 7.22. This is because in the thick plate, or in other components with lateral constraint such as grooves in shafts, the surrounding elastic material restrains notch deformation in the thickness direction, $\varepsilon_z \approx 0$, resulting in a stress, σ_z, in that direction. The value of Poisson's ratio, v, in Fig. 7.22 approaches 0.5 for large notch root plastic deformations. For the plane strain conditions, a smaller notch strain range and a larger notch stress range in the loading direction result, as compared with the plane stress condition.

Plane stress and plane strain conditions represent two bounding conditions which occur at the notch root. A modified stress–strain relation based on plasticity equations [22] can be used with Neuber's rule for plane strain conditions. A plane strain version of the strain energy density rule has also been developed [23]. The equations are, however, more complex and the interested reader is referred to references 22 and 23. Multiaxial aspects of notch effects on fatigue behavior are discussed in Chapter 10.

If one required an intermediate formula between the linear rule for plane strain and Neuber's rule for plane stress, it might be

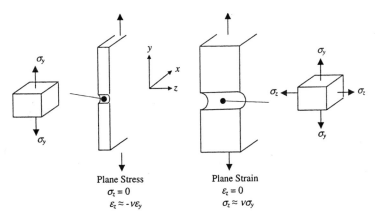

Plane Stress
$\sigma_z = 0$
$\varepsilon_z \approx -v\varepsilon_y$

Plane Strain
$\varepsilon_z = 0$
$\sigma_z \approx v\sigma_y$

Figure 7.22 Schematic representation of plane stress and plane strain conditions at the notch root.

$$\Delta\varepsilon = K_t \, \Delta e \left(\frac{K_t}{K_\sigma}\right)^m \tag{7.30}$$

where $m = 1$ for plane stress, $m = 0$ for plane strain, and $0 < m < 1$ for intermediate situations. By considering the elastic constraint at the notch, we can estimate a suitable value for K_ε. It is usually between K_t and K_t^2/K_σ. As mentioned previously, the Neuber assumption is good for thin sheets (plane stress) and is usually conservative by predicting strains higher than measured strains in almost all other cases. Using K_f rather than K_t in Neuber's relation for cyclic loading reduces the degree of conservatism. This rule, however, is always more conservative than the linear rule. Strain energy density or Glinka's rule has also been modified for geometries and loadings with intermediate stress/strain states [24].

7.3.5 Example of Life Estimation Using the Strain–Life Approach

This example applies the local strain approach to the same keyhole notched part, Fig. 7.13, with $K_t = 3$ and made of RQC-100 steel used as the example for the S–N approach in Section 7.2.4. We define a new problem and want to find:

    (a) notch stress and strain from a 53.4 kN (12 kip) monotonic load,

    (b) notch stress and strain after unloading from the monotonic load in part (a) to zero,

    (c) notch stress and strain amplitudes from constant amplitude alternating loads between 4.45 kN (1 kip) and 44.5 kN (10 kip), and

    (d) the expected fatigue life to the formation of cracks on the order of 1 mm from the loading in part (c).

The relevant properties quoted from Table A.2 are:

$E = 207$ GPa	$S_y = 883$ MPa	$K = 1172$ MPa	$n = 0.06$
	$S_y' = 600$ MPa	$K' = 1434$ MPa	$n' = 0.14$
$\sigma_f' = 1240$ MPa	$\varepsilon_f' = 0.66$	$b = -0.07$	$c = -0.69$

The part used for this example, a plate 9.5 mm (0.375 in.) thick with a notch radius of 4.75 mm (0.188 in.), behaves in an intermediate manner between plane stress and plane strain, which has been verified by strain measurement [25]. We use all three methods of notch strain analysis discussed in the previous sections in this example to illustrate their use.

    (a) From Example 7.2.4, the relation between nominal stress, S, and the applied load, P, is given by $S = P/A + Mc/I = 11.25\,P$, where P is in kN and S is in MPa. The nominal stress is therefore calculated as

$$S = 11.25\ P = 11.25\ (53.4) = 600\ \text{MPa}$$

The nominal monotonic behavior is elastic, since $S_{max} = 600$ MPa is about two-thirds of the yield strength, $S_y = 883$ MPa. It should be noted that the nominal behavior often starts to become inelastic at a stress level lower than the yield strength (usually about 0.75 to 0.9 S_y) since most materials are not elastic-perfectly plastic. However, the monotonic notch root behavior is inelastic since $K_t S_{max} = 3 \times 600 = 1800$ MPa $> S_y = 883$ MPa.

To find the notch root stress and strain using Neuber's rule, we have to solve the following two simultaneous equations:

Eq. 7.20: $$\varepsilon\sigma = \frac{(K_t S)^2}{E} = \frac{(3 \times 600)^2}{207\ 000} = 15.65$$

Eq. 7.16: $$\varepsilon = \frac{\sigma}{E} + \left(\frac{\sigma}{K}\right)^{1/n} = \frac{\sigma}{207\ 000} + \left(\frac{\sigma}{1172}\right)^{1/0.06}$$

The solution can be obtained by a trial-and-error, numerical, or graphical method (Shown in Fig. 7.23), resulting in $\sigma = 903$ MPa and $\varepsilon = 0.0173$.

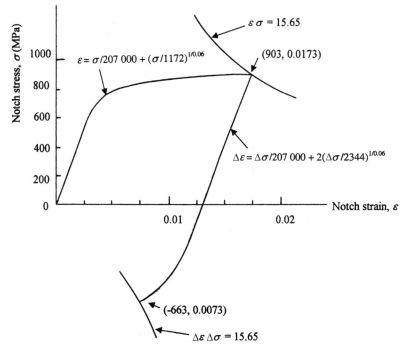

Figure 7.23 Determination of the strain at the root of a notch according to Neuber's rule; material: RQC-100; nominal stress 600 MPa followed by unloading to zero; $K_t = 3$.

If the strain energy density rule is used:

$$\text{Eq. 7.28:} \quad \frac{\sigma^2}{E} + \frac{2\sigma}{n+1}\left(\frac{\sigma}{K}\right)^{1/n} = \frac{(K_t S)^2}{E}$$

or

$$\frac{\sigma^2}{207\ 000} + \frac{2\sigma}{0.06+1}\left(\frac{\sigma}{1172}\right)^{1/0.06} = \frac{(3 \times 600)^2}{207\ 000} = 15.65$$

which gives $\sigma = 872\ \text{MPa}$. Substituting this value into the stress–strain equation results in

$$\text{Eq. 7.16:} \quad \varepsilon = \frac{\sigma}{E} + \left(\frac{\sigma}{K}\right)^{1/n} = \frac{872}{207\ 000} + \left(\frac{872}{1172}\right)^{1/0.06} = \underline{0.0115}$$

These values are smaller than those predicted by Neuber's rule, as expected. If we used the linear rule, notch root strain and stress would be calculated as

$$\text{Eq. 7.17:} \quad \varepsilon = K_t e = \frac{K_t S}{E} = \frac{3 \times 600}{207\ 000} = \underline{0.0087}$$

and

$$\text{Eq. 7.16:} \quad 0.0087 = \frac{\sigma}{207\ 000} + \left(\frac{\sigma}{1172}\right)^{1/0.06} \quad \text{or} \quad \sigma = \underline{849\ \text{MPa}}$$

(*b*)   For unloading, we can assume Masing behavior with a factor of 2 expansion of the monotonic stress–strain curve. Unloading takes place from $S_1 = S_{max} = 600\ \text{MPa}$ to $S_2 = 0$, or $\Delta S = 600\ \text{MPa}$. Therefore, from Neuber's rule we obtain:

$$\text{Eq. 7.23:} \quad \Delta\varepsilon\,\Delta\sigma = \frac{(K_t \Delta S)^2}{E} = \frac{(3 \times 600)^2}{207\ 000} = 15.65$$

and

$$\Delta\varepsilon = \frac{\Delta\sigma}{E} + 2\left(\frac{\Delta\sigma}{2K}\right)^{1/n} = \frac{\Delta\sigma}{207\ 000} + 2\left(\frac{\Delta\sigma}{2344}\right)^{1/0.06}$$

resulting in

$$\Delta\sigma = 1566\ \text{MPa} \quad \text{and} \quad \Delta\varepsilon = 0.0100$$

Residual stress, σ_{min}, and strain, ε_{min}, after unloading are calculated as

$$\sigma_{min} = \sigma_{max} - \Delta\sigma = 903 - 1566 = \underline{-663 \text{ MPa}}$$

and

$$\varepsilon_{min} = \varepsilon_{max} - \Delta\varepsilon = 0.0173 - 0.0100 = \underline{0.0073}$$

The residual notch stress and strain after unloading are shown in Fig. 7.23. Using the strain energy density rule:

$$\frac{(\Delta\sigma)^2}{E} + \frac{4\Delta\sigma}{n+1}\left(\frac{\Delta\sigma}{2K}\right)^{1/n} = \frac{(K_t \Delta S)^2}{E}$$

or

$$\frac{(\Delta\sigma)^2}{207\ 000} + \frac{4\Delta\sigma}{0.06+1}\left(\frac{\Delta\sigma}{2 \times 1172}\right)^{1/0.06} = \frac{(3 \times 600)^2}{207\ 000} = 15.65$$

resulting in

$$\Delta\sigma = 1524 \text{ MPa}$$

and

$$\Delta\varepsilon = \frac{\Delta\sigma}{E} + 2\left(\frac{\Delta\sigma}{2K}\right)^{1/n} = \frac{1524}{207\ 000} + 2\left(\frac{1524}{2344}\right)^{1/0.06} = 0.0089$$

Therefore

$$\sigma_{min} = \sigma_{max} - \Delta\sigma = 872 - 1524 = \underline{-652 \text{ MPa}}$$

and

$$\varepsilon_{min} = \varepsilon_{max} - \Delta\varepsilon = 0.0115 - 0.0089 = \underline{0.0026}$$

The linear rule does not result in a notch strain after unloading to zero load. Notch stress after unloading is calculated as

$$0.0087 = \frac{\Delta\sigma}{207\ 000} + 2\left(\frac{\Delta\sigma}{2344}\right)^{1/0.06} \quad \text{or} \quad \Delta\sigma = 1515 \text{ MPa}$$

Therefore

$$\sigma_{min} = \sigma_{max} - \Delta\sigma = 849 - 1515 = \underline{-666 \text{ MPa}}$$

As mentioned earlier, a factor of 2 expansion of the monotonic stress–strain curve was used as the unloading stress–strain path. This is a reasonable approximation in the absence of the actual unloading stress–strain curve of the material.

(c) For cyclic loading, the maximum, minimum, and amplitude of the nominal stress are calculated as

$$S_{max} = 11.25 \, P_{max} = 11.25 \, (44.5) = 500 \text{ MPa}$$

$$S_{min} = 11.25 \, P_{min} = 11.25 \, (4.45) = 50 \text{ MPa}$$

$$S_a = (S_{max} - S_{min})/2 = (500 - 50)/2 = 225 \text{ MPa}$$

The nominal behavior for cyclic loading is also elastic since S_{max}, $|S_{min}|$, and S_a are smaller than $S'_y = 600$ MPa.

We first need to calculate notch root stress and strain at the maximum load. Using Neuber's rule, we use the fatigue notch factor, $K_f = 2.82$, which was calculated in Example 7.2.4:

$$\varepsilon_{max} \sigma_{max} = \frac{(K_f S_{max})^2}{E} = \frac{(2.82 \times 500)^2}{207 \, 000} = 9.6$$

$$\varepsilon_{max} = \frac{\sigma_{max}}{E} + \left(\frac{\sigma_{max}}{K'}\right)^{1/n'} = \frac{\sigma_{max}}{207 \, 000} + \left(\frac{\sigma_{max}}{1434}\right)^{1/0.14}$$

This results in $\sigma_{max} = 745$ MPa, and $\varepsilon_{max} = 0.0129$. We assume a Masing behavior with a factor of 2 expansion of the stable cyclic stress–strain curve as the unloading deformation path for the material. Unloading takes place from 500 to 50 MPa, or $\Delta S = 450$ MPa. Therefore, from Neuber's rule we obtain

$$\text{Eq. 7.23:} \quad \Delta\varepsilon\Delta\sigma = \frac{(K_f \Delta S)^2}{E} = \frac{(2.82 \times 450)^2}{207 \, 000} = 7.78$$

and

$$\text{Eq. 5.13:} \quad \Delta\varepsilon = \frac{\Delta\sigma}{E} + 2\left(\frac{\Delta\sigma}{2K'}\right)^{1/n'} = \frac{\Delta\sigma}{207 \, 000} + 2\left(\frac{\Delta\sigma}{2868}\right)^{1/0.14}$$

resulting in

$$\Delta\sigma = 1082 \text{ MPa, and } \Delta\varepsilon = 0.0072$$

Therefore

$$\sigma_{min} = \sigma_{max} - \Delta\sigma = 745 - 1082 = -337 \text{ MPa}$$

$$\sigma_m = (\sigma_{max} + \sigma_{min})/2 = (745 - 337)/2 = 204 \text{ MPa}$$

and

$$\varepsilon_a = \Delta\varepsilon/2 = 0.0072/2 = \underline{0.0036}, \quad \sigma_a = \Delta\sigma/2 = 1082/2 = \underline{541 \text{ MPa}}$$

Using the strain energy density rule:

$$\frac{\sigma^2_{max}}{E} + \frac{2\sigma_{max}}{n' + 1}\left(\frac{\sigma_{max}}{K'}\right)^{1/n'} = \frac{(K_t S_{max})^2}{E}$$

or

$$\frac{\sigma^2_{max}}{207\,000} + \frac{2\sigma_{max}}{0.14 + 1}\left(\frac{\sigma_{max}}{1434}\right)^{1/0.14} = \frac{(3 \times 500)^2}{207\,000} = 10.9$$

which gives $\sigma_{max} = 712$ MPa. Substituting this value into the cyclic stress–strain equation results in

$$\varepsilon_{max} = \frac{\sigma_{max}}{E} + \left(\frac{\sigma_{max}}{K'}\right)^{1/n'} = \frac{712}{207\,000} + \left(\frac{712}{1434}\right)^{1/0.14} = 0.0102$$

For unloading:

Eq. 7.29: $\quad \dfrac{(\Delta\sigma)^2}{E} + \dfrac{4\Delta\sigma}{n' + 1}\left(\dfrac{\Delta\sigma}{2K'}\right)^{1/n'} = \dfrac{(K_t \Delta S)^2}{E}$

or

$$\frac{(\Delta\sigma)^2}{207\,000} + \frac{4\Delta\sigma}{0.14 + 1}\left(\frac{\Delta\sigma}{2 \times 1434}\right)^{1/0.14} = \frac{(3 \times 450)^2}{207\,000} = 8.8$$

resulting in $\Delta\sigma = 1070$ MPa and

Eq. 5.13: $\quad \Delta\varepsilon = \dfrac{\Delta\sigma}{E} + 2\left(\dfrac{\Delta\sigma}{2K'}\right)^{1/n'} = \dfrac{1070}{207\,000} + 2\left(\dfrac{1070}{2868}\right)^{1/0.14} = 0.0069$

Therefore

$$\sigma_{min} = \sigma_{max} - \Delta\sigma = 712 - 1070 = -358 \text{ MPa},$$

$$\sigma_m = (\sigma_{max} + \sigma_{min})/2 = (712 - 358)/2 = \underline{177 \text{ MPa}}$$

and

$$\varepsilon_a = \Delta\varepsilon/2 = 0.0069/2 = \underline{0.0035}, \quad \sigma_a = \Delta\sigma/2 = 1070/2 = \underline{535 \text{ MPa}}$$

If we use the linear rule:

$$\varepsilon_{max} = \frac{K_t S_{max}}{E} = \frac{3 \times 500}{207\ 000} = 0.0072$$

and

$$0.0072 = \frac{\sigma_{max}}{207\ 000} + \left(\frac{\sigma_{max}}{1434}\right)^{1/0.14} \quad \text{or} \quad \sigma_{max} = 663 \text{ MPa}$$

Then

$$\Delta\varepsilon = K_t \Delta e = \frac{K_t \Delta S}{E} = \frac{3 \times 450}{207\ 000} = 0.0065$$

and

$$0.0065 = \frac{\Delta\sigma}{207\ 000} + 2\left(\frac{\Delta\sigma}{2868}\right)^{1/0.14} \quad \text{or} \quad \Delta\sigma = 1045 \text{ MPa}$$

Therefore,

$$\sigma_{min} = \sigma_{max} - \Delta\sigma = 663 - 1045 = -382 \text{ MPa},$$

$$\sigma_m = (\sigma_{max} + \sigma_{min})/2 = (663 - 382)/2 = \underline{141 \text{ MPa}}$$

and

$$\varepsilon_a = \Delta\varepsilon/2 = 0.0065/2 = \underline{0.0033}, \quad \sigma_a = \Delta\sigma/2 = 1045/2 = \underline{523 \text{ MPa}}$$

(d) To predict fatigue life to formation of a small crack on the order of 1 mm for cyclic loading in part (c), we can use the Smith-Watson-Topper mean stress parameter (Eq. 5.24) or Morrow's mean stress parameters (Eq.

5.22 or 5.23) from Chapter 5. Here we choose the Smith-Watson-Topper parameter:

$$\varepsilon_a \sigma_{max} = (\sigma_f')^2 \frac{(2N_f)^{2b}}{E} + \varepsilon_f' \sigma_f' (2N_f)^{b+c} = 7.43(2N_f)^{-0.14} + 818(2N_f)^{-0.76}$$

The results are summarized in the following table:

	Neuber's Rule	Strain Energy Density Rule	Linear Rule
(a) Monotonic loading to 53.4 kN			
Notch stress, MPa	903	872	849
Notch strain	0.0173	0.0115	0.0087
(b) Unloading from 53.4 kN to 0			
Notch residual stress, MPa	−663	−652	−666
Notch strain	0.0073	0.0026	0
(c) Cyclic loading between 4.45 and 44.5 kN			
Notch stress amplitude, MPa	541	535	523
Notch strain amplitude	0.0036	0.0035	0.0033
(d) Fatigue life for loading in (c)			
Notch mean stress, σ_m (MPa)	204	177	141
Maximum notch stress, σ_{max} (MPa)	745	712	663
Product of $\varepsilon_a \sigma_{max}$	2.68	2.49	2.16
Fatigue life, N_f (cycles)	5750	7400	12 750

It can be seen from this table that the most conservative fatigue life is obtained from Neuber's rule and the least conservative life from the linear rule, with the result from the strain energy density rule falling in between. A 2 to 1 difference in estimated median life is shown for the three models, which is quite small. If K_t rather than K_f had been used in the application of Neuber's rule for cyclic loading, the notch stress amplitude would be 561 MPa rather than 541 MPa, the notch strain amplitude would be 0.0039 rather than 0.0036, and the expected fatigue life would be 3950 cycles rather than 5750 cycles. Therefore, more conservative predictions are obtained by using K_t, as expected.

It should be noted that the notch stress amplitudes obtained from the notch stress–strain analysis models presented could also be used in conjunction with the S–N approach. In this case, the S–N line or equation for smooth behavior should be used, since the effect of the notch is already incorporated in the stress amplitude. Since the minimum nominal stress in Part (c) is 50 MPa, the fatigue lives predicted from the strain–life approach in this example can be directly compared to those predicted from the S–N approach in Section 7.2.4. From Fig. 7.16, for a nominal stress amplitude of $S_a = 225$ MPa, the predicted life using Basquin's equation for the notched part is about 11 000 cycles based on the Haigh diagram approach for long life and about 28 000 cycles based on the modified Goodman equation.

The computation of $\Delta\sigma$ and $\Delta\varepsilon$ is not required for the application of Neuber's rule if the fatigue life is taken not from an $\varepsilon-N$ curve, but from a curve of the product $\varepsilon_a\sigma_{max}$ plotted versus N_f. One need only find the life that corresponds to the product $\varepsilon_a\sigma_{max}$. This approach was proposed by Smith et al. [26,27]. Finally, the remaining life (crack growth life from formation of a crack to fracture) of the notched part in this example problem can be estimated on the basis of linear elastic fracture mechanics discussed in Chapter 6. This is done in the next section.

7.4 APPLICATIONS OF FRACTURE MECHANICS TO CRACK GROWTH AT NOTCHES

As mentioned in Section 7.3, deformation of the material at the notch root under cyclic loading is often inelastic. Thus, cracks often start (nucleate) from a notch under conditions of local plasticity. As the cracks continue to grow, they grow into the elastic stress–strain field of the notch, and with subsequent growth extend into the net section elastic stress–strain field of the specimen or component. After the crack exceeds a certain length, the notch geometry has little influence on the elastic stress–strain field of the advancing crack, as shown in Fig. 7.2. Figure 7.24 schematically shows a crack and the crack tip

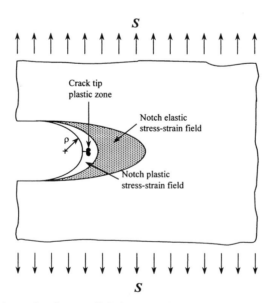

Figure 7.24 Schematic of a small fatigue crack emanating from a notch and the associated local elastic and plastic stress–strain fields.

plastic zone contained within the plastic stress–strain field of a notch as a result of an external applied load. Crack growth through a notch stress–strain field may represent a major or minor portion of the total fatigue life, depending on the conditions of the notch stress–strain field. These conditions include crack growth in the

Plastic stress–strain field of the notch.
Elastic stress–strain field of the notch.
Plastic and then the elastic stress–strain field of the notch.

Because the stress–strain field of a notch varies for each of these conditions, fatigue crack growth behavior also may vary as the crack grows through these different notch stress–strain fields. A schematic of the anomalous crack growth behavior from a notch for the conditions listed above is shown in Fig. 7.25. For cracks growing in the plastic stress–strain field of a notch, fatigue crack growth rates are initially high but decrease, i.e., crack growth deceleration, with increasing crack length as the crack grows further away from the notch root. For this condition, after the crack grows out of the notch plastic zone, it enters the elastic stress–strain field of the notch, yet its length may no longer be small, and the fatigue crack growth behavior tends to merge with that of the long crack growth curve. Fatigue crack growth behavior for cracks growing in the elastic stress–strain field of a notch when small is often similar to that presented and discussed for physically small cracks in Section 6.8. Crack growth behavior in the plastic and elastic stress–strain fields of the notch in general is a combination of the previous two conditions, assuming that the

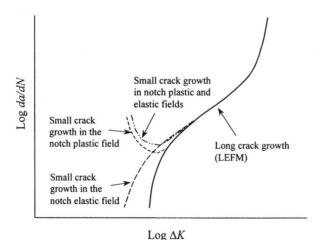

Figure 7.25 Schematic presentation of fatigue crack growth behavior from notches as a function of the stress–strain field at the notch.

crack is small, as the crack grows from the plastic to the elastic stress–strain field. Small cracks growing from notches that are influenced by the stress–strain field of the notch are generally referred to as "mechanically small" cracks.

As discussed above, fatigue crack growth from a notch is highly influenced by the stress–strain field of the notch. A crack that nucleates from a shallow or blunt notch will usually grow until failure (fracture) if the external applied cyclic load does not change. For this condition, fatigue behavior is often dominated by crack nucleation. Conversely, a crack may nucleate from a sharp notch rather quickly due to the elevated stresses at the notch root. However, it may stop growing once the crack grows through the notch-influenced region, i.e., crack arrest, or reach a minimum crack growth rate before continuing to grow. This is because the bulk or net nominal stress may not be high enough for the crack to grow, as ΔK is not high enough. For this condition, crack growth may dominate the fatigue behavior or total fatigue life. To illustrate the influence of the notch stress–strain field on a crack, consider a very wide sheet that contains a 5 mm and a 25 mm diameter circular hole loaded similar to that shown in Fig. 7.2. One can see that the stress decreases quite rapidly by moving away from the edge of the hole. However, the rate at which the stress decreases, $d\sigma_y/dx$, is a function of the hole radius. Table 7.1 summarizes the ratio of axial stress to nominal stress (σ_y/S) as one moves 0.1, 0.5, 1, and 2 mm away from the edge of the hole for both the 5 mm and 25 mm diameter holes. The ratio σ_y/S decreases much faster for a given distance away from the edge of the 5 mm hole in comparison to the 25 mm hole. Thus, if the distances (0.1, 0.5, 1, and 2 mm) represented the length of cracks emanating from the two different-size holes, the stress–strain field observed for the same crack length contained within each hole could be quite different. For instance, the stress for a 0.1 mm crack emanating from either hole is close to the maximum axial stress observed at the edge of the hole, which is $3S$ since $K_t = 3$. At 2 mm from the edge of the hole, however, the axial stress is only 30 percent greater than the nominal stress for the 5 mm hole, while for the 25 mm hole, the axial stress is still more than twice the nominal stress. For the 5 mm hole, the axial stress observed at the edge of the hole may be sufficient

TABLE 7.1 Ratio of σ_y/S as a Function of Distance from the Edge of a Circular Hole ($K_t = 3$)

Distance from Edge of Hole (mm)	σ_y/S	
	Hole dia. = 5 mm	Hole dia. = 25 mm
0	3.0	3.0
0.1	2.74	2.94
0.5	2.07	2.74
1	1.65	2.53
2	1.30	2.20

to cause a crack to nucleate, but the crack may become nonpropagating or reach a minimum growth rate as the crack grows into the lower local stressed region. For the 25 mm hole, a crack may nucleate and continue to grow or accelerate, as the local stress does not decrease as rapidly over a similar distance.

To quantify the influence of the stress–strain field of a notch on crack growth, consider the following example, adapted from reference 28. Assume that a very wide plate with a circular hole contains a pair of small through-thickness cracks, as shown in Fig. 7.26a. When the length of the fatigue crack, l, is small compared to the radius of the circle, c, the stress intensity factor, K_{small}, can be approximated by

$$K_{small} \approx 1.12(K_t S_{net})\sqrt{\pi l} \qquad (7.31)$$

where K_t is the stress concentration factor based on the net section stress, S_{net}, the value 1.12 is the free edge correction factor, as presented in Section 6.1.3, and the subscript "small" refers to a small crack under the influence of the elastic notch stress–strain field. Once the crack has grown a sufficient distance from the hole, the stress intensity factor solution takes on the form

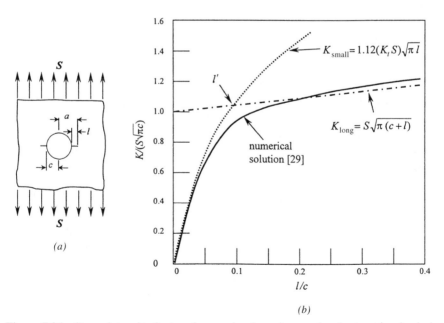

(a)

(b)

Figure 7.26 Stress intensity factors for a pair of cracks growing from a circular hole in a wide plate. (a) Nomenclature for cracks growing from notch. (b) Variation of normalized stress intensity factor ($K/S\sqrt{\pi c}$) versus normalized crack length (l/c).

$$K_{long} \approx S_{gross}\sqrt{\pi a} \qquad (7.32)$$

where $a = c + l$ and S_{gross} is based on the gross section area. Thus, for long cracks, the width or diameter of the hole acts as part of the crack. The subscript "long" refers to a long crack where the influence of the notch on the crack tip stress field is negligible. For a wide plate, $S_{net} \approx S_{gross}$, but for a narrow plate they are not equal. Figure 7.26b shows the variation of K based on Eqs. 7.31 and 7.32 as a function of the crack length, l, normalized by the radius of the hole, c. The finite element solution representing more realistic behavior, as predicted by Newman [29], shown by the solid curve in Fig. 7.26b, first follows K_{small}, falls below it, and then eventually approaches K_{long}. By equating Eqs. 7.31 and 7.32, an intersection point denoted l' can be approximated which identifies a transitional crack length where K_{small} can be used for crack lengths less than l' and K_{long} can be used for crack lengths greater than l'. The intersection point using the infinite plate solution is

$$l' = \frac{c}{(1.12K_t)^2 - 1} \qquad (7.33)$$

The stress intensity factor solutions presented in Eqs. 7.31 and 7.32 are valid for situations in which the notch tip plasticity effects are negligible. When this is the case, l' is usually a small fraction of the root radius, ρ, of the notch or stress concentration, where usually $0.1\rho \le l' \le 0.2\rho$. When the notch is circular, $\rho = c$. Using Eq. 7.33, the transitional crack length, l', associated with the 5 mm and 25 mm circular holes in the wide plate previously discussed, can now be calculated. For the 5 mm circular hole, $l' \approx 0.25$ mm, and for the 25 mm circular hole, $l' \approx 1.2$ mm.

Predicting fatigue crack growth from notches using LEFM is problematic, much the same as the situation presented for small crack behavior in Section 6.8. Even for conditions in which small-scale yielding prevails, fatigue crack growth rates for cracks emanating from notches tend to be higher than their long crack counterparts, as was shown in Fig. 7.25. A factor contributing to this behavior is the absence or reduction of tensile residual plastic deformation in the wake of the crack because the crack is small. It should be noted that the small tensile residual plastic deformation results from crack tip plasticity, not from inelasticity of the notch stress–strain field. Less residual plastic deformation in the crack wake means less crack closure, a greater ΔK_{eff}, and thus higher fatigue crack growth rates. When the crack is contained within the notch plasticity region, anomalous fatigue crack growth behavior also exists (shown in Fig. 7.25). For small cracks contained within the local elastic stress–strain field of a notch, extrapolation of the Paris equation to region I may provide a reasonable approach to life prediction, much like that presented in Section 6.8 for small cracks. For small cracks contained within the plastic stress–strain field of a notch, EPFM may provide at least a partial solution to the differences observed between the behavior of small cracks emanating from notches and long cracks. However, the decreased driving force with increasing crack length typically observed for these types of cracks is difficult to account for using LEFM or EPFM without the use of some type of crack closure model. Many attempts made to account for the anomalous fatigue crack growth from notches include a correction to the physical crack length [30–32]. While these models and corrections have proven useful, they have not provided a complete correlation between notch fatigue crack growth data and LEFM long crack growth data.

7.5   THE TWO-STAGE APPROACH TO FATIGUE LIFE ESTIMATION

The life of a structure is often considered in two stages:

Stage 1: Life to the formation of a crack on the order of 1 mm: "crack nucleation."

Stage 2: Life from this existing crack to fracture: "crack growth."

In some situations, stage 2 may be negligible in comparison to stage 1, e.g., the rotating bending of a small shaft with a mild shoulder root radius at a low load range. On the other hand, sometimes specifications require evidence of safety with given cracks, such as cracks approximately 1 mm in length. Stage 1 then becomes negligible in comparison to stage 2. Factors such as stress level, load history, residual stresses, material properties, and geometry contribute to the extent of each stage. Predictions of stage 1 are most commonly done using the ε–N method, which requires the two coefficients σ_f' and ε_f', the two exponents b and c, and the fatigue notch factor K_f, along with the cyclic stress–strain curve. Predictions of stage 2 require the stress intensity factor, K, for the given loading, geometry, and crack size, knowledge of the crack growth curve da/dN versus ΔK, the initial crack size, and knowledge of the fracture toughness. The two-stage approach also can be used with a damage-tolerant philosophy in which the prediction of fatigue crack growth life coupled with service inspection is important. Damage tolerance requires the interval of crack growth between an easily detectable crack size and a critical crack size to be longer than the interval between inspections.

In Section 7.3.5d, fatigue life calculations were made for stage 1 with a keyhole specimen (Fig. 7.13) using the strain–life approach. The following is a continuation of the example problem from Section 7.3.5d to predict the fatigue life associated with stage 2. The fatigue life associated with stage 2 (fatigue crack growth) can be predicted using information presented in Chapter 6 and again mentioned above. When the fatigue crack growth life associated with stage 2 is added to the results from Section 7.3.5d, this provides the "total fatigue life," which includes both fatigue crack nucleation life with a crack on the order of 1 mm (stage 1) and fatigue crack growth life (stage 2).

Example of Stage 2 and Total Fatigue Life Calculations   To use LEFM, the stress intensity factor solution for the specimen geometry is required. The stress intensity factor solution for the SAE keyhole specimen from Fig. 7.13 is given by [33]

$$K = (0.247)P \left[-54.98 + 417.28\left(\frac{a}{w}\right) - 939.76\left(\frac{a}{w}\right)^2 + 739.31\left(\frac{a}{w}\right)^3 \right] \quad (7.34)$$

where P is in kN and K is in MPa$\sqrt{m}$. This expression is applicable for a specimen thickness t equal to 9.5 mm, width w equal to 94 mm, and a/w between about 0.32 and 0.75, which corresponds to crack length a, as measured from the load line of 30 to 70 mm.

The fracture toughness, K_c, for RQC-100 steel for this thickness, along with the Paris equation (Eq. 6.19) for fatigue crack growth behavior with $R \approx 0$, were also obtained by Socie [33]. These values are

$$K_c \approx 165 \text{ MPa}\sqrt{m} \quad \text{and} \quad \frac{da}{dN} = 2.80 \times 10^{-12} \, (\Delta K)^{3.25} \quad (7.35)$$

where da/dN is in m/cycle and ΔK is in MPa$\sqrt{m}$. Although this crack growth rate equation was developed for $R = 0$, the mean stress effect ($R = 0.1$) observed in this example problem will not have much effect on crack growth compared to $R = 0$; thus, use of Eq. 7.35 is reasonable. The strain–life prediction models presented in Chapter 5 and in Section 7.3.5 are for crack nucleation lengths on the order of 1 mm. The example problem in Section 7.3.5 assumed a crack nucleation length of approximately 1 mm. The LEFM model must then assume an initial crack length comparable with this length; thus, let the initial crack length be 1 mm. Before we proceed with this example, we should check to see if this crack size emanating from the keyhole notch is greater than the transitional crack length l' presented in Section 7.4. If not, a more complex analysis involving the crack growing from the notch may be needed. Although the expression in Eq. 7.33 is for a crack emanating from both sides of a hole in a very wide plate, it can be used as a first approximation for the transitional crack length in this example problem. A more precise solution would take into account the geometry factor associated with the stress intensity factor solution of Eq. 7.34. Also, it should be recognized that use of Eq. 7.33 requires negligible notch tip plasticity, yet the notch root behavior was shown to be inelastic in Section 7.4.5. However, as a first approximation:

$$l' = \frac{c}{(1.12K_t)^2 - 1} = \frac{9.5/2}{[1.12(3)]^2 - 1} = 0.46 \text{ mm} \approx 0.5 \text{ mm}$$

Therefore, the crack length used to predict the fatigue life for stage 1 (1 mm) is greater than the estimated transitional crack length and can be assumed to be outside the influence of the notch stress–strain field. Thus, Eq. 7.34 can be used. The initial crack length is the notch length from the load line $(25.4 + 9.5/2) + 1$ mm. This gives an initial total crack length, a_i, of 31.15 mm and initial a/w of 0.33. The initial ΔK for this crack length with $\Delta P = P_{max} - P_{min} = 44.5 - 4.45 \approx 40$ kN from Eq. 7.34 is 69 MPa$\sqrt{m}$. This value is well above ΔK_{th}, and thus the Paris equation is applicable from that standpoint. Calculation of the plane stress initial plastic zone size, $2r_y$, from Eq. 6.10

$$2r_y = \frac{1}{\pi}\left(\frac{K_{max}}{S_y}\right)^2 \tag{6.11}$$

gives $2r_y = 2.4$ mm, or $r_y = 1.2$ mm, using $K_{max} = 77$ MPa and $S_y = 883$ MPa. While the initial ΔK is approaching region III of the sigmoidal curve $(da/dN \approx 2.7 \times 10^{-6}$ m/cycle) where the Paris equation gives nonconservative life predictions, we use the LEFM model and the Paris equation to obtain a first approximation.

The critical crack length at fracture can be approximated by setting K from Eq. 7.34 equal to K_c and solving for a/w using an iterative or trial-and-error procedure. For P_{max} equal to 44.5 kN, the critical a/w value is 0.58, or $a_c = 54$ mm.

The fatigue crack growth life, N_f, which we define here as N_g, can be found by integrating the Paris equation (Eq. 6.19):

$$\frac{da}{dN} = A(\Delta K)^n = A(K_{max} - K_{min})^n \tag{7.36}$$

or

$$N_g = N_f = \int_0^{N_f} dN = \int_{a_i}^{a_c} \frac{da}{A(K_{max} - K_{min})^n}$$

$$= \frac{1}{2.8 \times 10^{-12}[0.247(P_{max} - P_{min})]^{3.25}} \times \tag{7.37}$$

$$\int_{a_i}^{a_c} \frac{da}{\left[-54.98 + 417.28\left(\dfrac{a}{w}\right) - 939.76\left(\dfrac{a}{w}\right)^2 + 739.31\left(\dfrac{a}{w}\right)^3\right]^{3.25}}$$

Equation 7.37 must be integrated numerically between the limits a_i and a_c. Equations 6.20f or 6.20g cannot be used since K is not just a simple $\sqrt{a}$ function. Integration of Eq. 7.37 with the proper values of A, n, P_{max}, P_{min}, a_i, a_c, and w gives $N_g \approx 3000$ cycles, where N_g defines the fatigue crack growth life. This predicted fatigue crack growth life is one-half of the fatigue life calculated for stage 1, N_n, using Neuber's rule, and a smaller fraction compared to the strain energy density rule or the linear rule. Thus, for this example, stage 1, with crack nucleation to approximately 1 mm, has a greater contribution to the total fatigue life, N_t, where $N_t = N_n + N_g$. Now let us assume a keyhole diameter of 5 mm instead of 9.5 mm where the initial notch length from the load line is still 30.15 mm. Finite element results give $K_t = 4.1$ for this notch geometry, and from this $K_f = 3.7$. Using Neuber's rule and the Smith-Watson-Topper parameter, the fatigue life to approximately 1 mm based on $P_{max} = 44.5$ kN and $P_{min} = 4.45$ kN ($\Delta P \approx 40$ kN) results in $N_n \approx 1500$ cycles. The fatigue crack growth life, N_g, will not change (~ 3000 cycles), as the initial and final crack lengths do not change unless the transitional crack length (l') is greater than 1 mm for this notch size, which is not the case. Fatigue lives for both crack nucleation and crack growth are given in Table 7.2 for the two notch sizes described above and for two load ranges (ΔP). For the more blunt notch ($d = 9.5$ mm) with a smaller K_t and K_f, the nucleation life to a crack length of approximately 1 mm is a larger portion of the total fatigue life. This difference becomes even greater for the smaller load range ($\Delta P = 27$ kN), where the nucleation life is more than 90 percent of the total calculated fatigue life. For the sharper notch ($d = 5$ mm) with a larger K_t and K_f, the crack growth life is roughly two times longer than the nucleation life for the higher load range, while the nucleation life and crack growth life are similar for the lower load range. Thus, the fatigue crack nucleation and fatigue crack growth lives are dependent on both the notch geometry and the load level. From Table 7.2, fatigue crack nucleation life dominates the total fatigue life at the low load with the 9.5 mm diameter hole, while the fatigue crack growth life is a greater percentage of the total fatigue life at the high load with the 5 mm diameter hole. For the other two conditions presented in Table 7.2, the contribution of each stage falls between the two previous extremes. This load influence behavior is consistent with the schematic presented in Fig. 4.7.

7.6 SUMMARY

Notches cannot be avoided in many structures and machines, and understanding their effect is of key importance in the study of fatigue. The fatigue strength

TABLE 7.2 Fatigue Lives Based on Different Notch Radii and Load Ranges

ΔP (kN)	$K_t = 3$ ($d = 9.5$ mm)			$K_t = 4.1$ ($d = 5$ mm)		
	N_n	N_g	N_t	N_n	N_g	N_t
40	6 000	3 000	9 000	1 500	3 000	4 500
27	111 000	11 000	122 000	13 000	11 000	24 000

N_n denotes fatigue nucleation life of cracks on the order of 1 mm using Neuber's rule and the Smith-Watson-Topper parameter.
N_g denotes fatigue crack growth life of cracks from a_i to a_c by integrating Eq. 7.37.

of notched members is affected not only by the stress concentration factor, but also by the stress gradient, mean or residual stress at the notch, local yielding, and development and growth of cracks.

The ratio of smooth to notched fatigue strength is called the fatigue notch factor, K_f. This factor depends on the stress concentration factor, K_t, the notch sensitivity of the material, q, and the mean and alternating stress levels. For zero mean stress and long life, K_f is

$$K_f = 1 + q(K_t - 1) \qquad (7.8)$$

where empirical formulas for q have been developed by Neuber and Peterson.

In the S–N approach for fully reversed fatigue behavior (no mean or residual stress) of notched members, the S–N line can be approximated by connecting the ultimate tensile strength or true fracture strength at 1 cycle to the notched fatigue strength (S_f/K_f) at long life defined between 10^6 and 10^8 cycles, as appropriate. If a mean stress exists, the fatigue strength of the notched member at long life is estimated from a Haigh diagram or by using the modified Goodman equation:

$$\frac{S_a}{(S_f/K_f)} + \frac{S_m}{S_u} = 1 \qquad (7.12)$$

The strain–life approach for fatigue of notched members involves two steps. First, notch root stresses and strains are found by experimental, finite element, or analytical methods. A life prediction is then made using the notch root stresses and strains based on the strain–life equations from Chapter 5. Analytical models of notch stress/strain include the linear rule, Neuber's rule, and the strain energy density or Glinka's rule. The linear rule assumes the strain concentration factor, K_ε, to be equal to the theoretical stress concentration factor, K_t, which for nominal elastic behavior results in

$$\Delta\varepsilon = K_t \frac{\Delta S}{E}$$

This rule agrees with measurements in plane strain situations. Neuber's rule has been the most widely used model. It assumes the geometrical mean of the stress and strain concentration factors under plastic deformation conditions to be equal to the theoretical stress concentration factor $\sqrt{K_\sigma K_\varepsilon} = K_t$. This rule generally agrees with measurements in plane stress situations and is more conservative than the linear rule. To reduce the degree of conservatism with Neuber's rule, the fatigue notch factor, K_f, is often used in place of K_t. For nominal elastic behavior, Neuber's rule can be expressed as

$$\frac{(\Delta\sigma)^2}{E} + 2\Delta\sigma \left(\frac{\Delta\sigma}{2K'}\right)^{1/n'} = \frac{(K_f \Delta S)^2}{E} \qquad (7.24)$$

The strain energy density or Glinka's rule is based on the assumption that the strain energy density at the notch root is nearly the same for linear elastic and elastic-plastic notch behaviors. For nominal elastic behavior, Glinka's rule is expressed as

$$\frac{(\Delta\sigma)^2}{E} + \frac{4\Delta\sigma}{n'+1}\left(\frac{\Delta\sigma}{2K'}\right)^{1/n'} = \frac{(K_t\,\Delta S)^2}{E} \tag{7.29}$$

The local stress–strain field of a notch has a large influence on crack growth behavior. For cracks that nucleate from a shallow or blunt notch, the fatigue behavior is often dominated by crack nucleation. Cracks that nucleate from a sharp notch often nucleate rather quickly due to the elevated local stresses, and crack growth often dominates the fatigue behavior in this case. For essentially elastic notch conditions, a transitional crack length, l', can be estimated above which the contribution of the local stress–strain field of a notch to fatigue crack growth becomes small.

The two-stage approach can be used to estimate the total fatigue life that includes both crack nucleation (stage 1) and crack growth (stage 2). Strain–life (ε–N) methods are used for stage 1, while stage 2 uses fracture mechanics and its application to fatigue crack growth.

7.7 DOS AND DON'TS IN DESIGN

1. Do provide for smooth flow of stresses and forces by streamlining the contours of parts and consider design improvements like those shown in Fig. 7.6 to mitigate stress concentrations.
2. Don't permit sharp scratches or grooves with radii smaller than 0.25 mm (0.010 in.) on hard metals unless you know that they will not be subject to tensile stresses.
3. Don't worry about the exact numerical value of elastic stress concentration factors, but be aware of their trends.
4. Do recognize that the theoretical stress concentration factor, K_t, depends only on geometry and mode of loading, but that it can be used to relate notch stress to nominal stress for only linear elastic notch deformation behavior.
5. Do consider the effects of stress state and stress gradients at notches.
6. Do recognize that the fatigue strength of notched parts depends not only on the part geometry and loading but also on material notch sensitivity. The stronger the material, the higher the notch sensitivity.
7. Do expect mean stress to have more effect on the fatigue behavior of notched parts than on that of smooth parts.

8. Do recognize that when significant notch plastic deformation exists, the strain–life approach to life prediction is usually superior to the *S–N* approach.

9. Don't expect an elaborate, complex analysis to obviate the need for tests and service monitoring of machines or components.

10. Do recognize that fatigue crack growth from a notch may represent a major or minor portion of the total fatigue life.

11. Do recognize, using LEFM principles, that cracks growing from notches below some transitional crack length often display anomalous behavior compared to long cracks.

12. Do consider the two stages of fatigue life in estimating the total fatigue life of a component or structure: the formation of a crack on the order of 1 mm by the ε–N approach and the growth of the crack to failure by the LEFM approach.

REFERENCES

1. G. Kirsch, "Die Theorie Der Elastizitaet und Die Beduerfnisse Der Festigkeitslehre," ZVDI, 1898, mentioned by S. Timoshenko in *History of Strength of Materials,* McGraw-Hill Book Co., New York, 1953.

2. H. J. Grover, *Fatigue of Aircraft Structures,* NAVAIR OI-IA-13, U.S. Department of the Navy, Washington, DC, 1966.

3. M. Creagar, "The Elastic Stress-Field Near the Tip of a Blunt Notch," Ph.D. diss., Lehigh University, 1966.

4. W. D. Pilkey, *Peterson's Stress Concentration Factors,* 2nd ed., John Wiley and Sons, New York, 1997.

5. P. E. Erickson and W. F. Riley, "Minimizing Stress Concentrations Around Circular Holes in Uniaxially Loaded Plates," *Exp. Mech.,* Vol. 18, No. 3, 1978, p. 97.

6. J. Y. Mann, *Fatigue of Materials,* Melbourne University Press, Melbourne, Australia, 1967.

7. H. Neuber, *Kerbspannungstehre,* Springer, Berlin, 1958; *Translation Theory of Notch Stresses,* U.S. Office of Technical Services, Washington, DC, 1961.

8. P. Kuhn and H. F. Hardrath, "An Engineering Method for Estimating Notch Size Effect in Fatigue," NACA TN 2805, 1952.

9. R. E. Peterson, *Stress Concentration Factors,* John Wiley and Sons, New York, 1974.

10. R. C. Rice, ed., *SAE Fatigue Design Handbook,* 3rd ed., AE-22, SAE, Warrendale, PA, 1997.

11. H. O. Fuchs, "A Set of Fatigue Failure Criteria," *Trans. ASME, J. Basic Eng.,* Vol. 87, 1965, p. 333.

12. H. O. Fuchs, "Regional Tensile Stress as a Measure of the Fatigue Strength of Notched Parts," *Proceedings of the 1971 International Conference on the Mechanical Behavior of Materials,* Kyoto, Japan, 1972, p. 478.

13. "Technical Report on Low Cycle Fatigue Properties: Ferrous and Nonferrous Materials," SAE J1099, revised June 1998, *SAE Handbook,* Vol. 1, SAE, Warrendale, PA, 1999, p. 3.154.

14. D. V. Nelson and H. O. Fuchs, "Predictions of Cumulative Fatigue Damage Using Condensed Load Histories," *Fatigue Under Complex Loading: Analysis and Experiments,* AE-6, R. M. Wetzel, ed., SAE, Warrendale, PA, 1977, p. 163.

15. V. M. Faires, *Design of Machine Elements,* Macmillan, London, 1965.

16. L. Tucker and S. Bussa, "The SAE Cumulative Fatigue Damage Test Program," *Fatigue Under Complex Loading: Analysis and Experiments,* AE-6, R. M. Wetzel, ed., SAE, Warrendale, PA, 1977, p. 1.

17. S. Kotani, K. Koibuchi, and K. Kasai, "The Effect of Notches on Cyclic Stress–Strain Behavior and Fatigue Crack Initiation," *Proceedings of the Second International Conference on the Mechanical Behavior of Materials,* Boston, 1976, p. 606.

18. H. Neuber, "Theory of Stress Concentration for Shear-Strained Prismatical Bodies with Arbitrary Nonlinear Stress Strain Laws," *Trans. ASME, J. Appl. Mech.,* Vol. 28, 1961, p. 544.

19. T. Seeger and P. Heuler, "Generalized Application of Neuber's Rule," *J. Testing and Evaluation,* Vol. 8, No. 4, 1980, p. 199.

20. T. H. Topper, R. M. Wetzel, and J. D. Morrow, "Neuber's Rule Applied to Fatigue of Notched Specimens," *J. Materials, JMSLA,* Vol. 4, No. 1, 1969, p. 200.

21. K. Molski and G. Glinka, "A Method of Elastic-Plastic Stress and Strain Calculation at a Notch Root," *Mat. Sci. Eng.,* Vol. 50, 1981, p. 93.

22. N. E. Dowling, "Performance of Metal-Foil Strain Gages During Large Cyclic Strains," *Exp. Mech.,* Vol. 17, 1977, p. 193.

23. G. Glinka, "Relation between the Strain Energy Density Distribution and Elastic-Plastic Stress–Strain Fields near Cracks and Notches and Fatigue Life Calculation," *Low Cycle Fatigue,* ASTM STP 942, H. D. Soloman, G. R. Halford, L. Kaisand, and B. N. Leis, eds., ASTM, West Conshohocken, PA, 1988, p. 1022.

24. W. N. Sharpe, Jr., C. H. Yang, and R. L. Tregoning, "An Evaluation of the Neuber and Glinka Relations for Monotonic Loading," *ASME J. Applied Mech.,* Vol. 59, 1992, p. S50.

25. N. E. Dowling, W. R. Brose, and W. K. Wilson, "Notched Member Life Predictions by the Local Strain Approach," *Fatigue Under Complex Loading: Analysis and Experiments,* AE-6, R. M. Wetzel, ed., SAE, Warrendale, PA, 1977, p. 55.

26. K. N. Smith, P. Watson, and T. H. Topper, "A Stress–Strain Function for the Fatigue of Metals," *J. Mater.,* Vol. 5, No. 4, 1970, p. 767.

27. K. N. Smith, M. El Haddad, and J. F. Martin, "Fatigue Life and Crack Propagation Analyses of Welded Components Containing Residual Stresses, *J. Test. Eval.,* Vol. 5, No. 4, 1977, p. 327.

28. N. E. Dowling, "Notched Member Fatigue Life Predictions Combining Crack Initiation and Propagation," *Fatigue Eng. Mater. Struct.,* Vol. 2, 1979, p. 129.

29. J. C. Newman, "An Improved Method of Collocation for the Stress Analysis of Cracked Plates with Various Shaped Boundaries," *NASA Tech. Note D-6376,* Langley, VA, 1971.

30. M. H. El Haddad, T. H. Topper, and K. N. Smith, "Prediction of Non-Propagating Cracks," *Eng. Fract. Mech.,* Vol. 11, 1979, p. 573.

31. R. A. Smith and K. J. Miller, "Prediction of Fatigue Regimes in Notched Components," *Int. J. Mech. Sci.*, Vol. 20, 1978, p. 201.

32. M. H. El Haddad, N. E. Dowling, T. H. Topper, and K. N. Smith, "J Integral Applications for Short Fatigue Cracks at Notches," *Int. J. Fract.*, Vol. 16, 1980, p. 15.

33. D. F. Socie, "Estimating Fatigue Crack Initiation and Propagation Lives in Notched Plates under Variable Load Histories," Ph.D. diss., University of Illinois, Urbana, 1977.

PROBLEMS

1. Strain gages are often used to obtain strain–time histories and to obtain elastic stresses. If a strain gage is mounted in the loading direction (y direction) in Fig. 7.2 at a distance $x/r = 2.0$ (one radius away from the notch) and measured 0.002 strain for 7075-T6 aluminum, what is the stress, S? How does this value compare with a uniaxial state of stress at a point obtained from $\sigma = E\varepsilon$? What is the significance of the difference? Assume $E = 70$ GPa and $\nu = 0.33$.

2. Derive the relationship between K_t based on gross area and K_t based on net area for the finite-width plate of Fig. 7.1.

3. A stepped shaft, Fig. 7.3, has $D = 55$ mm, $d = 50$ mm, and $r = 2$ mm. Determine and compare K_t for (a) axial loading, (b) bending, and (c) torsion. Check other values in Fig. 7.3 and see if a conclusion can be drawn about the effect of type of loading on K_t for a given geometry.

4. A threaded bolt has a series of stress concentrations at each thread. Using the streamline analogy, how does K_t at the thread adjacent to the shank compare with K_t at intermediate threads?

5. With reference to K_t for an elliptical notch, how important are cracks aligned parallel to a uniaxial tensile stress?

6. For the stepped shaft of Problem 3, estimate the fatigue notch factor, K_f, at long life and $R = -1$ for axial loading for the following steels: (a) Man-Ten, (b) RQC- 100, (c) 4340 with $S_u = 745$ MPa, (d) 4340 with $S_u = 1260$ MPa, (e) 4340 with $S_u = 1530$ MPa, (f) 4340 with $S_u = 1950$ MPa. Comment on the notch sensitivity of the six steels.

7. How can the fatigue strength for the shaft geometry of Problem 3 be improved? List as many ways as you can think of to make this improvement.

8. The stepped shaft in Problem 3 is made of a steel alloy with an ultimate tensile strength of 700 MPa and is subjected to constant amplitude rotating bending. Estimate the magnitude of the bending moment which can be applied such that failure does not occur in 10^6 cycles.

9. Superimpose approximate fully reversed S–N curves for the notched specimen and the six steels in Problem 6.

10. Repeat Problem 8 for a fatigue life of 10^4 cycles.

11. Show that the criteria for the alternating tensile stress, $S_a = S_{cat}$, plot on the Haigh diagram as

$$S_a = S_{cat} \quad \text{for } R \geq 0$$

and

$$S_a + S_m = 2 S_{cat} \quad \text{for } R \leq 0$$

where S_a and S_m are the applied alternating and mean stress, respectively.

12. A stepped circular rod of Man-Ten steel with diameters of 50 and 40 mm has a root radius of 2 mm at the stepped section. The rod is to be subjected to axial cyclic loading. Using a Haigh diagram, determine the following for an approximate median fatigue life of 10^6 cycles:

    (a) What fully reversed alternating force, P_a, can be applied?
    (b) What is the maximum value of P_a if proper compressive residual stresses are present at the notch root? What is the magnitude of the compressive residual stress needed to obtain this maximum alternating stress?
    (c) What value of P_a can be applied if the residual stress calculated in (b) is tensile?

13. Same as Problem 12, except that the materials are as follows:

    (a) RQC-100 steel.
    (b) 4340 steel with $S_u = 827$ MPa.
    (c) 4340 steel with $S_u = 1240$ MPa.
    (d) 4340 steel with $S_u = 1468$ MPa.
    (e) 4142 steel with $S_u = 1930$ MPa.
    (f) 2024-T3 aluminum.
    (g) 5456-H3 aluminum.
    (h) 7075-T6 aluminum.

14. In Problem 12, estimate the values of P_a for parts (a), (b), and (c) for median fatigue life of 5×10^4 cycles.

15. Repeat Problem 14 with the materials given in Problem 13a to 13h.

16. A flat plate of Q & T 4340 steel with $S_u = 1250$ MPa is 80 mm wide, 300 mm long, and 10 mm thick and has a central through hole of 20 mm diameter. A constant amplitude, repeated axial force varying from 0 to

400 kN in tension consistently causes premature fatigue failures. It is desired to increase the life to 10^6 cycles without changing the minimum hole size and material. Three ways of increasing the fatigue life would be to change (a) thickness, (b) width, and (c) hole geometry. Provide three quantitative solutions using suggestions (a) to (c) separately such that fatigue failures should not occur in 10^6 cycles.

17. The notched member shown below has a strain gage reading of 0.002 at the notch root when a load of 40 kN is applied. Yielding for this material occurs at a strain of 0.0025. The load is increased such that the strain gage reads 0.0065 when the load is 80 kN. Determine K_t. Also determine K_ε, and compare it with values calculated using the linear rule, the Neuber rule, and the strain energy density rule. $E = 200$ GPa, $K = 950$ MPa, $n = 0.1$, and $\nu = 0.3$.

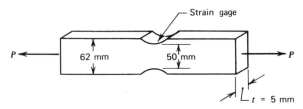

18. Rework the example problem in Section 7.3.5 for K_t equal to (a) 2.5, (b) 1.5, and (c) 1.0.

19. RQC-100 hot-rolled steel sheet was machined to the shape as shown in the following figure, with opposite notches, Fig. 7.4a. The member was subjected to a fully reversed alternating force, $P_a = 220$ kN. The notch root radius, r, is 3 mm.

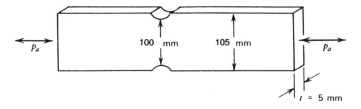

(a) Using strain–life fatigue concepts, determine the expected number of cycles to the appearance of a small crack of length $\Delta a \approx 1$ mm.

(b) Determine a reasonable number of cycles required to grow this crack from $\Delta a \approx 1$ mm to fracture (note: $a_i = \Delta a + 2.5$ mm).

(c) Compare (a) and (b) and discuss the significance and variability of each calculation.

(d) Repeat (a) to (c) for a notch root radius of $r = 30$ mm.

20. The same as Problem 19, except that the material is Q & T 4142 steel with $S_u = 1930$ MPa.

21. A stepped rotating shaft with diameters of 40 and 30 mm and a notch root radius at the step of 2 mm is to be subjected to a fully reversed, constant amplitude bending moment M. The desired life of the component is 90 000 cycles. The shaft is to be made of 1020 HR steel. Using the notch strain analysis method, estimate the allowable bending moment using the following:

   (*a*) Linear rule.

   (*b*) Nueber's rule.

   (*c*) Strain energy density rule.

   Discuss the difference in results and what value of M you would recommend as the final value.

22. The same as Problem 21, except that 2024-T3 aluminum is used.

23. For the hole radii compared in Table 7.1, plot on the same graph the stress distribution (σ_y/S) for each hole versus the distance from the edge of the hole. Extend this plot to a distance 5 mm from the hole edge. Comment on how the stress distributions could influence life calculations for a 0.1 mm and a 5 mm crack if the notch stress–strain field due to an external load is elastic.

24. A wide sheet contains a 10 mm diameter circular hole.

   (*a*) Using Eq. 7.33, estimate l'.

   (*b*) Repeat the calculation if the hole is elliptical where the major axis is $2b = 10$ mm and the radius of curvature is $r = 2$ mm.

   (*c*) Compare and comment on your findings from (*a*) and (*b*).

25. A component containing the notches compared in Table 7.1 is made of a material with a long crack threshold stress intensity factor range $\Delta K_{th} = 6$ MPa$\sqrt{m}$ for $R = 0$ and a yield strength $S_y = 450$ MPa. Each notch contains a pair of 0.4 mm small through-thickness cracks, as shown in Fig. 7.26*a*. If a stress range $\Delta S = 80$ MPa with $R = 0$ is applied to the component:

   (*a*) Determine if the cracks will grow from the 5 mm diameter hole.

   (*b*) Determine if the cracks will grow from the 25 mm diameter hole.

   (*c*) Comment on the applicability of LEFM to each case and discuss any shortcomings or limitations, if any, in the analysis performed.

CHAPTER 8

RESIDUAL STRESSES AND THEIR EFFECTS ON FATIGUE RESISTANCE

In Chapter 7 it was shown that the fatigue notch factor, K_f, can be about the same as, greater than, or much smaller than the elastic stress concentration factor, K_t, depending on the mean stress, S_m. To improve fatigue resistance, we should try to avoid tensile mean stress and have compressive mean stress. This can often be achieved by using residual stresses.

Residual stresses have also been called "self-equilibrating stresses" because they are in equilibrium within a part, without any external load. They are called "residual stresses" because they remain from a previous operation. Residual stresses exist in most manufactured parts, and their potential to improve or ruin components subjected to millions of load cycles can hardly be overestimated.

8.1 EXAMPLES

Figure 8.1 shows the $S-N$ behavior of a Ni-Cr alloy steel subjected to rotating bending with three different surface conditions involving smooth, notched, and notched shot-peened specimens [1]. K_t for the notched specimens was 1.76 and K_f at 10^7 cycles was also about 1.76, implying that the notched specimens were fully notch sensitive at this life. However, with the notched shot-peened specimens, the fatigue resistance is essentially the same as that of the smooth specimens, as they both occupy the same upper scatter band. Thus, the notch becomes perfectly harmless after it is shot-peened due to the desirable residual compressive stresses created at the notch surface from the shot-peening.

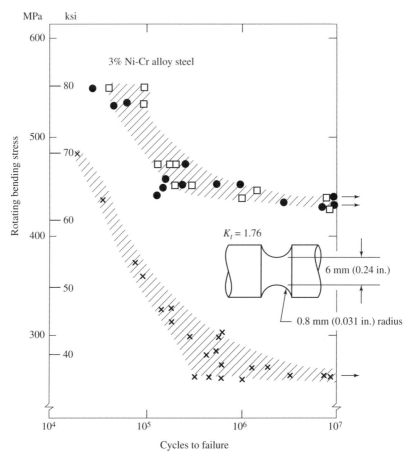

Figure 8.1 *S–N* behavior for smooth, notched unpeened, and notched peened specimens [1] (reprinted with permission of Pergamon Press). (●) smooth, polished, (x) notched, (□) notched, shot-peened.

Table 8.1 shows the effect of residual stresses produced by prestretching (tensile overload) on the fatigue strength and fatigue notch factor of specimens of 4340 steel with two different notches heat treated to two different hardness levels [2]. The prestretching caused localized inelastic tensile yielding at the notch, resulting in compressive residual stresses at the notch when the load was removed. The data in Table 8.1 compare the fatigue strengths at 2×10^6 cycles in rotating bending of round specimens with V-grooves. For the smooth specimens ($K_t = 1$) the fatigue strengths were 400 MPa (58 ksi) for the steel of 900 MPa (130 ksi) ultimate tensile strength and 630 MPa (91 ksi) for that of 1700 MPa (250 ksi) ultimate tensile strength. Stretching (tensile overload) with localized notch yielding increased the fatigue strength of the specimens made of the lower tensile strength steel by 130 percent for the sharper notch

TABLE 8.1   Fatigue Strengths and Fatigue Notch Factors with and without Residual Stresses

S_u, MPa	900			1700		
(K_t)	(1)	(2.15)	(3.2)	(1)	(2.15)	(3.2)
Without stretching						
S_f, MPa	400	205	160	630	240	190
(K_f)	(1)	(1.95)	(2.5)	(1)	(2.6)	(3.3)
With stretching						
S_f, MPa	390	390	370	635	620	610
(K_f)	(1.03)	(1.03)	(1.08)	(0.99)	(1.02)	(1.03)

Source: Courtesy of H. O. Fuchs.

($K_t = 3.2$) and by 90 percent for the more rounded notch ($K_t = 2.15$). The increases for the higher tensile strength steel specimens were 220 and 160 percent, respectively. The damaging effect of the notches without stretching was greater for the higher tensile strength steel. The residual stresses eliminated the notch effect almost completely. Without stretching, the worst notched specimens were those made of the lower tensile strength steel with the sharp notch, and the best notched specimens were those made of the higher tensile strength steel with the more rounded notch. It is interesting to note that, with the residual stresses induced by stretching, the worst notched specimens became much stronger than the best notched specimens without residual stresses. One can also see that the effects were not caused by work hardening of the steel, since the smooth specimens were essentially unaffected by the stretching operation, which had a great influence on the fatigue strength of the notched specimens.

Methods other than stretching and shot-peening are commonly used to produce desirable compressive residual surface stresses in metals that enhance fatigue resistance. On the other hand, some processes produce undesirable surface tensile residual stresses that decrease fatigue resistance. These methods and their fatigue effects are discussed in Section 8.2.

8.2   PRODUCTION OF RESIDUAL STRESSES AND FATIGUE RESISTANCE

The many methods of inducing residual stresses in parts can be divided into four main groups: mechanical methods, thermal methods, plating, and machining. The methods may be applicable to almost all metals, but in some cases they may be restricted to specific metal families.

8.2.1   Mechanical Methods

Mechanical methods of inducing residual stresses rely on applying external loads that produce localized inelastic deformation. Upon removal of the external loading, elastic "springback" occurs that produces both tensile and compressive residual stresses. Both tensile and compressive residual stresses must be present in order to satisfy all equations of internal force and moment equilibrium, i.e., $\Sigma F = \Sigma M = 0$. Figure 8.2 shows this process for inelastic bending of a beam behaving in an elastic-perfectly plastic manner. In Fig. 8.2a the beam is loaded inelastically by the moment, M_0, that results in the inelastic stress distribution shown. The depth of yielding depends upon the applied moment, geometry, and yield strength, S_y. In Fig. 8.2a the top fibers are in compression and the bottom fibers are in tension while the external moment is applied. Figure 8.2b shows the elastic springback stress distribution by unloading, i.e., removing the moment, M_0. The summation of the inelastic loading stress distribution and the elastic unloading stress distribution equals the remaining residual stress distribution with no external moment. This resultant summation is shown in Fig. 8.2c. Quantitative calculations are possible

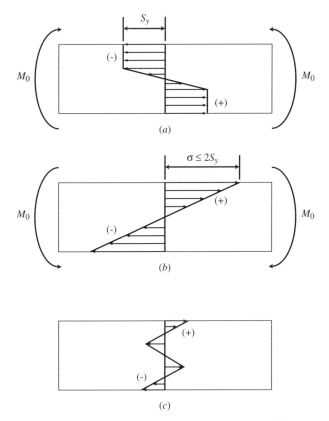

Figure 8.2   Residual stress formation from inelastic bending. (a) Stress distribution during inelastic loading. (b) Elastic recovery during unloading. (c) Resultant residual stress distribution after loading /unloading.

for this elastic-perfectly plastic material behavior by relating M_0 to the depth of yielding upon loading, using M_0 loading $= M_0$ unloading, the elastic bending stress equation $\sigma = My/I$ for unloading, and equations of equilibrium $\Sigma F = \Sigma M = 0$. This has been left as a homework problem. Quantitative calculations can also be made with other elastic-plastic behavior, but they become more complex. The maximum unloading stress must be $\leq 2S_y$ for the assumed elastic-perfectly plastic material. This implies that the largest residual stress cannot exceed S_y. A key factor to note in Fig. 8.2c is that the sign of the residual stress at each surface is opposite to the sign of the applied stress during inelastic loading in Fig. 8.2a. Thus, if a surface region yields in tension during loading, after unloading the residual surface stress will be in compression, which is desirable. Conversely, if a surface yields in compression upon loading, after unloading the residual surface stress will be in tension, which is undesirable. The beam in Fig. 8.2c will have better fatigue resistance at the bottom fibers than at the top fibers. Thus, straightening of parts by bending is usually detrimental due to the undesirable tensile residual stresses that form in regions overloaded in compression. Also, note that the residual stress distribution in Fig. 8.2c satisfies $\Sigma F = \Sigma M = 0$. If the material were not elastic-perfectly plastic, the residual stress distribution in Fig. 8.2c would be nonlinear but qualitatively similar to that shown.

Another example of forming residual stresses by mechanical means is the stretching (tensile overload) of the notched specimen in Fig. 8.3. Again, the material is assumed to be elastic-perfectly plastic. Figure 8.3a shows the non-uniform tensile stress distribution formed during the inelastic loading with P_0. The depth of tensile yielding at the notch depends upon the applied load, geometry, and the yield strength, S_y. The nonuniformity of stress is due to the stress concentration effect at the notch. Figure 8.3b shows the nonuniform elastic springback stress distribution by unloading, i.e., removing the load P_0. The maximum unloading stress is between S_y and $2S_y$. Summation of the inelastic loading stresses and the elastic unloading stresses results in the residual stress distribution shown in Fig. 8.3c. The maximum residual stress must be $\leq S_y$. Again, note that at the notch, the resultant residual stress sign is opposite to that during the original inelastic loading. Thus, tensile overloads with notches result in desirable residual compressive stresses at the notch, while compressive overloads with notches result in undesirable tensile residual stresses at the notch. The beneficial fatigue behavior presented in Table 8.1 is due to compressive residual stresses formed at the notch due to the pre-stretching.

The autofrettage of gun barrels [3] is an example of deliberate tensile overloading to enhance fatigue resistance, with desirable residual compressive stresses. Tensile stresses in the thick-walled tube are higher near the bore than on the outside, and one can yield the bore by high internal pressure without yielding the entire tube wall. When the pressure is removed, the elastic contraction of the tube diameter produces compressive residual stresses in the metal near the bore. Tensile overloading also is useful when cracks are

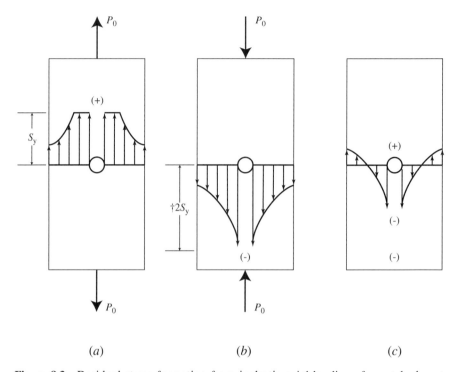

Figure 8.3   Residual stress formation from inelastic axial loading of a notched part. (*a*) Stress distribution during inelastic loading. (*b*) Elastic recovery during unloading. (*c*) Resultant residual stress distribution after loading/unloading.

present or suspected. If the cracks are small, the overload retards their growth. If the cracks are too large, the overload serves as a proof test by breaking parts that have low residual static strength.

The most widely used mechanical processes for producing beneficial compressive surface residual stresses for enhancing long and intermediate fatigue life are shot-peening and surface rolling. Both use local plastic deformation, one by the pressure of the impact of small balls, the other by the pressure of narrow rolls. Surface rolling is widely used in the production of threads. It is very economical as a forming operation for bolts and screws, as well as beneficial for fatigue resistance. Heywood [4] has reported 50 percent greater fatigue strength for rolled threads compared to cut or ground threads made of high-strength steel. For steels of lower hardness the improvement is less. For less than 550 MPa (80 ksi) ultimate tensile strength there is no improvement, according to Faires [5]. For internal threads, rolling is more expensive than cutting, but because of the improved resistance against fatigue and corrosion fatigue, rolling is standard procedure for some sucker-rod couplings. Bellow and Faulkner [6,7] obtained the results shown in Table 8.2 with AISI 8635 steel, quenched and tempered between R_c 16 and R_c 23. Rolling, as shown in

TABLE 8.2 Fatigue Strength at 10^5 Cycles, 0.2 Hz, AISI 8635 Steel [6,7]

Environment	Rolled Threads S_{Nf} − MPa (ksi)	Cut Threads S_{Nf} − MPa (ksi)	% Increase from Rolling
Air	510 (74)	303 (44)	68
3.5% NaCl	414 (60)	290 (42)	43
H_2S + CH_3COOH + 5% NaCl	317 (46)	<276 (40)	>15

Fig. 8.4, is also used to produce desired compressive residual stresses in fillets for components such as crankshafts, axles, gear teeth, turbine blades, and between the shank and head of bolts.

Shot-peening has been used successfully with steels including stainless and maraging steels, ductile iron, and aluminum, titanium, and nickel base alloys. Small balls (shot) that range from 0.18 to 3.35 mm (0.007 to 0.132 in.) with 14 different size specifications [8] are thrown or shot at high velocities against the work pieces. They produce surface dimples and would produce considerable plastic stretching of the skin of the part if this were not restrained by the elastic core. Compressive stresses are thus produced in the skin. The depth of the compressive layer and the roughness of the dimpled surface are determined by the material of the work piece and by the intensity of peening, which depends on shot size, material, velocity or flow rate of the shot, and time of exposure. The magnitude of the compressive residual stress depends mainly on the material of the work piece. The intensity is specified in Almen numbers [9,10]. Excessive intensities may produce excessive surface roughness and excessive tensile stresses in the core of the work piece. Insufficient intensities may fail to provide enough protection against fatigue failures. Recommended shot-peening intensities that have been used successfully on a variety of parts, along with other shot-peening suggestions, can be found in reference 11.

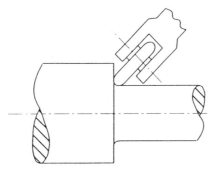

Figure 8.4 Introduction of compressive residual surface stresses at the fillet radius by localized cold rolling.

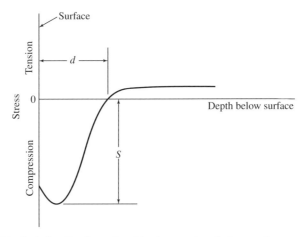

Figure 8.5   Typical distribution of residual stress just below a shot-peened surface.

A typical stress distribution produced by shot-peening is shown in Fig. 8.5. The depth of the compressive residual stress, distance d, ranges from about 0.025 to 0.5 mm (0.001 to 0.02 in.). The relation of the stress peak to material hardness obtained by Brodrick [12] is shown in Fig. 8.6. Shot-peening is used on many parts, ranging from small blades for chain saws to large crankshafts for diesel locomotives. Application to high-performance gears and springs is almost universal. Figure 8.7, for carburized gears, shows a 10-fold fatigue life

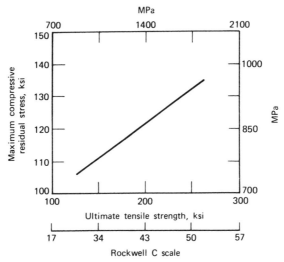

Figure 8.6   Compressive residual stress produced by shot-peening versus tensile strength of steel [12].

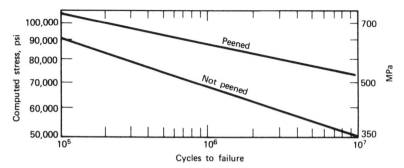

Figure 8.7 Effect of shot-peening on fatigue behavior of carburized gears [13] (reprinted with permission of McGraw-Hill Book Co.).

increase and a fatigue strength increase of 45 percent at 10 million cycles, as obtained by Straub [13].

Residual stresses are especially valuable when used with higher tensile strength metals because the full potential of greater yield strength can be used only if the damaging effect of notches and scratches can be overcome. Compressive residual stresses can do just that, as indicated in Fig. 8.8 from data by Brodrick [12]. Here it is seen that shot-peening increased the fatigue

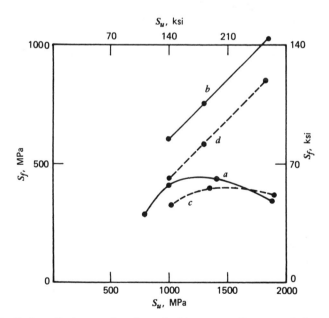

Figure 8.8 Fatigue limit as a function of ultimate tensile strength for peened and unpeened specimens [12]. (*a*) Shaft not peened. (*b*) Shaft peened. (*c*) Scratched plate not peened. (*d*) Scratched plate peened.

limit by a factor of 1.35 to 1.5 for $S_u \approx 1000$ MPa (145 ksi) and by a factor of 2.25 to 3 for $S_u \approx 1800$ MPa (260 ksi). Adequate depth of the compressively stressed layer is important because cracks can grow about as fast in higher tensile strength metals as in lower tensile strength metals, and the critical crack length is smaller. The compressed layer must be deep enough to be able to stop the cracks. However, due to the compressive layer, fatigue crack nucleation sites and growth may sometimes be shifted to subsurface tensile residual stress regions.

Other mechanical processes that achieve improvement of fatigue strength by compressive residual stresses include coining around holes, expansion of holes, and hammer-peening of welds.

8.2.2   Thermal Methods

Manufacturing procedures used in forming parts induce a wide variety of microstructure, surface finish, and residual stress. Many of these manufacturing procedures involve thermal processes such as casting, forging, hot-rolling, extrusion, injection molding, welding, brazing, quench and tempering, temper stress relief, flame or induction hardening, carburizing, and nitriding. Residual stresses from these thermal processes may be beneficial or detrimental. Thermal processes are perhaps the oldest means of improving the fatigue resistance of components. Surface hardening of steel is the chief example. If it is properly done, it leaves components with a surface skin (case) that is hard and in compression. The compressive residual stress can reach the yield strength of the hardened skin; it very effectively prevents the formation and growth of cracks and thus permits us to realize the gain in fatigue strength that we would expect from the increased hardness, as shown in Fig. 8.8. Surface hardening can be accomplished by induction hardening, carburizing, nitriding, severe quenching of carbon steel, or similar methods. Figure 8.9 shows residual stress distributions (axial, hoop, and radial directions) in an SAE 1045 40 mm (1.57 in.) diameter induction hardened steel shaft with a case hardness of about R_c 55 and a core hardness of about R_c 10 [14]. The horizontal axis is the normalized shaft radius such that the center of the shaft is at the left edge and the free surface is at the right edge. Both axial and hoop residual stresses are compressive in the surface region and decrease in magnitude toward the center. The transition from compression to tension for the axial and hoop residual stresses occurs in the same region as the microstructure and hardness transitions. For the data shown, at a normalized radius of about 0.75 or less, the three residual stresses are all tension, which creates an undesirable tri-axial tensile state of stress. This is a region where subsurface fatigue cracks can nucleate and grow. This can be very common in long life situations where applied stresses are low. High applied stresses may relax the surface compressive residual stresses and shift the fatigue failure to the surface. Thus, both surface and subsurface fatigue failures have occurred in induction hardened shafts. However, induction hardened shafts with either surface or subsurface

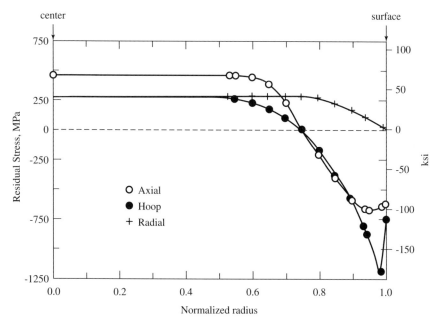

Figure 8.9 Typical residual stress distributions of SAE 1045 induction hardened 40 mm (1.57 in.) diameter steel shaft specimens as measured by x-ray diffraction [14] (copyright ASTM; reprinted with permission).

fatigue failure sites have significantly greater long and intermediate fatigue life resistance than noninduction hardened shafts. The case depth and residual stress magnitudes and depths for induction hardened shafts can have a wide range of values, depending on the material and the induction hardening proce- dure. Carburizing and nitriding produce residual stresses similar to those discussed above, except that the surface compressive residual stresses and case depth are not as deep.

Figure 8.10 shows the fatigue limit of several shallow quenched steels with a severe 60° V notch radius of 0.64 mm (0.025 in.) producing a $K_t = 3.6$ versus the measured axial compressive residual stresses at the notch surface [15,16]. The fatigue limit is almost exactly equal to the measured axial compressive residual stress, where the linear line in Fig. 8.10 represents exact equality between the fatigue limit and the compressive residual stress and the curved line represents the actual test results. This again indicates the importance of compressive residual stress at notches in high-strength metals that are capable of achieving high compressive residual stresses. The depths of the compressed layers are adequate for arresting cracks. Thermal treatments can also produce the opposite effect. The heat applied in welding can produce tensile stresses up to the yield strength of the material. They reduce fatigue strength and exacerbate the effects of notches and cracks. Many examples of the effects

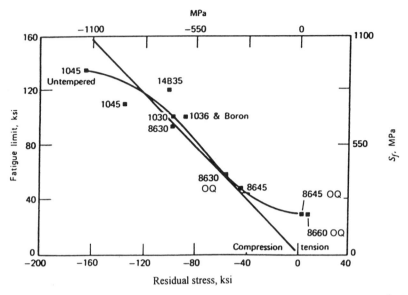

Figure 8.10 Effect of residual stress on fatigue limit of notched steel bars [15,16] (reprinted with permission from the Society of Automotive Engineers). OQ, oil quenched; all others, water quenched.

of residual stresses involving mechanical and thermal procedures are given in references 17 and 18.

8.2.3 Plating

Plating by electrolytic means can involve soft plating materials such as cadmium, zinc, tin, lead, or copper or harder plating materials such as chromium and nickel. Plating of parts is done to increase corrosion resistance and to improve the esthetic appearance. In addition, chromium plating is used to increase wear resistance and to build up the size of worn and undersized parts. Electroplating with chromium or nickel creates significant tensile residual stresses in the plating material and microcracking. Both of these situations contribute to significant reduction in the fatigue resistance of chromium- or nickel-plated parts. These reductions are greatest in higher-strength steels at longer and intermediate lives and depend upon the plating thickness. With lower-strength steels or under low-cycle fatigue, significant plasticity can occur from external loading that relaxes the residual stresses. In addition, during electroplating, hydrogen can be introduced into the base metal, causing susceptibility to hydrogen embrittlement. This problem is best circumvented by thermal stress relief of the chromium-plated parts, usually above about 400°C (750°F), which drives out the undesirable hydrogen and relaxes some of the residual stresses.

Figure 8.11 shows the influence of chrome plating on the fatigue resistance of 4130 steel heat treated to 965–1100 MPa (140–160 ksi) ultimate tensile strength [19]. At 10^6 cycles the fatigue strength is decreased by a factor of about 1.4 relative to the control specimens. However, with superimposed shot-peening or shot-peening with polishing, fatigue strength is brought back to about 90 percent of the control condition. This example clearly emphasizes the important concept that superposition of desirable processing techniques can eliminate much of the detrimental fatigue aspects of undesirable processes. Thus, methods that produce desirable compressive surface stresses such as shot-peening, nitriding, or surface rolling can be used to nullify many of the detrimental fatigue aspects of chromium or nickel plating. This has been done successfully both before and after electroplating.

The softer electrolytically deposited materials, such as cadmium, zinc, tin, lead, and copper, often have only a small influence on fatigue resistance in air environments, but they may contribute to improved fatigue resistance in corrosive environments. Galvanizing (hot-dip zinc coating), however, significantly degrades the air fatigue resistance, particularly with higher-strength steels. This degradation has been attributed to the greater susceptibility of cracks due to plating [20].

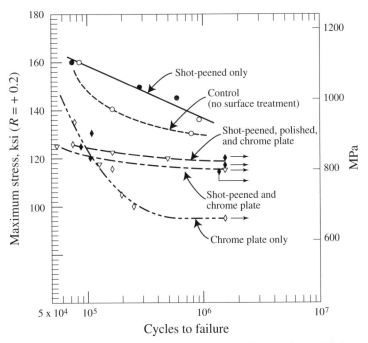

Figure 8.11 Influence of chromium plating and chromium plating with shot-peening on the axial loaded fatigue behavior of 4130 steel [19] (reprinted with permission from the Society of Automotive Engineers).

Anodizing treatments produce a hard oxide surface coating that protects against corrosion and abrasion. These treatments often have only a small influence on the fatigue resistance of many aluminum alloys, but exceptions with significant reductions in fatigue resistance for some aluminum alloys have also occurred [21]. Alcladding uses a high-purity aluminum surface coating on aluminum alloys to increase corrosion resistance. In general, this causes a decrease in fatigue resistance in air; however, in corrosive environments the method can result in increased fatigue resistance.

Summarizing the effects of plating on fatigue resistance, chromium and nickel platings are detrimental due to high tensile residual surface stresses and plating microcracking. Other softer platings often have a much smaller influence. However, under corrosive fatigue conditions, the platings may not be as detrimental and can be beneficial. Processes used before or after plating that produce surface compressive residual stresses can reduce or eliminate many of the detrimental plating fatigue effects.

8.2.4 Machining

Machining operations such as turning, milling, planing, and broaching and abrasion operations such as grinding, polishing, and honing can significantly affect fatigue resistance. These methods all involve surface operations in which fatigue cracks can nucleate and grow. They can involve four major factors that affect fatigue resistance, namely, surface finish, cold working, possible phase transformations, and residual stresses. All four of these factors contribute to fatigue resistance; residual stresses may be the most dominant factor. Regardless of which factors are most dominant, the greatest effects of machining on fatigue resistance occur at longer and intermediate lives.

During machining, three major regions are involved as one moves from the surface to the interior: regions of plastically deformed material, elastically deformed material, and undeformed material. Residual stresses are dominant in the plastically and elastically deformed surface and near surface material. Significant local heating may occur, particularly within the plastically deformed material. Residual stress depth, sign, and magnitude, as well as surface finish, are dependent upon cutting velocity, tool pressure, feed, tool geometry/wear, and cooling. As indicated in Fig. 4.15, machined surfaces for steel have lower fatigue resistance than reference mirror polished specimens, and this resistance is lower for the higher ultimate tensile strength materials. However, with the simplified model of Fig. 4.15, only one general relation exists. In reality, significant variation in machined surface fatigue resistance exists for different metals and machining operations. For the many different machining operations and metals available, surface residual stresses are usually tensile, with subsurface residual compressive stresses. However, the opposite also occurs in some machining/metal combinations. The surface depth of tensile residual stresses is often small, about 0.02–0.2 mm (about 0.001–0.01 in.); hence, polishing can remove some, most, or all of the residual stresses. By controlling cutting tool

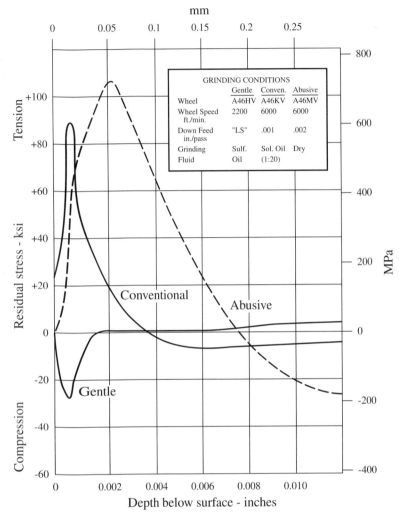

Figure 8.12 Residual stresses in Q&T 4340 steel from abusive, conventional or gentle grinding [19] (reprinted with permission from the Society of Automotive Engineers).

150 MPa (100 − 80 = 20 ksi). Residual stress determination and relaxation from simple to complex applied load histories are best obtained using the local strain analysis described in Section 7.3.

In constant amplitude, long-life fatigue, when millions of cycles must be endured, there is little danger or hope of decreasing the beneficial or harmful residual stresses in high-strength materials. In the intermediate life range we get intermediate results. In practice, shot-peening is very successfully used not only for parts that must withstand millions of cycles, such as valve springs,

speed, geometry, velocity, and feed, along with cooling methods,
of residual stresses, surface finishes, cold work, and transformatio
machining [22].

Grinding also produces wide variation in residual stresses and f.
tance. Figure 4.15 shows significantly less grinding effect on the f
of steels than produced by the machining methods described abov
tional or abusive grinding, using high-speed, high feed, water as a lu
no lubricant, introduces shallow, but high-magnitude, surface tensi
stresses. Gentle grinding with low speed, low feed, and oil as a
can provide shallow, low-magnitude compressive residual surfac
Residual stress distributions for gentle, conventional, and abusive
are shown in Fig. 8.12 for 4340 Q&T steel with R_c 50 [19]. These
residual stress distributions caused decreased fatigue resistance for
tional or abusive grinding and increased fatigue resistance for gentle

Polishing and honing are performed with lower speed, pressure, an
depth, and hence impart fewer residual stresses to parts and a small
on fatigue resistance from residual stresses. However, surface finis
are still very significant.

8.3 RELAXATION OF RESIDUAL STRESSES

Residual stresses produce the same effects that mean stresses of th
amount and distribution would produce. Thus, similitude exists betweer
stress and residual stress, and $S-N$, $\varepsilon-N$, and $da/dN-\Delta K$ methods can b
for both mean and residual stresses. However, there is a difference. The
stresses persist as long as the mean load remains. The residual stresses p
as long as the sum of residual stress and applied stress does not excee
pertinent yield strength, S_y or S'_y, of the material. Thus, they are more bene
(and potentially more harmful) when applied to hard metals with high-
strengths. In softer metals such as mild steel, the residual stresses can be
easily decreased by yielding. This is one of the reasons that mild steel is
usually shot-peened and why it can be welded with fewer precautions
high-tensile strength metals. Greater care must be taken to minimize ter
residual stresses when high-tensile strength metals are welded.

If the sum of the residual stress and the maximum or minimum app
stress exceeds the yield strength, neglecting strain hardening, the result
residual stress will be equal to the yield strength minus the maximum
minimum applied stress. As an example, we may think of a shaft in rotati
bending. Using approximate round numbers, we assume that it has a yie
strength of 700 MPa (100 ksi) and a compressive residual stress of 400 M
(60 ksi) in the skin layer. We can say that an applied alternating stress
550 MPa (80 ksi) will reduce the residual stress to 150 MPa (20 ksi) becaus
the maximum compressive stress cannot be 400 + 550 = 950 MPa but onl
700 MPa. Removing the applied stress leaves a residual stress of 700 − 550 =

but also for parts such as axle shafts, which are expected to last for a few hundred thousand cycles of maximum stress range, and landing gears, which must withstand only a few thousand cycles of large stress ranges and many cycles of smaller stress ranges.

Note that loading in one direction only, as in springs and most gears, will not destroy beneficial residual stresses. Automobile leaf springs, for instance, are usually shot-peened. The load stress is tensile on the upper side. If the sum of residual stress and applied stress is always less than the yield strength, it will not produce yielding and will not decrease the residual stress on the peened side of the leaf. On the lower side of the leaf, the load stress is compressive. If that side were peened, the load stress and residual stress together would produce yielding. But fatigue failures will not normally develop on the compressively stressed side of the leaf. Therefore, it would not matter if the peening stresses diminished there. As a matter of fact, only the tension side is peened because tests have shown that peening the compressive side would cause an unnecessary expense. In springs, as in other parts that are loaded predominantly in one direction, an overload applied early in life is beneficial because it introduces desirable residual compressive stresses at the proper surface. Springs, hoists, and pressure vessels are strengthened by proof loading with a load higher than the highest expected service load.

In addition to the above mechanical loading situations that relax or alter residual stresses, thermal stress relief can also relax residual stresses. This relaxation can occur without applying external loads and is a function of material, temperature, time, and residual stress magnitude and distribution. For a given material, temperature is the dominant factor. At proper stress relief temperatures, residual stresses will relax with time in a decreasing exponential manner. Different materials will have different best stress relief temperatures and time at temperature. Approximate temperatures at which stress relief will be significant, as given by Zahavi and Torbilo [22], are:

400–450°C (750–840°F) for titanium alloys
450–500°C (840–930°F) for high-temperature resistance steels
650–800°C (1200–1475°F) for high-temperature resistance nickel alloys
800–900°C (1475–1650°F) for superalloys

8.4 MEASUREMENT OF RESIDUAL STRESSES

Residual stresses may be determined analytically, computationally with finite element analysis, and experimentally. The local strain approach given in Section 7.3 is an analytical method used with applied loading. For other mechanical, thermal, plating, or machining operations, analytical methods are complex and thus limited in use. Finite element analysis for determining residual stresses is becoming more popular due to its higher computational efficiency and better accuracy. However, the most common methods for determining surface and subsurface residual stresses rely on experimental methods. The determination of surface residual stresses is mostly nondestructive, while subsurface residual stress determination is mostly destructive. The Society for Experimental Mechanics (SEM) has published an excellent, comprehensive *Handbook of Measurement of Residual Stresses* [23] that describes in detail, with significant references, seven major experimental methods for determining residual stresses: hole-drilling and ring core, layer removal, sectioning, X-ray diffraction, neutron diffraction, ultrasonic, and magnetic methods. ASTM has standard test methods for the hole-drilling method [24] and for X-ray diffraction measurements [25,26]. Four of these experimental methods will now be briefly described.

The hole-drilling method involves drilling a small hole, typically 1.5 to 3 mm (0.06 to 0.12 in.) deep, through a three-element radial strain gage rosette attached to the part. The strain gage relaxation around the hole from the drilling is then measured and converted to biaxial residual stresses in the vicinity of the hole. Since the hole drilling is local and can be repaired, the hole-drilling method is considered only semidestructive.

Sectioning methods are used to measure subsurface residual stresses by removing a beam, ring, or prism specimen from a residual stressed part of concern. The surface is subjected to repetitive surface layer removal by electrochemical polishing, etching, or machining. The curvature changes or deflections of the specimen for removal of each layer are measured, and these measurements are then related to residual stress magnitudes. The thickness of the layer removed can be small, about 0.02 mm (0.0008 in.), if steep residual stress gradients exist, and thicker layers can be removed for smaller residual stress gradients. The method is destructive but is quick and reliable.

X-ray diffraction can be used nondestructively to measure surface residual stresses and destructively for subsurface residual stresses. Residual stresses cause crystal lattice distortion, and a measurement of the interplanar spacing of the crystal lattice indicates the magnitude of the residual stress. The angle of diffraction is the measured quantity. Since the X-ray beam penetrates to a depth of less than 0.025 mm (0.001 in.), only surface residual stresses are measured. However, by electrochemically polishing away thin layers of metal, subsurface residual stresses can be measured. Both portable and nonportable X-ray diffraction equipment is available for many diverse situations, making the X-ray diffraction method very popular. The typical precision of X-ray diffraction residual stress measurements can be as low as ±7 MPa (±1 ksi) or as high as ±35 MPa (±5 ksi).

The ultrasonic method for determining residual stresses assumes that the speed of sound traveling through metals is linearly proportional to the residual stress. Acoustic waves are transmitted into parts, and their velocity-related parameters are measured. Longitudinal, shear, and Rayleigh waves are used in this method. Measurement of residual stresses using ultrasonics has produced mixed results because of the many parameters in metals, such as hardness, microstructure, composition, discontinuities, and dislocation density, that can affect the results.

Despite the importance of residual stresses and the availability of methods to measure them, routine measurement of residual stresses is still the exception rather than the rule. Most users of residual stresses rely on careful control of processes such as carburizing, shot-peening, and coining.

8.5 STRESS INTENSITY FACTORS FOR RESIDUAL STRESSES

Residual stress effects on fatigue crack growth have been handled quantitatively with crack closure models or superposition of applied stress intensity factors and residual stress intensity factors. These two methods have been used with appropriate fatigue crack growth models involving the sigmoidal $da/dN–\Delta K$ behavior discussed in Chapter 6. Superposition of applied and residual stress intensity factors is appropriate due to the linear elastic models involved, and hence

$$K_T = K + K_{res} \qquad (8.1)$$

where K is the applied stress intensity factor associated with the applied stresses, K_{res} is the residual stress intensity factor associated with the residual stresses, and K_T is the total or sum of the applied and residual stress intensity factors under mode I conditions.

To determine K_{res}, the residual stress magnitude and profile without cracks must be known, obtained, or assumed. K_{res} can then be obtained by inserting a crack face at the desired location and then loading the inserted crack face with the residual stresses that exist normal to the plane of crack growth. Methods for obtaining K_{res} include weight functions, Green's function, boundary integration, alternating methods, and finite element methods. A few K_{res} solutions are available in the literature [27,28] and additional solutions can be generated using applicable computer programs. The accuracy of integration of the superposition model, da/dN, as a function of the appropriate ΔK_T, for fatigue crack growth is limited, since the residual stress pattern will change as the crack grows through the residual stress field, altering the original K_{res} solution. Residual stress relaxation can also occur due to cyclic plasticity, cyclic softening, and temperature stress relief, as discussed in Section 8.3. Also, the crack contains its own residual stress field at the crack tip within the plastic zone, $2r_y'$, along with residual deformation or crack closure in the wake of the growing crack. Integrating past the change in sign of the initial residual stress sign is not reasonable, since the original residual stress field would have changed significantly. Most of the fatigue crack growth would have been accounted for by then anyway.

Figure 8.13a shows a stress intensity factor schematic for a component subjected to both an applied stress intensity factor and a tensile residual stress intensity factor, and the superposition of both. From Chapter 6, the applied stress intensity factor range and the applied R ratio are

$$\Delta K = K_{max} - K_{min} \quad \text{and} \quad R = \frac{K_{min}}{K_{max}} \tag{8.2}$$

The total stress intensity factors are

$$K_{Tmax} = K_{max} + K_{res} \quad \text{and} \quad K_{Tmin} = K_{min} + K_{res}$$
$$\Delta K_T = K_{Tmax} - K_{Tmin} = (K_{max} + K_{res}) - (K_{min} + K_{res}) = \Delta K \tag{8.3}$$

Thus, it is important to note that the total stress intensity factor range, ΔK_T, is equal to the applied stress intensity factor range, ΔK. However,

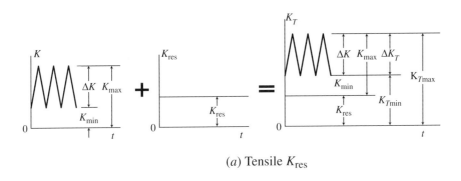

(a) Tensile K_{res}

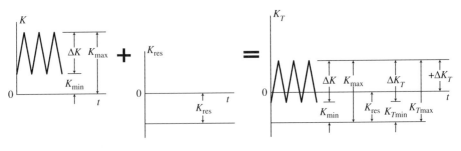

(b) Compressive K_{res}

Figure 8.13 Initial stress intensity factors for applied loading, residual stresses, and superposition. (a) Tensile K_{res}. (b) Compressive K_{res}.

$$R_T = \frac{K_{Tmin}}{K_{Tmax}} = \frac{K_{min} + K_{res}}{K_{max} + K_{res}} > R \tag{8.4}$$

Thus, the detrimental tensile residual stress effects are incorporated by using a higher R ratio in the integration of

$$da/dN = f(\Delta K_T) \tag{8.5}$$

The function, $f(\Delta K_T)$, could include the Paris, Forman, or Walker equations presented in Section 6.5, or other possible equations, by using $\Delta K_T = \Delta K$ and R_T rather than R.

   Figure 8.13b shows a schematic superposition of applied stress intensity factors with a negative K_{res} formed from compressive residual stresses. A negative K_{res} has no mathematical basis, but it characterizes the crack as being closed. If the applied K is greater than $|-K_{res}|$, then the crack will open. Under compressive K_{res} conditions, Eqs. 8.1 to 8.3 still hold. However ΔK_T and R_T, as given by Eqs. 8.3 and 8.4, may not be the appropriate total values to use in the integration, since ΔK_T may contain a negative portion, as shown in Fig. 8.13b. Here the appropriate stress intensity factor range is the positive stress intensity factor, $+\Delta K_T$, where

$$+\Delta K_T = K_{max} - |K_{res}| \quad \text{and} \quad R_T = 0 \tag{8.6}$$

Under compressive residual stresses, ΔK_T could be ≤ 0, indicating that the crack is closed and that no fatigue crack growth is expected. Also, fatigue crack growth should not occur with tensile residual stresses if $\Delta K_T \leq \Delta K_{th}$ for the appropriate R_T value of ΔK_{th}. For compressive residual stresses, fatigue crack growth should not occur if $+\Delta K_T \leq \Delta K_{th}$ with $R_T = 0$ and hence with the $R = 0$, ΔK_{th} value.

   As an example of the application of this analysis, Fig. 8.14a shows an assumed analytical residual stress field for a longitudinal welded wide plate containing a central crack of length $2a$ [27,28]. The residual stress is tensile in the central crack region, with a maximum value of σ_0, and eventually becomes compressive at $x/c > 1$ to satisfy the equilibrium equations, $\Sigma F = \Sigma M = 0$. The residual stress intensity factor, K_{res}, obtained using the Green's function method, can be determined with Fig. 8.14b from

$$K_{res} = \sigma_0 \sqrt{\pi a}\, f(a/c) \tag{8.7}$$

Figure 8.14b shows that $f(a/c)$, and hence K_{res}, decreases as the crack grows into the residual stress field. This expression plus the applied stress intensity factor, K, is incorporated into the integration of Eq. 8.5 for the fatigue life to grow a crack from one length to another. Since K_{res} is a complex expression such that α, from Eq. 6.2b, is not a constant, this implies that Eqs. 6.20f or

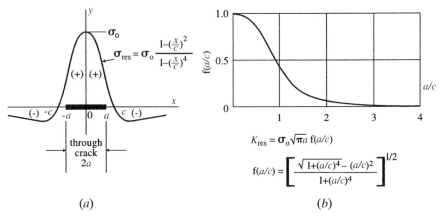

Figure 8.14 Residual stress distribution and residual stress intensity factor, K_{res}, for a longitudinal welded infinite plate containing a central through crack [27,28]. (a) Residual stress distribution, σ_{res}, across the plate width. (b) Residual stress intensity factor, K_{res} (reprinted with permission of P. C. Paris).

6.20g cannot be used in the integration and hence numerical integration must be used.

8.6 SUMMARY

Residual stresses in fatigue design are of the highest importance. Compressive residual stresses efficiently retard the formation and growth of cracks subjected to cyclic loading and thus enhance fatigue resistance. The opposite occurs for tensile residual stresses. Residual stresses are analogous to mean stresses and therefore can be incorporated into S–N, ε–N, and da/dN–ΔK fatigue life methodologies. With S–N and ε–N models, applied stresses and residual stresses, for a specific direction, can be added algebraically. With LEFM fatigue crack growth models, these stresses cannot be added algebraically, but separate stress intensity factors for both applied stresses and residual stresses can be added algebraically for a given mode of crack extension. Since residual stresses are in self-equilibrium, both tensile and compressive values must exist if residual stresses are present. These can be uniaxial, biaxial, or triaxial. During nonuniform inelastic loading, surface regions that yield in tension result in desirable surface compressive residual stresses when the load is removed. Surface regions that yield in compression during nonuniform inelastic loading result in undesirable surface tensile residual stresses when the load is removed. Thus, one single high-load can have a significant influence on fatigue resistance.

Residual stresses can be formed from many manufacturing methods involving mechanical, thermal, plating, and machining operations. These methods can be beneficial, detrimental, or have little influence on fatigue resistance.

Shot-peening and surface cold-rolling are the two most common mechanical methods for introducing surface compressive residual stresses. They can increase fatigue limits by more than a factor of 2, particularly in notched parts. Induction hardening, carburizing, and nitriding are common thermal methods for introducing surface compressive residual stresses for enhanced fatigue resistance. Hard plating such as chromium or nickel plating introduces surface tensile residual stresses and hence is detrimental to fatigue resistance. Soft plating such as cadmium and zinc plating produces small residual stresses and hence has only a small influence on fatigue resistance. Galvanizing is an exception and is detrimental. However, under corrosive environmental fatigue conditions, these plating methods can enhance fatigue resistance. Machining operations most often introduce surface tensile residual stresses and hence decrease fatigue resistance. Many of the operations that introduce undesirable surface tensile residual stresses, such as chrome plating or machining, can have the detrimental aspects reduced or eliminated by additional beneficial treatments such as shot-peening, cold-rolling, and others.

Residual stresses have greater influence in long and intermediate fatigue life than in low-cycle fatigue. This is particularly true for higher-strength metals. The reason for this behavior is that with low-cycle fatigue and/or lower strength metals, relaxation of residual stresses is more likely to occur due to localized cumulative plasticity from cyclic loading, in which the sum of applied stresses and residual stresses exceeds the yield strength. Relaxation of residual stresses can also result from thermal stress relief.

The magnitude and distribution of residual stresses are most commonly obtained using experimental methods such as X-ray diffraction, hole-drilling, or sectioning. Analytical or computational methods are often complicated. However, the local strain method presented in Section 7.3 can be used to obtain local residual stresses from cyclic loading.

8.7 DOS AND DON'TS IN DESIGN

1. Do consider the beneficial and harmful effects of residual stresses for all long-life (high-cycle) and intermediate-life fatigue applications and remember that the greatest influence and importance of residual stresses is at notches.
2. Don't expect much help from residual stresses in very low-cycle (fewer than 10^3 cycles) applications due to residual stress relaxation.
3. Do realize that residual stresses are in self-equilibrium; hence, both compressive and tensile residual stresses must exist if residual stresses are present.
4. Don't permit surface tensile residual stresses in high tensile strength parts.
5. Don't forget the importance of sufficient compressive residual stress magnitude and depth.

6. Do remember that grinding and welding can produce very harmful tensile residual stresses and that tensile overloading, peening, and surface hardening can produce very beneficial surface compressive residual stresses.

7. Don't overlook the fact that straightening will introduce tensile and compressive residual surface stresses on opposite sides, and that except for continued one-way bending, this method will be detrimental to fatigue resistance.

8. Do note that residual stresses and mean stresses are similar and, hence, that S–N, ε–N, and da/dN–ΔK fatigue models can be used with both. For LEFM application to fatigue crack growth, superposition of applied and residual stress intensity factors, not superposition of stresses, is required.

9. Do consult with the people who will process your parts and discuss your requirements for residual stresses with them.

REFERENCES

1. W. J. Harris, *Metallic Fatigue*, Pergamon Press, London, 1961.

2. H. O. Fuchs, "Regional Tensile Stress as a Measure of the Fatigue Strength of Notched Parts," *Mechanical Behavior of Materials, Proceedings of the International Conference*, Vol. II, Society of Materials Science, Kyoto, Japan, 1972, p. 478.

3. T. E. Davidson, D. P. Kendall, and A. N. Reiner, "Residual Stresses in Thick-Walled Cylinders Resulting from Mechanically Induced Overstrain," *Exp. Mech.*, Vol. 3, No. 11, 1963, p. 253.

4. R. B. Heywood, *Designing Against Fatigue of Metals*, Reinhold, New York, 1962.

5. V. M. Faires, *Design of Machine Elements*, 4th ed., Macmillian, London, 1965.

6. D. G. Bellow and M. G. Faulkner, "Development of an Improved Internal Thread for the Petroleum Industry," *Closed Loop*, MTS Systems Corp., Vol. 8, No. 1, 1978, p. 3.

7. D. G. Bellow and M. G. Faulkner, "Salt Water and Hydrogen Sulfide Corrosion Fatigue of Work-Hardened Threaded Elements," *J. Test. Eval.*, Vol. 4, No. 2, 1976, p. 141.

8. "Cast Shot and Grit Size Specifications for Peening and Cleaning" SAE Standard J444, *SAE Handbook*, Vol. 1, SAE, Warrendale, PA, 1998, p. 8.14.

9. "Test Strip, Holder and Gage for Shot-Peening," SAE Standard J442, *SAE Handbook*, Vol. 1, SAE, Warrendale, PA, 1998, p. 8.06.

10. H. O. Fuchs, "Shotpeening Effects and Specifications," *Metals*, ASTM STP 196, ASTM, West Conshohocken, PA, 1962, p. 22.

11. "Shot-Peening," *Metals Handbook*, Vol. 5, 9th ed., American Society of Metals, Metals Park, OH, 1982, p. 138.

12. R. F. Brodrick, "Protective Shot-Peening of Propellers," Wright Air Development Center Technical Report TR55-56, 1955.

13. J. C. Straub, "Shot-Peening," *Metals Engineering, Design,* 2nd ed., O. J. Horger, ed., McGraw-Hill Book Co., New York, 1965, p. 258.

14. H. Y. Zhang, R. I. Stephens, and G. Glinka, "Subsurface Fatigue Crack Initiation and Propagation Behavior of Induction Hardened Shafts Under the Effect of Residual and Applied Bending Stresses," *Fatigue and Fracture Mechanics,* 30th Vol., P. C. Paris and K. L. Jerina, eds., ASTM STP 1360, ASTM, West Conshohocken, PA, 1999, p. 240.

15. D. V. Nelson, R. E. Ricklefs, and W. P. Evans, "The Role of Residual Stresses in Increasing Long Life Fatigue Strength of Notched Machine Members," *Achievement of High Fatigue Resistance Metals and Alloys,* ASTM STP 467, ASTM, West Conshohocken, PA, 1970, p. 228.

16. R. B. Liss, C. G. Massieon, and A. S. McKloskey, "The Development of Heat Treat Stresses and Their Effect on Fatigue Strength of Hardened Steel," SAE Paper No. 650517, SAE, Warrendale, PA, 1965.

17. J. O. Almen and P. H. Black, *Residual Stresses and Fatigue in Metals,* McGraw-Hill Book Co., New York, 1963.

18. O. J. Horger, ed., *Metals Engineering, Design,* 2nd ed., McGraw-Hill Book Co., New York, 1965.

19. R. C. Rice, ed., *SAE Fatigue Design Handbook,* 3rd ed., *AE-22,* Society of Automotive Engineers, Warrendale, PA, 1997.

20. N. E. Frost, K. J. Marsh, and L. P. Pook, *Metal Fatigue,* Clarendon Press, Oxford, 1974.

21. P. G. Forrest, *Fatigue of Metals,* Pergamon Press, London, 1962.

22. E. Zahavi and V. Torbilo, *Fatigue Design and Life Expectancy of Machine Parts,* CRC Press, Boca Raton, FL, 1996.

23. J. Lu, M. James, and G. Roy, *Handbook of Measurement of Residual Stresses,* Society for Experimental Mechanics, Bethel, CT, 1996.

24. "Standard Test Method for Determining Residual Stresses by the Hole-Drilling Strain-Gage Method," ASTM E837, Vol. 03.01, ASTM, West Conshohocken, PA, 2000, p. 670.

25. "Standard Test Method for Verifying the Alignment of X-Ray Diffraction Instrumentation for Residual Stress Measurement," ASTM E915, Vol. 03.01, ASTM, West Conshohocken, PA, 2000, p. 702.

26. "Standard Test Method for Determining the Effective Elastic Parameter for X-Ray Diffraction Measurements of Residual Stresses," ASTM E1426, Vol. 03.01, ASTM, West Conshohocken, PA, 2000, p. 916.

27. Y. Murakami, ed. in chief, *Stress Intensity Factors Handbook,* Vol. 2, Pergamon Press, 1987.

28. H. Tada, P. C. Paris, and G. R. Irwin, *The Stress Analysis of Cracks Handbook,* 2nd ed., Paris Productions, St Louis, MO, 1985.

PROBLEMS

1. A flat specimen similar to, say, the Almen strip is shot-peened on one side as shown. Qualitatively sketch a proper residual stress distribution along section A-B, such that all equations of equilibrium are qualitatively satisfied. This assumes that $\Sigma F = \Sigma M = 0$.

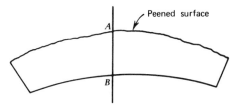

2. Each day 5000 heavy beer barrels are rolled down the sloping cantilever beam to point A, where a crane picks them up for shipment. About once a year the cantilever beam breaks. It has been suggested that compressive residual stresses might increase the life of the beam. Which way would you overload the beam so that surface compressive residual stresses would be formed at the proper locations? Would this be a good solution if failure occurred once a week? Why?

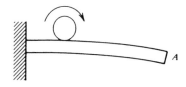

3. A 100×150 mm rectangular beam is subjected to pure bending as shown. $M = 1.4\ M_0$, where M_0 is the moment that just causes the first fiber to yield. Assume an ideally elastic-perfectly plastic material with the given stress–strain properties in both tension and compression. Quantitatively determine the resultant residual stress distribution if the moment is completely removed. Make a final plot of these residual stresses on rectangular coordinate paper.

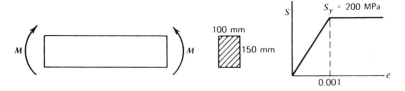

4. For an axially loaded finite width plate with edge notches on each side (Fig. 7.4a), qualitatively sketch the resulting residual stress pattern if an inelastic compressive overload is applied and then removed. Assume no buckling. What effect could this have on fatigue resistance?

5. Figure 8.8 shows significant shaft fatigue limit improvement due to shot-peening. Based on these fatigue limits, estimate the magnitude of the shot-

peened near surface residual stresses for S_u = 1000 MPa and 1800 MPa. Use similitude between residual and mean stresses and use a mean stress model from Chapter 4. Compare your results with those shown in Fig. 8.6. Comment on your findings.

6. Assume that the RQC-100 steel specimens in Problem 8 of Chapter 5 are reground using (a) conventional and (b) gentle grinding, and the residual stress distributions shown in Fig. 8.12 are formed at the surface regions. Repeat Problem 8 of Chapter 5 for the two new RQC-100 surface conditions using similitude between residual and mean stresses. Use a mean stress model from Chapter 5. Comment on your results.

7. Using similitude between residual and mean stresses, estimate the magnitude of the residual stresses produced at the notch of Fig. 8.1 based on 10^6 to 10^7 cycles. Consider the nominal S–N (modified Goodman or Haigh diagrams) and local ε–N approach with a mean stress model. Which method provides a reasonable answer? What would you do to determine if your calculations are reasonable?

8. Repeat Example 7.2.4 using the S–N method, assuming that the drilled hole (9.5 mm or 0.375 in. diameter) has been treated by expansion or shot-peening to have a compressive residual stress of 700 MPa (100 ksi) near the surface. Comment on the difference in results with and without expansion or shot-peening.

9. A wide plate is formed by welding together two identical plates using a longitudinal butt weld. A longitudinal (axial) load resulting in S_{max} = 200 MPa with R = 0 is applied. An initial through-thickness center crack with a crack length of $2a$ = 2 mm is perpendicular to the weld bead and the applied load. Material properties for R = 0 are da/dN = $10^{-12}(\Delta K)^3$, with da/dN in m/cycle and ΔK in MPa$\sqrt{m}$, ΔK_{th} =5 MPa$\sqrt{m}$, and K_c = 75 MPa$\sqrt{m}$. Assuming S_y and S_y' are large enough to use LEFM principles, estimate the initial crack growth rate if

(a) No residual stresses exist in the crack region.

(b) Tensile residual stresses with σ_0 = 300 MPa, as shown in Fig. 8.14, exist with parameter $2c$ = 4 mm.

(c) Can you estimate the number of cycles to fracture for parts (a) and (b)? If so, how many cycles?

(d) What stress, S_{max}, can be applied so that no fatigue crack growth occurs for parts (a) and (b)?

CHAPTER 9

FATIGUE FROM VARIABLE AMPLITUDE LOADING

9.1 SPECTRUM LOADS AND CUMULATIVE DAMAGE

Service load histories are usually variable amplitude, as shown in Fig. 9.1*a–c* [1]. Figure 9.1*d* shows details to a larger time scale for the beginning of the spectrum shown in Fig. 9.1*b*. The loads on an aircraft wing or a tractor shovel are far from constant amplitude. Realistic representation of service loads is a key ingredient of successful fatigue analysis or design. It is therefore important to accurately measure the applied loads on an existing component or structure or to predict loads on a component or structure that does not yet exist.

To measure the load history, transducers (most commonly electrical resistance strain gages) are attached to the critical areas of the component. These critical areas are often identified either analytically by FEA or experimentally. The acquired data from the transducers are usually recorded and stored by a computer or by other devices. The recorded data may be filtered to isolate the primary loads from noise, and then are often summarized or compressed by cycle counting methods in order to simplify the fatigue damage computations. Cycle counting methods are discussed in Section 9.5. Prediction of the load history of a component or structure that does not yet exist is discussed in Section 9.9. Detailed discussions of measurement and prediction of service load histories are presented in [2].

If closed-loop electrohydraulic test systems are used, real-life load histories can be applied directly to small test specimens, components, subassemblies, and even entire products. Historically, complex load histories were often, and still are sometimes, replaced in test programs by more simplified loadings,

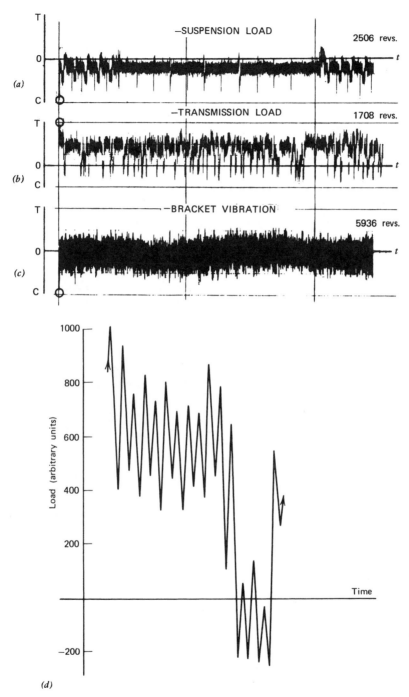

Figure 9.1 Load-time display of service histories [1] (reprinted with permission of the Society of Automotive Engineers). (*a*) Suspension load history. (*b*) Transmission load history. (*c*) Bracket vibration load history. (*d*) Beginning of the transmission load history shown in (*b*) at a larger time scale.

such as the block programs shown in Figs. 9.2*a* and 9.2*b* or by constant amplitude tests. More discussion of fatigue testing is provided in Section 2.3.

Techniques for analysis and testing have been developed to predict whether variable amplitude service loads will produce acceptable or unacceptable fatigue lives. Data from constant amplitude tests are the basis for the analysis. The analysis may be simple, based on nominal stresses and the assumption that damage is linear with the number of cycles. Or it may be more complex, for instance, in considering the early stages of fatigue by notch strain analysis and the later stages by crack growth analysis. Which approach is most appropriate in a particular instance depends on several criteria, which are discussed in this chapter.

In tests with constant amplitude load cycles, the cumulative effect of all the cycles eventually produces fatigue failure (unless the load is below the prevailing fatigue limit). When loading and unloading do not occur in constant amplitude cycles but in a variable amplitude manner, the cumulative effect of these events may also produce fatigue failure. The term "cumulative damage" refers to the fatigue effects of loading events other than constant amplitude cycles. The term "spectrum," as used in fatigue literature, often means a series of fatigue loading events other than uniformly repeated cycles. Sometimes spectrum means a listing, as, for instance, in Table 9.1 for the suspension service history in Fig. 9.1*a*. Other parameters, such as maximum and minimum loads, are also used to define the classifications or "boxes" in which the counts

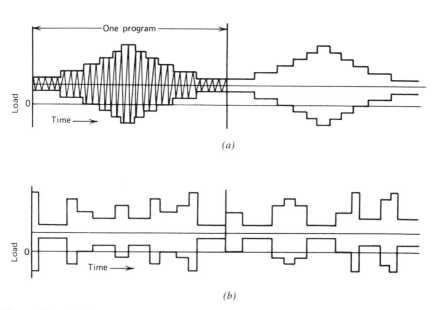

Figure 9.2 Block program load spectra. (*a*) Programmed six load level test. (*b*) Random block program loading.

TABLE 9.1 Example of the Number of Cycles at Various Stress Range and Mean Stress Combinations from the Suspension Load History in Fig. 9.1a [1]

Stress Range (MPa)	Mean Stress (MPa)													
	−500	−450	−400	−350	−300	−250	−200	−150	−100	−50	0	50	100	150
200	8	16	24	8	6	66	358	104	6	10	6	6	4	2
250	2	16	46	10	4	72	564	118	20	10	10	2	2	6
300	2	8	22	6		36	312	90	14	8	4			
350			4	6		14	148	36	20	6	2			
400			2	10	6	4	64	12	10	6				
450				4		2	20	10	4			2		
500				2	2		8	12				4		
550					4	2	8	2		2				
600				2			2							
650				2	4	4	2		2					
700				4	2									
750			2											
800				2										
850				4	2									
900				4	4									
950				2	4	2								
1000			2		2									
1050					2									
1100														
1150			2											
1200														
1250														
1300				2										
1350														

Reprinted with permission from the Society of Automotive Engineers.

of cycles are listed. The boxes in Table 9.1 contain the cycle counts for different combinations of stress ranges and means.

9.2   DAMAGE QUANTIFICATION AND THE CONCEPTS OF DAMAGE FRACTION AND ACCUMULATION

One approach to variable load histories uses the concept of "damage," defined as the fraction of life (also referred to as "cycle ratio") used up by an event or a series of events. These fractions are added together; when their sum reaches 1.0 or 100 percent, we expect and predict failure.

After a crack has been started, one can define damage by the growth of that crack. According to this definition, loading events that produce zero crack growth would cause zero damage. A loading event that extends the crack by $\Delta a = 0.01$ mm would cause damage, which might be defined as $D = \Delta a/a_c$, where a_c is the critical crack length for the highest expected load and D is damage fraction. However, the same loading event that produced 0.01 mm crack growth at an earlier stage may produce 1 mm crack growth at a later stage.

In addition to the life fraction (or cycle ratio) and crack length or crack population, many other measures have been used to quantify fatigue damage. These include metallurgical parameters, mechanical parameters, and physical measures. Metallurgical parameters include the size or number of dislocations and the spacing or intensity of slipbands. These measures represent the physical nature of fatigue damage explicitly, but their quantification usually involves destructive evaluation techniques. Mechanical parameters directly reflect damage through changes in the mechanical responses of the material such as hardness, stress, strain, stiffness, and strain energy as the material degrades with fatigue damage accumulation. Physical measures indirectly quantify the fatigue damage and consist of mainly nondestructive techniques such as X-radiography, acoustic emission, ultrasonic techniques, magnetic field methods, potential drop, and eddy current techniques. A detailed review and discussion of the damage-quantifying parameters and their use in cumulative fatigue damage assessment is given in [3]. The most common measure of damage, however, is the life fraction or cycle ratio, and this is the quantifying measure we will use in this book.

9.3   CUMULATIVE DAMAGE THEORIES

9.3.1   Palmgren-Miner Linear Damage Rule

The damage caused by one cycle is defined as $D = 1/N_f$, where N_f is the number of repetitions of this same cycle that equals the median life to failure.

The damage produced by n such cycles is then $nD = n/N_f$. Figure 9.3 shows two blocks of constant amplitude stress cycles, and the corresponding S–N curve, with fatigue lives at stress amplitudes S_{a1} and S_{a2} denoted by N_{f1} and N_{f2}, respectively. The damaging effect of n_1 cycles at S_{a1} stress amplitude is assumed to be $n_1 D_1 = n_1/N_{f1}$, while the damaging effect of n_2 cycles at S_{a2} stress amplitude is assumed to be $n_2 D_2 = n_2/N_{f2}$. Similarly, the cycle ratio or damage caused by n_i cycles at S_{ai} stress amplitude is $n_i D_i = n_i/N_{fi}$. Failure is predicted when the sum of all ratios becomes 1.0 or 100 percent. The relation

$$\sum \frac{n_i}{N_{fi}} = \frac{n_1}{N_{f1}} + \frac{n_2}{N_{f2}} + \ldots = 1 \tag{9.1}$$

expresses the linear damage rule, proposed by Palmgren [4] for prediction of ball bearing life and later by Miner [5] for prediction of aircraft fatigue life. Equation 9.1 is also used with other fatigue curves such as load–life or ε–N curves.

Applied to ball bearings, the linear damage rule works in the following way: Tests show the median life of bearings model X operating at high frequency to be 2×10^8 cycles under a 1 kN load and 3×10^7 cycles under a 2 kN load. How many cycles can we expect the bearing to last if the load is 1 kN 90 percent of the time and 2 kN 10 percent of the time?

If the total number of applied cycles is n, the number of cycles at the 1 kN and 2 kN loads are $n_1 = 0.9n$ and $n_2 = 0.1n$, respectively. The total damage done will be the sum of the damages done by the 1 kN load, n_1/N_{f1}, plus the damage done by the 2 kN load, n_2/N_{f2}:

$$D = \frac{n_1}{N_{f1}} + \frac{n_2}{N_{f2}} = \frac{0.9n}{2 \times 10^8} + \frac{0.1n}{3 \times 10^7} = 7.83 \times 10^{-9}\, n = 1$$

and the expected life is $1/(7.83 \times 10^{-9}) = 1.3 \times 10^8$ cycles.

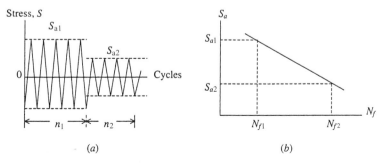

Figure 9.3 Constant amplitude stress blocks and S–N curve. (a) Constant amplitude stress blocks. (b) S–N curve.

The assumption of linear damage is open to many objections. For example, the sequence and interaction of events may have major influences on the fatigue life. This is discussed in Section 9.4. Also, the rate of damage accumulation may be a function of the load amplitude such that at low load levels most of the life is involved in crack nucleation, while at high load levels most of the life is spent in crack growth. Experimental evidence under completely reversed loading conditions for both smooth and notched specimens often indicates that $\Sigma n_i/N_{fi} \neq 1$ for a low-to-high or a high-to-low loading sequence. Even though the linear damage rule ignores these effects, it is commonly used because none of the other proposed methods achieves better agreement with data from many different tests.

Palmgren and Miner were well aware of these shortcomings. They decided to use an average damage, and their method is the simplest and still the most widely used approach to predicting fatigue life to the appearance of cracks and in many cases to predicting the total fatigue life to fracture. Another approach [6] applies the linear damage law to two separate stages of the fatigue process, namely, crack nucleation and crack growth.

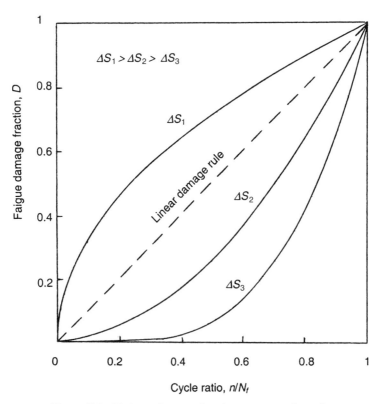

Figure 9.4   Fatigue damage fraction versus cycle ratio.

9.3.2 Nonlinear Damage Theories

To remedy the deficiencies associated with the linear damage assumption, many nonlinear cumulative fatigue damage rules have been proposed. Fatemi and Yang have reviewed and classified most of these rules into several categories in a recent review paper [7]. These theories account for the nonlinear nature of fatigue damage accumulation by using nonlinear relations such as $D = \Sigma(n_i/N_{fi})^{\alpha_i}$, where the power α_i depends on the load level, proposed by Marco and Starkey [8], rather than the linear relation in Eq. 9.1. Figure 9.4 shows the linear damage rule and the aforementioned nonlinear rule at three stress levels in a plot of fatigue damage versus cycle ratio. This figure indicates that according to this nonlinear rule, a cycle ratio of $n/N_f = 0.6$, for example, produces damage fractions of $D = 0.77$, 0.36, and 0.13 at ΔS_1, ΔS_2, and ΔS_3 stress levels, respectively. The damage fraction is of course 0.6, independent of the stress level, according to the linear damage rule.

Though many nonlinear damage models have been developed, unfortunately none can encompass many of the complicating factors encountered during complex variable amplitude loading. Consequently, the Palmgren-Miner linear damage rule is still dominantly used in fatigue analysis or design in spite of its many shortcomings.

9.4 LOAD INTERACTION AND SEQUENCE EFFECTS

Sequence effects exist both in the early stages (crack nucleation and microcrack growth) and in the later stages (macrocrack growth) of fatigue. The same principles govern both stages. We begin by considering the early stages.

It has been shown [9] that the fatigue strength of smooth specimens is reduced more than indicated by the linear damage rule if a few cycles of fully reversed high-stress amplitude are applied before testing with lower-stress amplitudes. This effect, however, is very small compared to the sequence effects on notched parts. On notched parts the sequence effects can be very strong. We recall Table 8.1, which shows that the fatigue strength of notched specimens can be more than tripled by a single high tensile overload. The fatigue life might be 10 times or even 100 times as great if that overload is applied at the beginning of the sequence rather than at the end. We also recall from Chapter 8 that this effect can be explained by residual stress effects.

Figure 9.5, taken from Crews [10], shows another sequence effect. Here the life of a specimen with a hole was 460 000 cycles of low load after 9.5 cycles of high load but only 63 000 cycles after 10 cycles of high load. This difference is also explained by the residual stresses remaining from the high loads. When the high load cycles ended on a tension peak, the effect was beneficial to fatigue life; when they ended on a compression reversal, the effect was harmful.

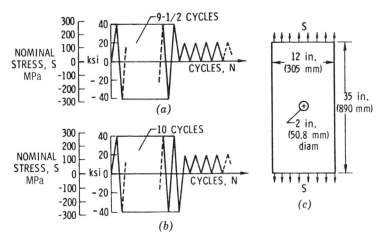

Figure 9.5 Effect of prior loading on fatigue life [10] (copyright ASTM; reprinted with permission). (*a*) Beneficial prior loading. (*b*) Detrimental prior loading. (*c*) Notched specimen.

In Figs. 9.6 and 9.7, reprinted from Stephens et al. [11], sequence effects from different load patterns are shown in terms of fatigue crack growth. They confirm the results that were given above in terms of life to failure, N_f. Yielding and the resulting residual stresses and crack closure near the crack tip are the main causes of these effects. Here, even more than in the early stages of fatigue, the details of the sequence, the directions of the last overloads, are of prime importance. Even a small tensile load can produce a plastic zone at the tip of the crack, forming compressive residual stresses and crack closure that may retard the growth of cracks. Compressive loads have a different effect. They do not open the cracks, and they must be of greater magnitude before they produce yielding. Then they form tensile residual stresses at the crack tip and thus accelerate the growth of cracks.

We see that sequence effects can be very important and that they depend not on the number of cycles but on the exact details of the load history. This requires, of course, detailed, step-by-step analysis. Many service histories, however, are such that sequence effects either cancel each other or are entirely unpredictable. In those cases, nothing is gained by a detailed analysis, and the simpler prediction methods should be used because they provide equally close predictions and do not create the illusion of higher accuracy. We do not yet have quantitative rules that tell when sequence effects must be considered in predicting fatigue life. A few qualitative rules based on experience can be stated as follows:

1. If the sequence of service loads is completely unknown, one must decide whether to assume significant sequence effects or not.

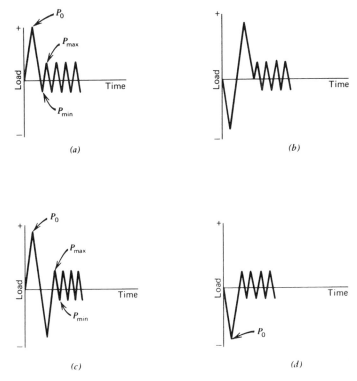

Figure 9.6 Four different overload patterns: (*a*) tension, (*b*) compression-tension, (*c*) tension-compression, (*d*) compression [11] (reprinted by permission of the American Society for Testing and Materials).

2. If the loading is random, with a normal (or Gaussian) probability distribution with widely varying amplitudes at similar frequencies (i.e., a narrow frequency band), there will be no definable sequence. Figure 9.1c is an example.
3. If the loading history shows infrequent high loads in one direction, as for instance in the ground–air–ground cycle of aircraft, one should expect sequence effects.
4. Infrequent tensile overloads produce retardation of crack growth or crack arrest. Compressive overloads large enough to produce yielding can produce the opposite effect.

For the histories shown in Fig. 9.1, predictions that included sequence effects were not significantly better than predictions that neglected sequence effects [12,13]. The negligible influence of load sequence here is explained by the short intervals between the large amplitudes, which produce by far the greatest damage. If we had fewer large amplitudes and many more small amplitudes

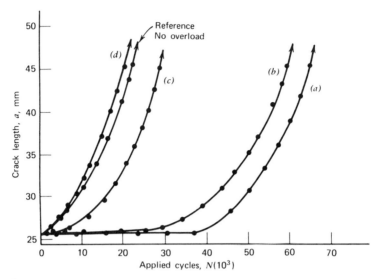

Figure 9.7 Crack growth following different overload patterns in 7075-T6 [11] (reprinted by permission of the American Society for Testing and Materials). (*a*) Tension overload. (*b*) Compression-tension overload. (*c*) Tension-compression overload. (*d*) Compression overload. $K_{\text{overload}} = 45.3 \text{ MPa}\sqrt{m}$. $K_{\text{max}} = 19.7 \text{ MPa}\sqrt{m}$.

(as in aircraft wings), the damage done by the small amplitudes and the sequence effects would be significant. In the preceding discussion, "significant" means larger than the uncertainty inherent in our calculations.

When the future load history is known, one can follow it reversal by reversal, either by test programs applied to the part or analytically [12,14,15]. Less sophisticated methods may achieve equally accurate predictions, depending on the certainty with which the future load history and the fatigue properties of the parts are known. In general, the future load history is not known precisely. Therefore, it would make little sense to try to allow for sequence effects and other interactions of the loading events. In some cases, either the future load history can be predicted fairly well or it is prescribed by the customer as the basis for the analysis and for acceptance tests.

If sequence effects can be neglected, the predictive calculations can be simpler. Sequence effects must be expected when, between many minor ranges, there are a few major deviations that always end in the same way, coming back to the minor ranges either always from high compression or always from high tension. Figure 9.1*b* shows a history that could be expected to show sequence effects because between the 1600 minor ranges there are about 100 major deviations that always come back from high compression. Calculations with and without considering sequence effects were made for two materials and three load levels, and these were compared with fatigue test results for a notched specimen [14]. There were no significant differences for the softer material, Man-Ten, or for the lowest load level, which produced fatigue lives of about 40 000 blocks of 1708 reversals. For the harder material, RQC-100, at higher load levels the results were as follows:

Test lives	22–30	269–460	Blocks
Calculated lives without sequence effects	69	1300	Blocks
Calculated lives with sequence effects	10	170	Blocks

These numbers are for lives to the appearance of obvious cracks taken as 2.5 mm long.

The methods used to incorporate load sequence effects require knowledge of the monotonic and cyclic stress–strain curves in addition to the fatigue properties. Load interaction models are usually used with a computer program.

9.5 CYCLE COUNTING METHODS

The purpose of all cycle counting methods is to compare the effect of variable amplitude load histories to fatigue data and curves obtained with simple constant amplitude load cycles. The application of the linear damage rule, $\Sigma n_i/N_{fi} = 1$, requires that we know the condition (mean and amplitude of stress or strain) to which the damaging event should be compared. Different counting methods can change the resulting predictions by an order of magnitude [16]. A very simple example is shown in Fig. 9.8. The history in Fig. 9.8a could be analyzed as cycles, as shown in Fig. 9.8b, which uses only the segment from point 1 to point 11, which is the beginning of a repetition of the same events. In this method of counting we find one cycle with each of the following five pairs of extreme values:

100/200   100/300   −200/+200   −200/−100   −300/−100

Another counting method produces the five pairs shown in Fig. 9.8c:

−300/300   100/200 twice   −200/−100 twice

The two methods result in very different calculations, and the results of the latter method (Fig. 9.8c) correlate much better with experimental results.

All good counting methods must count a cycle with the range from the highest peak to the lowest valley and must try to count other cycles in a manner that maximizes the ranges that are counted. This rule can be justified either by assuming that damage is a function of the magnitude of the hysteresis loop or by considering that in fatigue (as in many other fields) intermediate fluctuations are less important than the overall differences between high points

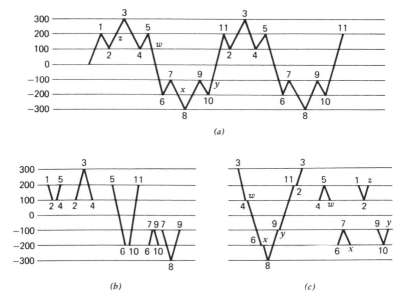

Figure 9.8 Two ways of counting cycles.

and low points. In addition, all good counting methods count every part of every overall range once and only once. They also count smaller ranges down to some predetermined threshold once and only once. Several counting methods that achieve these objectives are well documented in the literature and will now be discussed.

9.5.1 Rainflow Method

This is the most popular and probably the best method of cycle counting; it was first proposed by Matsuishi and Endo [17]. With the load–time, stress–time, or strain–time history plotted such that the time axis is vertically downward, these authors thought of the lines going horizontally from a reversal to a succeeding range as rain flowing down a pagoda roof represented by the history of peaks and valleys. Therefore, the method was called "rainflow counting."

The operation of the rainflow method is shown in Fig. 9.9 for a history consisting of four peaks and four valleys (Fig. 9.9a). The rules are:

1. Rearrange the history to start with either the highest peak or the lowest valley.
2. Starting from the highest peak (or the lowest valley), go down to the next reversal. The rainflow runs down and continues unless either the magnitude of the following peak (or the following valley, if we started

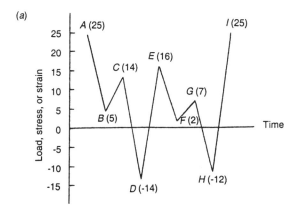

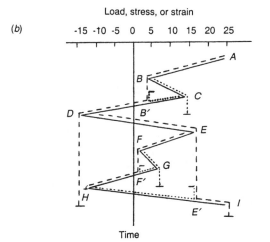

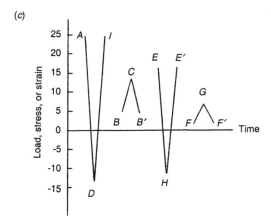

Figure 9.9 Example of rainflow counting. (*a*) Load, stress, or strain history. (*b*) Illustration of rainflow counting. (*c*) Resulting count.

from the lowest valley) is equal to or larger than the peak (or valley) from which it initiated, or a previous rainflow is encountered.

3. Repeat the same procedure for the next reversal and continue these steps to the end.

4. Repeat the procedure for all the ranges and parts of a range that were not used in previous steps.

This procedure is illustrated in Fig. 9.9b. For this history, the largest peak is at point A and the history starts with this peak. Therefore, we start at peak A and go down to the next reversal at point B. Since the next peak at point C is not larger than the starting peak at point A, we go down to point B' and proceed to the next reversal at point D. The following two peaks at points E and G are also not larger than the starting peak at point A. Therefore, we continue to point H at the end of the history. This results in counting a half cycle with range A–D from point A to point D. We now go to the next reversal, which is the valley at point B, and go down to the next reversal at point C. Since the following valley at point D is larger than the one we started from, we stop at point C and count a half cycle with range B–C. The next reversal is the peak at point C. Starting at this point, we must stop at point B' since a previous rainflow from point A is encountered, resulting in a half cycle with range C–B'. At the next reversal, which is the valley at point D, we go down to the next reversal at point E. Since the following valleys at points F and H are smaller than the starting valley at point D, we proceed to point E' and stop at point I at the end of the history. The half cycle from this count, therefore, has the range D–A (or D–I). This procedure is repeated until the loading history is exhausted.

It should be noted that every part of the load history is counted only once. Also, the counted half cycles always occur in pairs of equal magnitude, resulting in full cycles. This is the reason for rearranging the history to start with a peak or valley having the largest magnitude. In the example history, pairs of half cycles A–D and D–I, B–C and C–B', E–H and H–E', and F–G and G–F' form full cycles A–D–I, B–C–B', E–H–E', and F–G–F', shown in Fig. 9.9c. The resulting range and mean values are as follows:

Cycle	Maximum	Minimum	Range	Mean
A–D–I	25	−14	39	5.5
B–C–B'	14	5	9	9.5
E–H–E'	16	−12	28	2
F–G–F'	7	2	5	4.5

An advantage of rainflow counting is when it is combined with a notch strain analysis, as shown in Fig. 9.10 [18]. The load history in Fig. 9.10a is applied to the notched component in Fig. 9.10b, resulting in the notch strain history shown in Fig. 9.10c. This notch strain history is then applied to the

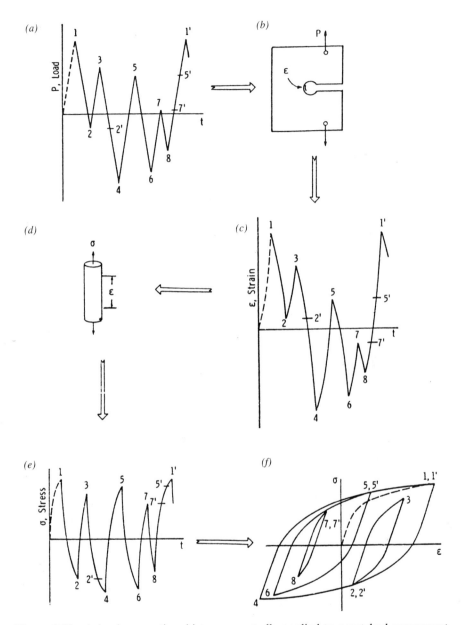

Figure 9.10 A load versus time history repeatedly applied to a notched component and the notch stress–strain response [18] (reprinted with permission of the Society of Automotive Engineers). (*a*) Load–time history. (*b*) Notched component. (*c*) Notch strain history. (*d*) Notch strain history applied to a smooth specimen. (*e*) Notch stress response obtained from a smooth specimen. (*f*) Notch stress–strain hysteresis loops.

smooth axial specimen shown in Fig. 9.10*d,* resulting in the stress response shown in Fig. 9.10*e* and the stress–strain hysteresis loops shown in Fig. 9.10*f.* The stress response can also be obtained from the cyclic stress–strain curve or equation, if available. Note that rainflow counting results in closed hysteresis loops, with each closed loop representing a counted cycle. Therefore, the closed hysteresis loops can also be used to perform the cycle counting. The tips of the largest hysteresis loop are at the largest tensile and compressive loads in the load history (points 1 and 4). Also, note that the notch strain–time history (Fig. 9.10*c*) is quite different from the corresponding notch stress–time history (Fig. 9.10*e*). During each segment of the loading, the material "remembers" its prior deformation (this is called "material memory"). For example, after unloading from point 1 to point 2, loading from point 2 to point 3 follows the hysteresis loop shown in Fig. 9.10*f.* But upon unloading from point 3 to point 4, the unloading deformation path continues to point 4 along the same hysteresis path from point 1 to point 2. The damage from each counted cycle (or each closed hysteresis loop) can be computed from the strain amplitude and mean stress for that cycle as soon as it has been identified in the counting procedure. The corresponding reversal points can then be discarded. More details of the fatigue life prediction analysis using the strain–life approach in conjunction with the rainflow counting method are presented and illustrated by an example problem in Section 9.7.

A computer program that performs rainflow cycle counting can be found in [2]. Such a program applied to a complex history such as that in Fig. 9.1*a* results in the table of ranges and means shown in Table 9.1. Data acquisition systems for real-time rainflow counting of strain gage signals are commercially available. More details of the rainflow counting procedure can also be found in [19]. Other methods that accomplish the same purpose as the rainflow cycle counting method have been shown in a number of sources [1,2,14].

9.5.2 Other Cycle Counting Methods

Variations of rainflow counting and similar methods such as range-pair and racetrack counting methods have also been used for cycle counting. All of these methods result in identical counts if the load history begins and ends with its maximum peak or with its minimum valley. In addition to variations of the rainflow method, other cycle counting methods such as level-crossing and peak counting methods are also in use. As mentioned earlier, the rainflow method is currently the most popular. However, a brief discussion of several of the other methods is also included here.

Range-Pair Method   Figure 9.11 shows the operation of this method [20,21]. Cycles containing small ranges are counted first, and their reversal points (peaks and valleys) are eliminated from further consideration. In Fig. 9.11*a* there are 20 reversals. Fourteen of them are counted and eliminated by counting the seven pairs indicated by crosshatching. This leaves the six reversals

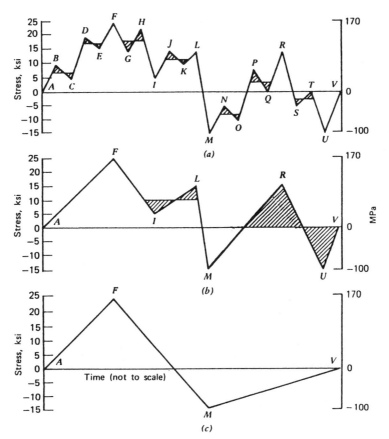

Figure 9.11 Hayes' method of counting [20,21]. (*a*) Counting smaller cycles and discarding the corresponding reversal points. (*b*) Further counting of smaller cycles and discarding the corresponding reversal points. (*c*) Final cycle left from range-pair counting.

of Fig. 9.11*b*. Looking only at Fig. 9.11*b*, we see three peaks and three valleys as follows:

	25		14		16		(0)
(0)		5		−14		−12	

They would be counted as cycles (shown by maximum and minimum loads separated by a slant) as follows:

5/14 (crosshatched)   16/−12 (crosshatched)

leaving 25/−14, which is shown in Fig. 9.11c. The result of a range-pair count is a table of the occurrence of ranges and, if desired, of their mean values.

Racetrack Counting   The racetrack method of counting cycles is shown in Fig. 9.12 [20]. The original history in Fig. 9.12a is condensed to the history in Fig. 9.12c. The method of eliminating smaller ranges is indicated in Fig. 9.12b. A "racetrack" of width S is defined, bounded by "fences" that have the same profile as the original history. Only those reversal points at which a "racer"

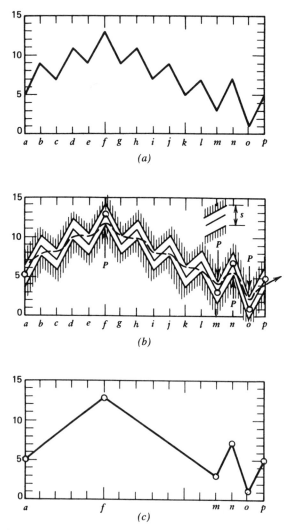

Figure 9.12   The racetrack method of listing reversals [20]. (a) An irregular history. (b) Screening through width S. (c) The resulting condensed history.

would have to change from upward to downward, as at f and n, or vice versa, as at m and o, are counted. The width, S, of the track determines the number of reversals that will be counted. The object of this method is to condense a long, complex history of reversals or a long, complex chart of peaks and valleys to make it more useful. The racetrack method can be used in two modes. The first produces a histogram, spectrum, or listing in which magnitudes of overall ranges and frequencies of their occurrence are given. The second mode produces a condensed history in which essential peaks and valleys are listed in their original sequence. The condensed history includes the sequence of events, which may be important if yielding produces residual stresses that remain active for many succeeding reversals. This method is useful for condensing histories to those few events, perhaps the 10 percent of events that do most of the damage, which usually account for more than 90 percent of all calculated damage [12,16]. The condensed histories accelerate testing and computation and permit focusing of attention on a few significant events. A computer program for the racetrack method was published by Fuchs et al. [16].

Level-Crossing Method   Figure 9.13 illustrates the application and results of the level-crossing counting method [19]. First, a reference load is chosen and the load history is divided into preset load increments. In Fig. 9.13a the load reference is at m and the load history has been divided by unit load increments. A count is then recorded each time a positively sloped portion

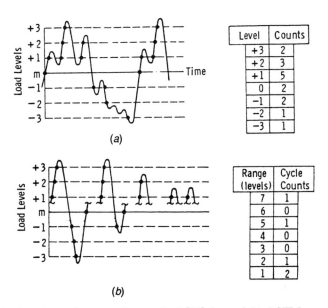

(a)

(b)

Figure 9.13   Level-crossing counting method [19] (copyright ASTM; reprinted with permission). (a) Load history and the resulting load level count. (b) Construction of the most damaging cycles and the resulting load range cycle counts.

of the load history crosses the preset increment above the reference load and each time a negatively sloped portion of the load history crosses the preset limit below the reference load. Reference load crossings are counted on the positively sloped portion of the load history. The recorded count for the history in Fig. 9.13a is tabulated next to the load history. To obtain the most fatigue-damaging cycles, the largest possible cycle is constructed from the recorded count, followed by the second largest possible cycle from the remaining available counts, and so on, until all the counts are used. Reversal points are assumed to occur halfway between increment levels. This process and its cycle count results are shown in Fig. 9.13b. The resulting cycles shown in Fig. 9.13b can be applied in any desired order. Therefore, this method does not account for load sequence effects, which may be significant in fatigue damage analysis, as discussed in Section 9.4. Variations of the level-crossing counting method to eliminate small-amplitude cycles, or to derive a cycle count different from that shown in Fig. 9.13b based on the level-crossing count in Fig. 9.13a, also exist.

Peak Counting Method In this method, a reference load level is first chosen for the load history. The peaks (or maximum load values) above and valleys (or minimum load values) below the reference load level are then identified and counted, as shown in Fig. 9.14a [19]. The most damaging fatigue cycle is

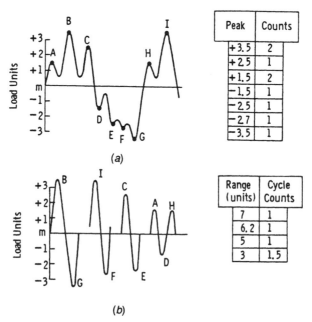

Figure 9.14 Peak counting method [19] (copyright ASTM; reprinted with permission). (a) Load history and the resulting peak and valley counts. (b) Resulting cycle count.

obtained by constructing the largest possible cycle using the highest peak and lowest valley. This is followed by constructing the next largest cycle from the largest peak and valley of the remaining counts, and this process is continued until all the peak counts are used. The final cycles resulting from this process are shown in Fig. 9.14*b*. Similar to the level-crossing counting method, the cycles resulting from peak counting method are in no particular order and, therefore, load sequence effects are not accounted for by this method. A variation of the peak counting method counts all peaks and valleys without regard to the reference load. Other variations also exist to eliminate small-amplitude loadings or to derive a cycle count different from that shown in Fig. 9.14*b*.

More details of the cycle counting methods mentioned as well as others can be found in [1,2,19]. It should also be mentioned that cycle counting is not the only method that can be used to assess cumulative damage. Statistical measures of the events in a measured or expected load history may be useful. For example, it has been shown [22] that the damage produced by random loadings is proportional to the "root-mean sixth power" of the load ranges. In a statistical approach the details, which may be all-important in fatigue analysis, are submerged. Root-mean-square measures have also been used successfully for predicting crack growth with the spectrum loading in Fig. 4.1*c* [23].

9.6 LIFE ESTIMATION USING THE STRESS–LIFE APPROACH

This method neglects sequence effects, and an *S–N* curve is used as the primary input. If an experimental load–life or *S–N* curve of the part is available, it is the best input. Many complicating factors such as stress concentrations, joints, surface conditions, and so on are included in this curve. If an experimental *S–N* curve of only the material is available, it must be modified to account for mean or residual stresses, stress concentrations at the critical location, and other factors such as surface finish, as explained in Chapters 4, 7, and 8. For the most preliminary of estimates it may be necessary to estimate an *S–N* curve for the material, as discussed in Chapter 4. It should be noted, however, that materials and service conditions for which a fatigue limit may exist under constant amplitude cycling, may no longer exhibit a fatigue limit under variable amplitude loading if the largest loads are above the fatigue limit. In this case, the *S–N* line in a log-log plot which is usually represented by the Basquin equation can be extrapolated to below the fatigue limit [24].

The next step is to count load ranges and mean loads of the given load history by rainflow or another cycle counting method, as discussed in Section 9.5. As mentioned earlier, in many cases the 10 percent of all overall ranges that are greatest will do more than 90 percent of the damage [12,16]. These greatest overall ranges can easily be picked out either from a computer printout

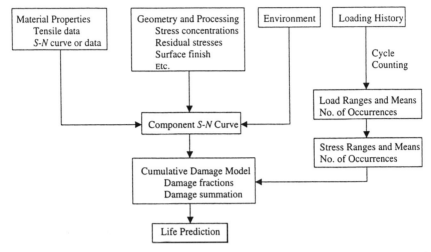

Figure 9.15   Sequential steps in predicting fatigue life based on the *S–N* approach.

or by eye from a plot of peaks and valleys. Figure 9.12 shows how to do this, neglecting obviously minor load cycles.

Next, the load ranges and means must be converted to nominal stress ranges and means. Finally, the damage expected from each of the stress ranges is calculated from the *S–N* curve as $1/N_f$, and the damages are added. The ratio of 1/(sum of damages) is the number of times we expect the given history to be endured until failure occurs. The entire procedure is summarized in a flow chart in Fig. 9.15, which is similar to the procedure for the analysis part of the fatigue design flow chart shown in Fig. 2.1 and illustrated by the following example.

A round shaft made of RQC-100 steel is repeatedly subjected to the block of nominal axial stress history shown in Fig. 9.16. (*a*) If the shaft is smooth, with a polished surface finish, how many blocks of this stress history can be applied before failure is expected? (*b*) Repeat part (*a*) if the shaft has a circumferencial notch with a notch root radius of 1 mm and a stress concentration factor of $K_t = 2$. The applied stress history for the notched shaft and the given K_t are based on the net nominal stress.

The loading history shown in Fig. 9.16 consists of three constant amplitude load segments summarized as follows:

Load Segment	Minimum Stress, S_{min} (MPa)	Maximum Stress, S_{max} (MPa)	Stress Amplitude, S_a (MPa)	Mean Stress, S_m (MPa)	Applied Cycles, n
1	−500	500	500	0	3
2	−500	650	575	75	1
3	0	650	325	325	10

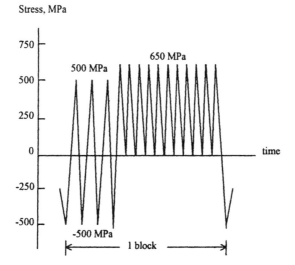

Figure 9.16 Nominal axial stress history applied to a shaft.

Note that load segment 2 may not be apparent at first glance from the load history in Fig. 9.16, but in fact, as will be seen later, it is the most damaging cycle in the load history. A rainflow count will identify this cycle.

(*a*) Using the material fatigue properties from Table A.2, the *S–N* curve for RQC-100 steel is given by

$$S_{Nf} = \sigma_f' \, (2N_f)^b = 1240 \, (2N_f)^{-0.07} \tag{9.2}$$

where S_{Nf} is the fully reversed ($R = -1$) fatigue strength at $2N_f$ reversals. The *S–N* curve for fully reversed loading is shown in Fig. 9.17. Since the loadings in the second and third segments are not completely reversed, we can account for the mean stress by using one of the mean stress correction parameters discussed in Section 4.3. Here we use the modified Goodman equation (Eq. 4.8):

$$\frac{S_a}{S_{Nf}} + \frac{S_m}{S_u} = 1$$

with $S_u = 931$ MPa from Table A.2. The equivalent fully reversed stress amplitude, S_{Nf}, at the 75 MPa and 325 MPa mean stress levels is now computed:

At $S_m = 75$ MPa: $\quad \dfrac{S_a}{S_{Nf}} + \dfrac{75}{931} = 1 \quad$ or $\quad S_{Nf} = 1.088 \, S_a$

At $S_m = 325$ MPa: $\quad \dfrac{S_a}{S_{Nf}} + \dfrac{325}{931} = 1 \quad$ or $\quad S_{Nf} = 1.536 \, S_a$

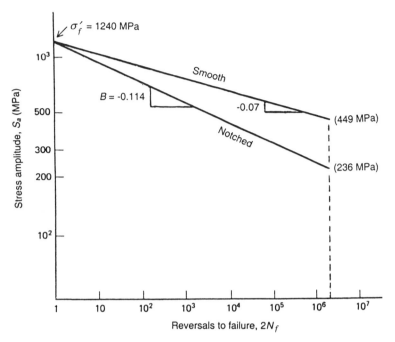

Figure 9.17 S–N lines for RQC-100 steel for completely reversed loading, $R = -1$, for smooth and notched conditions.

We can now obtain cycles to failure, N_f, at each equivalent fully reversed stress amplitude, S_{Nf}, from Eq. 9.2 and compute the damage fraction for each load segment, n/N_f, as summarized in the following table:

Load Segment	S_a (MPa)	S_{Nf} (MPa)	N_f	n	n/N_f
1	500	500	215 770	3	1.4×10^{-5}
2	575	625	8 815	1	1.13×10^{-4}
3	325	499	219 630	10	4.6×10^{-5}
					$\Sigma n_i/N_{fi} = 1.73 \times 10^{-4}$

The expected life is calculated as the reciprocal of $\Sigma n_i/N_{fi}$ or $1/(1.73 \times 10^{-4})$ = 5780 blocks.

(*b*) For the notched shaft, we first obtain the fatigue notch factor, K_f. Using Peterson's equation (Eq. 7.10 along with Eq. 7.11a):

$$a = 0.0254 \left(\frac{2070}{S_u}\right)^{1.8} = 0.0254 \left(\frac{2070}{931}\right)^{1.8} = 0.11 \text{ mm}$$

$$K_f = 1 + \frac{K_t - 1}{1 + a/r} = 1 + \frac{2 - 1}{1 + 0.11/1} = 1.90$$

The S–N line equation for fully reversed loading ($R = -1$) of the notched shaft is obtained by connecting $\sigma_f' = 1240$ MPa at 1 reversal to S_f/K_f at 2×10^6 reversals as follows:

$$S_f = 1240 \, (2 \times 10^6)^{-0.07} = 449 \text{ MPa} \quad \text{and} \quad S_f/K_f = 449/1.90 = 236 \text{ MPa}$$

Then

$$236 = 1240 \, (2 \times 10^6)^B \quad \text{or} \quad B = -0.114 \quad \text{and} \quad S_{Nf} = 1240 \, (2N_f)^{-0.114}$$

The S–N line for fully reversed loading ($R = -1$) of the notched shaft is also shown in Fig. 9.17. To obtain the equivalent completely reversed net alternating stress, S_{Nf}, for the second and third load segments where mean stress exists, we can again use the modified Goodman equation, as in part (a). The relations between the applied alternating stress, S_a, and the equivalent completely reversed stress, S_{Nf}, remain the same as in part (a). We now calculate damage fraction for each load segment as follows:

Load Segment	S_a (MPa)	S_{Nf} (MPa)	N_f	n	n/N_f
1	500	500	1442	3	2.1×10^{-3}
2	575	625	202	1	4.9×10^{-3}
3	325	499	1458	10	6.9×10^{-3}
					$\Sigma n_i/N_{fi} = 0.0139$

The expected life is calculated as $1/(0.0139) = 72$ blocks. It should be remembered that the life calculated is to the nucleation or formation of a crack between 0.25 and 5 mm in length. If this is on the order of the shaft diameter, the number of blocks calculated represents an estimate of the total life of the shaft. If this is much smaller than the shaft diameter, however, there may be substantial life involved in crack growth. In this case, fracture mechanics can be used to obtain the crack growth life of the shaft.

9.7 LIFE ESTIMATION USING THE STRAIN–LIFE APPROACH

When the load history contains large overloads, significant plastic deformation can exist, particularly at stress concentrations, and load sequence effects can be significant. In these cases, the strain–life approach is generally superior to the stress–life approach for cumulative fatigue damage analysis. However, when the load levels are relatively low such that the resulting strains are mainly elastic, the strain–life and stress–life approaches usually result in similar predictions.

Application of the strain–life approach requires the material strain–life curve or equation, as explained in Chapter 5. For a notched member, notch strain analysis, typically with an analytical model such as Neuber, Glinka, or linear rules, as explained in Section 7.3, along with the cyclic stress–strain curve of the material, are used to relate nominal stresses and strains to notch stresses and strains. The effect of mean or residual stresses is also accounted for by using one of the mean stress correction parameters, as discussed in Section 5.6. It should be noted, however, that in situations where mean stresses are large, some mean stress relaxation and creep can take place at the notch root. The effect of lower mean stress due to plastic deformation at the notch root is accounted for by the notch stress–strain analysis mentioned above. But since the material deformation model used is for cyclic stable behavior of the material, transient relaxation and creep are not included in this analysis. This does not, however, usually alter the results of life predictions significantly. Mean stress relaxation is also discussed in Section 5.6.

The load history is converted to a nominal stress history, which is then applied to the component. Using the material cyclic stress–strain relation, and for a notched member a notch strain analysis model (e.g., Neuber's rule), the applied nominal stress history results in the notch strain history and stress–strain hysteresis loops, as explained in Section 9.5.1 and shown in Fig. 9.10. The strain amplitude and mean stress can now be obtained for each rainflow counted cycle from the notch stress and strain history or from the hysteresis loops. When the strain amplitude and mean stress are known, the damage fraction corresponding to each cycle, $1/N_f$, is then computed, and the damages are added to predict failure. The procedure described above is illustrated by the following example problem.

The notched shaft in Part (b) of the example problem in Section 9.6 is subjected to repeated blocks of the variable amplitude net nominal stress in Fig. 9.9, where each unit in Fig. 9.9 is equivalent to 20 MPa. What is the expected life to the appearance of a small crack, according to the strain–life approach?

Cyclic stress–strain properties (K' and n') and strain–life properties (σ_f', b, ε_f', and c) for RQC-100 steel are given in Table A.2. The load history was rainflow counted in Section 9.5.1, as shown in Fig. 9.9c. To convert the applied net nominal stress history to the notch stress–strain history, we choose to use Neuber's rule with $K_f = 1.9$, as calculated for the example problem in Section 9.6.

Initial loading to point A in Fig. 9.9a, which is the highest load in the history, is at a net nominal stress of $S = 25 \times 20 = 500$ MPa. This stress level is 83 percent of the cyclic yield strength ($S_y' = 600$ MPa) and, therefore, the nominal behavior is elastic ($e = S/E$). Notch root stress and strain at point A are then calculated from Neuber's rule and the cyclic stress–strain equation as follows (using $E \approx 200$ GPa):

$$\varepsilon\,\sigma = \frac{(K_f\,S)^2}{E} = \frac{(1.9 \times 500)^2}{200\ 000} = 4.51$$

and

$$\varepsilon = \frac{\sigma}{E} + \left(\frac{\sigma}{K'}\right)^{1/n'} = \frac{\sigma}{200\ 000} + \left(\frac{\sigma}{1434}\right)^{1/0.14}$$

resulting in $\varepsilon = 0.0069$ and $\sigma = 653$ MPa. The initial loading to point A is shown by the dashed curve in Fig. 9.18. The notch deformation at point A is inelastic, as the stress level at this point is above the cyclic yield strength.

For loading from point A to point B, we use the hysteresis loop equation, with a nominal stress range of $\Delta S_{A\ to\ B} = (25 - 5) \times 20 = 400$ MPa, as follows:

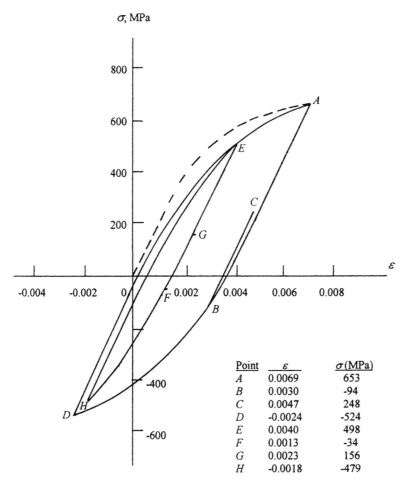

Point	ε	σ(MPa)
A	0.0069	653
B	0.0030	-94
C	0.0047	248
D	-0.0024	-524
E	0.0040	498
F	0.0013	-34
G	0.0023	156
H	-0.0018	-479

Figure 9.18 Hysteresis loops and reversal point stress and strain values for a notched shaft subjected to variable amplitude axial stressing.

$$\Delta\varepsilon\Delta\sigma = \frac{(K_f\Delta S)^2}{E} = \frac{(1.9 \times 400)^2}{200\ 000} = 2.89$$

and

$$\Delta\varepsilon = \frac{\Delta\sigma}{E} + 2\left(\frac{\Delta\sigma}{2K'}\right)^{1/n'} = \frac{\Delta\sigma}{200\ 000} + 2\left(\frac{\Delta\sigma}{2868}\right)^{1/0.14}$$

resulting in $\Delta\varepsilon = 0.0039$ and $\Delta\sigma = 747$ MPa. The loading from point A to point B, following the hysteresis loop, is shown in Fig. 9.18. Now we can compute notch root stress and strain at point B in the loading:

$$\sigma_{at\ B} = \sigma_{at\ A} - \Delta\sigma_{A\ to\ B} = 653 - 747 = -94\ \text{MPa}$$

and

$$\varepsilon_{at\ B} = \varepsilon_{at\ A} - \Delta\varepsilon_{A\ to\ B} = 0.0069 - 0.0039 = 0.0030$$

This process is continued throughout the loading, with the hysteresis loop equation remaining the same but the value of the right side of Neuber's equation changing as ΔS changes for each point in the history. Neuber's equations and notch stresses and strains for the remainder of the loading are summarized as follows:

From point B to point C: $\Delta S = 180$ MPa, $\Delta\varepsilon\Delta\sigma = 0.58$,
$\Delta\varepsilon = 0.0017$, and $\Delta\sigma = 342$ MPa
$\sigma_{at\ C} = -94 + 342 = 248$ MPa,
$\varepsilon_{at\ C} = 0.0030 + 0.0017 = 0.0047$

From point A to point D: $\Delta S = 780$ MPa, $\Delta\varepsilon\Delta\sigma = 10.98$,
$\Delta\varepsilon = 0.0093$, and $\Delta\sigma = 1177$ MPa
$\sigma_{at\ D} = 653 - 1177 = -524$ MPa,
$\varepsilon_{at\ D} = 0.0069 - 0.0093 = -0.0024$

From point D to point E: $\Delta S = 600$ MPa, $\Delta\varepsilon\Delta\sigma = 6.50$,
$\Delta\varepsilon = 0.0064$, and $\Delta\sigma = 1022$ MPa
$\sigma_{at\ E} = -524 + 1022 = 498$ MPa,
$\varepsilon_{at\ E} = -0.0024 + 0.0064 = 0.0040$

From point E to point F: $\Delta S = 280$ MPa, $\Delta\varepsilon\Delta\sigma = 1.42$,
$\Delta\varepsilon = 0.0027$, and $\Delta\sigma = 532$ MPa
$\sigma_{at\ F} = 498 - 532 = -34$ MPa,
$\varepsilon_{at\ F} = 0.0040 - 0.0027 = 0.0013$

From point F to point G: $\Delta S = 100$ MPa, $\Delta\varepsilon\Delta\sigma = 0.18$,
$\Delta\varepsilon = 0.0010$, and $\Delta\sigma = 190$ MPa
$\sigma_{at\ G} = -34 + 190 = 156$ MPa,
$\varepsilon_{at\ G} = 0.0013 + 0.0010 = 0.0023$

From point E to point H: $\Delta S = 560$ MPa, $\Delta\varepsilon\Delta\sigma = 5.66$,

$$\Delta\varepsilon = 0.0058, \text{ and } \Delta\sigma = 977 \text{ MPa}$$
$$\sigma_{\text{at } H} = 498 - 977 = -479 \text{ MPa},$$
$$\varepsilon_{\text{at } H} = 0.0040 - 0.0058 = -0.0018$$

Notch stress and strain values at each reversal point are also shown in Fig. 9.18. We can now use the strain–life equation with one of the mean stress correction parameters to calculate cycles to failure, N_f, for each combination of strain amplitude and mean stress. Here we choose the Smith-Watson-Topper parameter (Eq. 5.24), where $\sigma_{\max} = \sigma_a + \sigma_m$.

$$\varepsilon_a \; \sigma_{\max} = \frac{(\sigma_f')^2}{E}(2N_f)^{2b} + \sigma_f' \; \varepsilon_f' \; (2N_f)^{b+c} = \frac{(1240)^2}{200\,000}(2N_f)^{2(-0.07)}$$

$$+ (1240)\,(0.66)\,(2N_f)^{-0.07-0.69} = 7.69(2N_f)^{-0.14} + 818(2N_f)^{-0.76}$$

Cycle ratios, n/N_f, are then computed and summed, as shown in the following table:

Cycle	ΔS (MPa)	ε_a	$\sigma_{\max}$	N_f	n	n/N_f
$A–D–I$	780	0.00465	659	4 050	1	2.47×10^{-4}
$B–C–B'$	190	0.00085	247	∞	1	0
$E–H–E'$	560	0.00290	498	112 500	1	9×10^{-6}
$F–G–F'$	100	0.00050	156	∞	1	0
					$\Sigma n_i/N_{fi} =$	2.56×10^{-4}

The expected fatigue life is calculated as $1/(2.56 \times 10^{-4}) = 3900$ repetitions of the variable amplitude load block.

    If we use only the largest cycle for damage calculation (i.e., omitting the other three cycles), the expected life is calculated as $1/(2.47 \times 10^{-4}) = 4050$ repetitions. This differs by only about 4 percent from the previous answer. This indicates that 96 percent of the damage is caused by cycle $A–D–I$, and 4 percent of the damage is caused by cycle $E–H–E'$. Cycles $B–C–B'$ and $F–G–F'$ are nondamaging cycles, according to these calculations. Also, the mean stress levels for the two damaging cycles are $(653 - 524)/2 = 65$ MPa for the $A–D–I$ cycle and $(498 - 479)/2 = 10$ MPa for the $E–H–E'$ cycle. These mean stress levels are low compared to their corresponding stress amplitudes, and their effects on fatigue life are expected to be small. Therefore, we could have neglected the mean stress effect and used the strain–life equation (Eq. 5.14) rather than the Smith-Watson-Topper parameter in this problem.

9.8 CRACK GROWTH AND LIFE ESTIMATION MODELS

Life estimations associated with fatigue crack growth involve the growth of cracks from an initial length to some intermediate length or to the final length at fracture. It would be straightforward if one could use the crack growth data from tests with a uniform range of load or stress intensity factor, similar to the conditions presented in Chapter 6. However, fatigue crack growth depends not only on the range of stress intensity factor and stress ratio, but also very significantly on the previous load history, which may have left compressive or tensile residual stress fields. If compressive, the residual stress field tends to retard (delay) or arrest crack growth; if tensile, it tends to accelerate it. The major interest in fatigue crack growth life estimations has been motivated by the damage-tolerant design philosophy in which proper inspection periods must be predetermined.

Fatigue crack growth life estimation models for variable amplitude loading can be simple and straightforward or complex, requiring extensive computations. In some cases, the actual service load history can be approximated and simplified to give repeated applications of a multiple block loading sequence, as in Fig. 9.2, where each block represents a given load level. Estimation of the fatigue crack growth life can then be made, evaluating crack growth related to each block (load level) or, in some cases, the entire repeating load history. Another approach is the summation of crack increments based on each cycle. These methods of life estimation can be used on the assumption that crack growth for a given cycle or block is not influenced by the prior loading history (load sequence). This may be a reasonable assumption if the load history is highly irregular (random), where the frequency of overloads in both tension and compression is similar, or if the overloads are not too severe. However, many loading conditions exist, for example the upper and lower wing surfaces of an airplane, where high overloads can occur and are predominantly in one direction. These overloads may introduce load sequence effects that can significantly affect the overall fatigue crack growth life, much the same as for the fatigue crack nucleation life, as described in Section 9.4. In these cases, models incorporating load sequence effects have been developed to provide more accurate life estimations. Several common methods and models used for making fatigue crack growth life estimations for variable amplitude loading are described below.

Direct Summation   Perhaps the simplest model of damage accumulation based on fatigue crack growth under variable amplitude loading is the direct summation of damage caused by each cycle. This model implies the crack length after N cycles of load applications is

$$a_N = a_0 + \sum_{i=1}^{N} \Delta a_i = a_0 + \sum_{i=1}^{N} f(\Delta K_i) \tag{9.3}$$

where a_0 is the initial crack length. The crack growth increment, Δa_i, associated with each cycle of variable amplitude loading can be estimated from the constant amplitude fatigue crack growth curve, da/dN versus ΔK. The crack growth rate, da/dN, can be estimated from the stress intensity factor range, ΔK, and the stress ratio, R, associated with the given cycle. Cycle-by-cycle summation continues until fracture occurs or a predetermined crack length is reached, at which point the number of cycles accumulated is the estimated fatigue crack growth life. Using direct summation, the accumulation of fatigue damage can also be evaluated block by block. The first step usually involves evaluation of the variable amplitude load history, for example, that shown in Fig. 9.1a. Using an applicable cycle counting method, such as the rainflow or range-pair method, cycle counts for different combinations of stress ranges and means can be determined, similar to that shown in Table 9.1. Each of these "boxes" represent a block of identical constant amplitude cycles with a specific stress range, ΔS, and mean stress, S_m. An increment of crack growth can then be estimated for each block based on the constant amplitude fatigue crack growth curve associated with the stress range and mean stress for each block and summation continues, block by block, until failure occurs. While direct summation accounts for every cycle in the load history, it assumes that the crack growth increment, Δa, related to each cycle or block is not influenced by the prior cycles, i.e., load sequence effects are not taken into account. This assumption could lead to a reduced level of accuracy, depending on the load history being evaluated. However, many fatigue crack growth life estimation codes do incorporate the ability to use crack closure or other load interaction models using direct summation. These models are described later in this section.

Equivalent K Methods   For repeating histories or block load spectra, where load sequence effects are not expected to be significant, it is often possible to correlate crack growth rates and make life estimations using an equivalent stress intensity factor. Once the equivalent load(s) are developed, the analysis becomes one of constant amplitude loading. Use of an equivalent stress intensity factor is usually limited to spectra that exhibit a periodic behavior, i.e., where a repeating load history is relatively short and where crack extension during a single repeating load history is small. While equivalent K approaches are typically not as accurate as direct summation methods, their use has resulted in sufficient accuracy and substantial computational time savings.

One such method is based on the premise that an equivalent $R = 0$ (zero to tension) stress intensity factor range, ΔK_{eq}, can be determined that will cause the same amount of fatigue crack growth as in the variable amplitude load history, as given by

$$da/dN = f(\Delta K_{eq})_{\text{VA loading}} = f(\Delta K)_{\text{CA loading}} \tag{9.4}$$

Assuming that the fatigue crack growth behavior conforms to the Paris equation (Eq. 6.19), the equivalent stress intensity factor range, ΔK_{eq}, can be written as

$$\Delta K_{eq} = \Delta S_{eq} \sqrt{\pi a} \ \alpha = \left[\frac{\sum\limits_{i=1}^{N} (\Delta S'_i)^n}{N} \right]^{1/n} \sqrt{\pi a} \ \alpha \qquad (9.5)$$

where ΔS_{eq} is the equivalent $R = 0$ stress range for the repeating load history, $\Delta S'_i$ is the equivalent $R = 0$ stress range in the ith cycle within the variable amplitude load history, n is the slope of the Paris equation region II line, and N is the number of cycles contained within the variable amplitude load history. ΔS_i is obtained for each constant amplitude block (which can contain one or many cycles) within the variable amplitude load history. To account for mean stress effects associated with each constant amplitude block, the Walker approach can be used to calculate the equivalent $R = 0$ stress range where

$$\Delta S' = S_{max} (1 - R)^{\gamma} \qquad (9.6)$$

Once ΔK_{eq} has been determined, it can be substituted for ΔK in Eq. 6.19. The fatigue crack growth life can then be estimated using one of the suggested techniques presented in Section 6.4.

Another equivalent K method used is based on the assumption that the variation of the crack tip stress–strain field can be described in terms of the root-mean-square value of ΔK (ΔK_{rms}) where

$$\Delta K_{rms} = \Delta S_{rms} \sqrt{\pi a} \ \alpha = \sqrt{\frac{\sum\limits_{i=1}^{N} \Delta S_i^2}{N}} \ \sqrt{\pi a} \ \alpha \qquad (9.7)$$

This equation is similar to Eq. 9.5 with $n = 2$ and $\Delta S'_i$ replaced by ΔS_i. ΔS_{rms} is the nominal root mean-square-stress range for the variable amplitude load history, and ΔS_i is the nominal stress range in the ith cycle. ΔS_i can also be determined for a constant amplitude block if it exists in a load history as well as for a cycle. The root-mean-square stress ratio, R_{rms}, can be calculated from $R_{rms} = S_{min(rms)}/S_{max(rms)}$ where

$$S_{max(rms)} = \left[\frac{1}{N} \sum_{i=1}^{N} (S_{max(i)})^2 \right]^{1/2} \quad \text{and} \quad S_{min(rms)} = \left[\frac{1}{N} \sum_{i=1}^{N} (S_{min(i)})^2 \right]^{1/2} \qquad (9.8)$$

$S_{max(i)}$ and $S_{min(i)}$ are the maximum and minimum stresses observed in the ith cycle, with $S_{min(i)} = 0$ if the minimum stress applied in the ith cycle is in compression.

Once obtained, ΔK_{rms} and R_{rms} can then be used in one of the general power law expressions, for example the Walker equation (Eq. 6.22), where

$$\frac{da}{dN} = f(\Delta K_{rms}) = \frac{A(\Delta K_{rms})^n}{(1 - R_{rms})^{n(1-\lambda)}} \qquad (9.9)$$

It should be noted that for constant amplitude loading $\Delta K_{rms} = \Delta K$ and $R_{rms} = R$. If $R_{rms} = 0$, Eq. 9.9 reduces to the Paris equation (Eq. 6.19). Integration of Eq. 9.9, in a manner similar to that presented in Section 6.5, or by a numerical integration technique, will provide a fatigue crack growth life estimate.

Example Assume that the wide, thin plate in the example problem from Section 6.4.4 is subjected to repetitions of the variable amplitude nominal stress history in Fig. 9.9, where each unit in Fig. 9.9 is equivalent to 10 MPa. How many times can this history be repeated before fatigue fracture is estimated to occur using the two equivalent K methods described above?

From Section 6.4.4, $K_c = 104$ MPa$\sqrt{m}$, $A = 6.9 \times 10^{-12}$, and $n = 3$. The stress history has already been cycle counted using the rainflow method with the minimum-maximum pairs given in Section 9.5.1. Thus, four full cycles exist in this stress history ($N = 4$). The resulting maximum and minimum stresses, stress ratio, applied positive stress range, ΔS, and equivalent stress range, $\Delta S'$, for each cycle are as follows:

Cycle	S_{min} (MPa)	S_{max} (MPa)	R	ΔS (MPa)	$\Delta S'$ (MPa)
1	-140	250	0	250	250
2	50	140	0.36	90	112
3	-120	160	0	160	160
4	20	70	0.29	50	59

Note that when a negative stress is encountered in a stress cycle, it is set equal to zero in calculations of R, $\Delta S'$, and ΔS. $\Delta S'$ values were calculated assuming $\gamma = 0.5$, consistent with the value used in Section 6.5.1. Substituting $\Delta S'$ values for the four cycles into the bracket portion of Eq. 9.5 and using the Paris equation exponent $n = 3$, yields $\Delta S_{eq} = 175$ MPa. In a similar manner, using Eq. 9.7, $\Delta S_{rms} = 157$ MPa. R_{rms} is calculated to be $27/168 = 0.16$ where $S_{min(rms)}$ and $S_{max(rms)}$ from Eq. 9.8 are 27 MPa and 168 MPa, respectively. Again, when $S_{min(i)}$ is negative, it is set equal to zero in calculation of $S_{min(rms)}$. The remaining calculation involves determination of the final crack length at fracture, a_f, which corresponds to the most severe stress in the load history, that being $S_{max} = 250$ MPa. Thus, from Eq. 6.20c with $\alpha = 1.12$ and $K_c = 104$ MPa$\sqrt{m}$, we obtain $a_f = 44$ mm. Using the direct integration method from Section 6.4.4 with Eq. 6.20g and using Eq. 9.5, $N_{f(eq)}$ results in 110 000 cycles. The estimated number of load history repetitions to failure is then 110 000/4 = 27 500. Similarly, using methods from Section 6.5 and using Eq. 9.9, $N_{f(rms)}$ results in 117 000 cycles and the number of load history repetitions is 29 250. The difference in estimated fatigue crack growth life between the two equivalent K methods for this example is less than 10 percent. Because the load history is short and irregular, these results can be expected to be reasonable. It should be noted that the calculations made above were based on the assumption that LEFM principles were not violated. Calculation of the initial plastic zone radius results in $r_y = (K_{max}/S_y)^2/2\pi = (15.7/630)^2/2\pi \approx 0.1$ mm, while the plastic zone radius at fracture is $r_y = (104/630)^2/2\pi \approx 4.3$ mm. These values are both less than the general LEFM restriction where $r_y \leq a/4$. In addition, the maximum applied stress, $S_{max} = 250$ MPa, is well below the yield strength, $S_y = 630$ MPa. Thus, LEFM appears very reasonable for this problem.

The methods described above assume that crack growth is not affected by prior events in the load history. In some cases, this assumption may be very reasonable. However, in many cases this assumption could lead to significant error. As mentioned earlier, crack growth depends on the previous load history, which can retard or accelerate crack growth. Quantitative fatigue crack growth life estimations under variable amplitude conditions that account for load sequence effects are based primarily on three different models formulated in the early 1970s by Wheeler [25], Willenborg et al. [26], and Elber [27]. These models, along with their extensions and modifications, are still used extensively in fatigue crack growth life estimation programs and codes. Use of these models often requires substantial numerical calculations. A good knowledge of the load spectrum is needed to properly include interaction (load sequence) effects. The accuracy of the models has ranged from very good to very poor. However, variable amplitude crack growth estimations involving either load interaction or noninteraction models have not been developed to a point where simulated or actual field tests can be eliminated. A description of the three load interaction models follows. Substantial literature exists in which these models have been analyzed, modified, expanded, and used in detail. Their use has been most prevalent in the aerospace field.

Wheeler Model The Wheeler model, often referred to as a "yield zone or crack tip plasticity model," uses a modification of the basic functional relationship between constant amplitude crack growth rate and the stress intensity factor range

$$\frac{da}{dN} = f(\Delta K) \tag{9.10}$$

The Paris equation (Eq. 6.19), Forman equation (Eq. 6.21), or Walker equation (Eq. 6.22) could represent this functional relationship. The model modifies the basic summation of Eq. 9.3. Note that Eq. 9.3 implies that load sequence effects are omitted. The Wheeler modification introduces an empirical retardation parameter, C_i, that allows for retardation by reducing the crack growth rate following a tensile overload where

$$C_i = \left[\frac{r_{yi}}{(a_{OL} + r_{OL}) - a_i} \right]^m \tag{9.11}$$

and

r_{yi} = the plastic zone at the ith cycle after the tensile overload
a_{OL} = the crack length at a tensile overload application
r_{OL} = the plastic zone caused by the tensile overload
a_i = the crack length at the ith cycle after the tensile overload
m = an empirical shaping exponent

The terms are defined schematically in Fig. 9.19. Equation 9.11 is the proper retardation parameter as long as

$$a_i + r_{yi} < a_{OL} + r_{OL} \tag{9.12}$$

If

$$a_i + r_{yi} \geq a_{OL} + r_{OL} \tag{9.13}$$

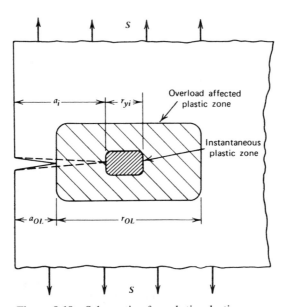

Figure 9.19 Schematic of crack tip plastic zones.

then $C_i = 1$. This implies that as long as the instantaneous plastic zone is within the tensile overload affected plastic zone, crack growth retardation will be present. When the instantaneous plastic zone just reaches or passes beyond the tensile overload affected plastic zone, retardation effects disappear. The resultant equations for the Wheeler crack length summation and crack growth rate during the ith cycle are

$$a_N = a_0 + \sum_{i=1}^{N} C_i f(\Delta K_i) \quad \text{and} \quad \left(\frac{da}{dN}\right)_i = C_i f(\Delta K) \tag{9.14}$$

Values for r_{yi} and r_{OL} for plane stress from Eq. 6.10 are

$$r_{yi} = \frac{1}{2\pi}\left(\frac{K_{maxi}}{S_y}\right)^2 \quad \text{and} \quad r_{OL} = \frac{1}{2\pi}\left(\frac{K_{OL}}{S_y}\right)^2 \tag{9.15}$$

Values of r_{yi} and r_{OL} for plane strain are one-third of those in Eq. 9.15. The exponent m shapes the retardation parameter, C_i, to correlate with test data. This value is obtained experimentally and can pertain to an entire known spectrum or can be varied to consider specific load ratios between a high load and a low load. Wheeler found values for m of 1.3 for D6AC steel and 3.4 for titanium in his experiments. If $m = 0$ then no retardation influence exists. The numerical computation scheme requires crack growth summation one load application at a time, similar to the direct summation model discussed earlier. While the Wheeler model takes into account retardation effects by the use of C_i, the model does not account for crack growth accelerations due to compressive overloads (underloads) or initial accelerations sometimes observed immediately following a tensile overload. An acceleration immediately following a tensile overload has been attributed to a decrease in the opening stress magnitude, S_{op}, due to the plastically deformed material ahead of the crack tip. This can result in a brief initial acceleration prior to a decrease in crack growth rate (retardation).

Willenborg Model The Willenborg model, also a yield zone or crack tip plasticity model, handles crack growth retardation by using an effective stress concept to reduce the applied stresses and hence the crack tip stress intensity factor. The model does not rely on empirically derived parameters except the constant amplitude crack growth parameters for a given material, such as in the Paris equation (Eq. 6.19), Forman equation (Eq. 6.21), or Walker equation (Eq. 6.22). An effective value of the stress intensity factor range is computed by assuming that a given or predetermined crack tip compressive residual stress is present after a tensile overload. The model assumes that retardation decays over a length equal to the plastic zone r_{OL}, from a tensile overload, as shown in Fig. 9.19. Any load greater than the preceding tensile overload creates a new retardation condition independent of all preceding conditions. Only tensile loads are counted and compression loads are neglected. Thus, similar to the Wheeler model, fatigue crack growth accelerations are not accounted for in this model. The model operates in terms of total crack length rather than crack length increments. Retardation decays to zero when the instantaneous crack length, a_i, plus its associated yield zone, r_{yi}, exceeds a_p where $a_p = a_{OL} + r_{OL}$. The model postulates that the compressive residual stress due to the tensile overload is the difference between the maximum stress occurring at a_i and the required stress necessary to terminate retardation. This implies that the retardation is due to a reduction in K_{max}. Thus, effective maximum and minimum stresses are determined on the basis of this difference and are accounted for by determining an effective load ratio, R_{eff}, and an effective stress intensity factor range, ΔK_{eff}. The fatigue crack growth rate, da/dN, can then be calculated from, say, the Forman equation (Eq. 6.21) based on effective values where

$$\frac{da}{dN} = \frac{A(\Delta K_{eff})^n}{(1 - R_{eff})K_c - \Delta K_{eff}} \tag{9.16}$$

The above model can be applied to individual cycles, load segments, entire blocks, or complete spectra using a number of available commercial or U.S. government-sponsored computer programs.

Elber Model The Elber model, referred to as a "crack closure model," is an empirically based crack closure model that uses the effective stress intensity factor range concept, as described in Section 6.7, to incorporate load interaction effects in variable amplitude fatigue crack growth life estimations. Studies indicate that the crack opening load varies in a random loading history; thus, ΔK_{eff}, defined as $\Delta K_{eff} = K_{max} - K_{op}$, can vary with each cycle. Note that ΔK_{eff} here is defined differently than for the Willenborg model of Eq. 9.16. Based on Elber's crack closure model, the effective stress intensity factor range, ΔK_{eff}, can be used in place of the nominal stress intensity factor range, ΔK. Using, for example, the Paris equation (Eq. 6.19), and replacing ΔK with ΔK_{eff} or ΔS with ΔS_{eff} results in

$$\frac{da}{dN} = A(\Delta K_{eff})^n = A(\Delta S_{eff}\sqrt{\pi a}\ \alpha)^n \tag{9.17}$$

where $\Delta S_{eff} = S_{max} - S_{op}$. Equation 9.17 can then be used in a numerical integration scheme to obtain cycle-by-cycle fatigue crack growth from some initial crack size to a final crack size. It should be noted that in order to perform this integration scheme, knowledge of the variation of K_{op} is required during the variable amplitude loading sequence. Methods used to obtain K_{op} include fixing or holding K_{op} constant during a given block of variable amplitude loading, or assuming that K_{op} is the same under variable amplitude loading and constant amplitude loading.

In addition to crack tip plasticity and crack closure, the transient fatigue crack growth behavior associated with load excursions has been attributed to other mechanisms, such as crack tip blunting, crack deflection, and strain hardening. A detailed review of these mechanisms can be found in [28]. A number of fatigue crack growth life estimation computer codes currently exist that have adopted the above-mentioned models or extensions and modifications of these models. Concepts associated with small crack behavior and crack growth from notches, while complex, have also been incorporated into some life estimation codes. Some of the more widely used U.S. government-sponsored computer codes include NASGRO [29], FASTRAN-II [30], MOD-GRO [31], and FLAGRO [32]. These and other life estimation codes typically require significant information regarding material properties. In many cases, fatigue crack growth lives can be reasonably estimated with these codes.

9.9   SIMULATING SERVICE HISTORIES IN THE LABORATORY AND DIGITAL PROTOTYPING

9.9.1   Laboratory Test Methods

According to a proverb, experience is the best teacher. Unfortunately, it is also a very expensive teacher. Much fatigue research has aimed at providing data and rules that distill the essence of experience so the designer can calculate sizes, shapes, and processing of the part and confidently expect them to last long enough without being too heavy, bulky, or expensive. For constant amplitude loading of well-known materials in well-known environments, this can be done. For variable amplitude loading, we still lack tools for such confident analysis and prediction. To avoid the high cost of learning from field failures, many companies and agencies have developed laboratory test methods that can approximate the results of field experience for variable amplitude load histories. Six different test methods are discussed below. They differ in philosophy and in cost. With sufficient experience in comparing laboratory data to field data, they all can perform very well. They all require knowledge of the expected history as primary input. Some agencies, such as the U.S. Air Force, provide a spectrum or history of use. For other customers, like drivers of cars, it is very difficult to decide on a "representative" load history. Additional aspects of fatigue testing of components and full-scale structures are discussed in Section 2.3.

In order of increasing physical complexity, the six methods discussed are (1) characteristic constant amplitude, (2) block testing, (3) condensed histories, (4) truncated histories, (5) complete histories, and (6) statistically simulated histories.

Testing with a Characteristic Constant Amplitude   A characteristic constant amplitude, based on experience, serves very well if it can be validated by field data. For automobile suspension springs, for instance, constant amplitude tests to a few hundred thousand cycles of maximum possible deflection reproduce the field history well enough. The reasons for this are that (1) the large amplitudes actually do most of the fatigue damage and (2) field use is so diverse that any one field history would be just as wrong as this constant amplitude test.

An "equivalent" constant amplitude has been defined as the constant amplitude that produces the same median life, in number of reversals, as the real history. Because it is difficult to know the median life for the real history and because important large ranges will be omitted to obtain equal numbers of reversals, we see little merit in an equivalent constant amplitude test.

Block Testing   The real history is replaced by a number of "blocks" of constant maximum and minimum load or deflection, as shown in Fig. 9.2. The series of blocks form a program. In principle, the program contains the same number of reversals as the history, and its blocks approximate the distribution of peaks and valleys. In practice, large numbers of small ranges, well below the fatigue limit, are omitted. Six to 10 blocks provide adequate approximations. The sequence of blocks is important. A random or pseudorandom sequence will minimize undesirable sequence effects. High-amplitude ranges that occur only seldom are added once in every nth program. It has been found that a test should contain at least a dozen repetitions of the program in order to represent a real history. This method of testing has been widely used in the aircraft industry.

Testing with Condensed Histories   Current electronic-hydraulic controls are capable of producing practically any prescribed sequence of loads, deflection, or strains. To reduce testing cost and time, a condensed version of the real history may be used. Condensed histories include selected peaks and valleys of the real history in their real sequence and omit many smaller peaks and valleys. It has been shown that with only 2 percent of the reversals, properly selected, the fatigue life to the appearance of 2.5 mm deep cracks is practically the same as that obtained with all the reversals for the histories shown in Fig. 9.1 [16,33]. The selection of the most significant peaks and valleys is done by racetrack counting or by editing a rainflow count to retain only the largest ranges.

Truncated Histories   Histories can be truncated by omitting all ranges smaller than a given level. If this is done after rainflow counting, it amounts to the same as condensed histories. If it is done without regard to overall ranges, it may lead to serious errors because a small range (say, 10 percent of the maximum range) may easily add 5 percent to the highest peak or lowest valley and thus increase the damage 30 percent or more. The same magnitude, discarded by racetrack or rainflow counting, might have contributed only one-millionth (0.0001 percent) of the damage of the maximum range. (The above numbers assumed an $S-N$ slope of $-1/6$, or damage to be proportional to the sixth power of amplitudes.)

Complete Histories   Suitable test machines can apply the record of a load history on computer disk or other devices to the test specimen, component, or structure over and over again. This produces a good test, but it may be unnecessarily expensive in time in view of the capabilities of condensed histories.

Statistically Simulated Histories   Repeating a recorded history over and over again may not be the best way to test parts. It may, for instance, omit or exaggerate some sequence effects. To overcome this problem, one can arrange truly random input to the test machine with prescribed parameters, such as the distribution of peaks and valleys or of amplitudes and means. Other parameters that might be prescribed are RMS value, maximum value, and harmonic component spectrum, for instance, "white Gaussian noise."

9.9.2 Digital Prototyping

Digital prototyping and computer simulation techniques for fatigue analysis, as introduced in Section 2.5, are much more recent and rapidly progressing. An important element of digital prototyping is dynamic simulation of the structure. This involves building a computer model and exciting it by representative service loads, such as a digital road profile for an automobile. The output from the excited computer model can then be used for fatigue damage assessment and life predictions.

Building the computer or digital model consists of representing the structure by multibody elements (such as a control arm of an automobile represented by rigid or flexible mass elements) and connecting them by appropriate force elements (such as springs or damping elements) or appropriate constraints (such as spherical joints). In building a model, there is always a trade-off between a complex, expensive model with better solution accuracy and a simpler, less expensive model with poorer solution accuracy. The accuracy of the simulation results, however, depends not only on model complexity, but also on setting appropriate solution parameters such as error tolerance.

Commercial software including both multibody dynamic and finite element analyses are now commonly used for dynamic simulation. Multibody dynamic analysis software is generally used to obtain the dynamic response of the structure, such as displacements and reaction forces. The dynamic response is then used as the input to finite element analysis software to obtain stresses and strains at the critical locations, which in turn are used for fatigue life predictions. More details on dynamic modeling and simulation can be found in [2] and from individual software vendors.

The Society of Automotive Engineers Fatigue Design and Evaluation Committee created a task group on digital prototyping for structural durability in

1998. An All Terrain Vehicle (ATV) was chosen for system simulation and fatigue life predictions. The scope of this project includes many aspects, such as dynamic simulation, modal analysis, rig testing, FEA, real-time simulation, and fatigue life predictions. More details of the SAE Digital Prototype Task Group can be found in [34].

9.10 SUMMARY

Service load histories are usually variable amplitude, and their realistic representation is a key ingredient of successful fatigue design. Transducers in the critical locations of the component or structure are usually used to measure the load or strain history. To compare fatigue behavior from variable amplitude histories to fatigue curves obtained with simple constant amplitude loading, a cycle counting method is needed. Good cycle counting methods must count a cycle with the range from the highest peak to the lowest valley and must try to count other cycles in a manner that maximizes the ranges that are counted. The rainflow method is the most popular method of cycle counting. However, variations of rainflow and other counting methods are also in use.

To evaluate damage from each cycle in a variable amplitude load or strain history, a quantifying measure is required. The most common measure is the life fraction or cycle ratio for crack nucleation and the crack length for crack growth. Life fraction defines the damage caused by one cycle as $D = 1/N_f$, where N_f is the number of repetitions of this same cycle that equals the median fatigue life to failure. Once the damage from each cycle has been calculated, it is accumulated or summed over the entire load history. The linear damage rule, also referred to as the "Palmgren-Miner rule," is the simplest rule and is often used. This rule assumes a linear summation of damage and predicts failure when $\Sigma n_i/N_{fi} = 1$. The assumption of linear damage ignores load sequence and interaction effects which may have a major influence on the fatigue life. Even though many nonlinear damage models have been developed, they do not generally encompass many of the complicating factors encountered during complex variable amplitude loading. As a result, the linear damage rule remains the most widely used rule for cumulative fatigue damage analysis.

The stress–life approach neglects load sequence effects in service load histories. However, it is simple and the $S–N$ curve, which is used as input to life predictions, can include many of the complex factors influencing fatigue behavior. The strain–life approach accounts for load sequence effects and is generally advantageous for cumulative damage analysis of notch members, in which significant plasticity usually exists. This approach uses the material cyclic stress–strain curve and a notch strain analysis method to obtain notch root stresses and strains, which are then used for life predictions. Both the $S–N$ and $\varepsilon–N$ approaches are used to predict life to crack formation or nucleation, with a crack length on the order of 1 mm.

Fatigue crack growth life estimations for variable amplitude loading can be made by either neglecting or taking into account load sequence effects. If service load histories can be approximated to give repeated applications of a loading sequence and if peak loads have minor effects on crack growth and retardation, a reasonable and simple approach may be one in which load sequence effects are neglected. However, in many cases, fatigue crack growth depends on load history and sequence effects that must be taken into account for the development of reliable models. Current fatigue crack growth life estimation codes that account for load sequence effects use crack tip plasticity or crack closure models. In most cases, computer software is necessary to make fatigue crack growth life estimations for variable amplitude loading.

To avoid the high cost of field failures, laboratory and/or field test methods may still be required to complement fatigue analysis and life predictions from variable amplitude load histories. Laboratory and/or field test methods with different degrees of physical complexity and cost can be used to approximate the results of actual service experience. Digital prototyping and computer simulation techniques for fatigue analysis have also been developed.

9.11 DOS AND DON'TS IN DESIGN

1. Do learn as much as possible from service failures. They may validate or invalidate your assumptions about service loads.

2. Do determine (by test or by agreement) the service load histories for which you are designing.

3. Do ask yourself whether sequence effects are likely to be important. Infrequent one-sided overloads are expected to produce sequence effects. For many service histories, however, sequence effects either cancel each other or are entirely unpredictable.

4. Don't ignore the fact that often only a few events in the load history produce most of the damage. Properly condensed load histories containing these events can significantly reduce analysis as well as testing time.

5. Don't forget that only a few overloads in the load history can significantly affect the fatigue behavior in a notched or cracked component by producing beneficial compressive or detrimental tensile residual stresses at the notch root or at the crack tip.

6. Do allow margins for error in keeping with the certainty or uncertainty of your assumptions.

7. Don't forget that fatigue analysis and predictions alone, for complex variable amplitude load histories and service conditions, may not be sufficient to avoid costly field failures. Laboratory and/or field tests may also be necessary to approximate the results of field experience and validate the predictions.

8. Do place prototypes or early production machines in severe service and follow them very carefully to obtain load histories and to detect weak spots.

REFERENCES

1. R. M. Wetzel, ed., *Fatigue Under Complex Loading: Analysis and Experiments,* AE-6, SAE, Warrendale, PA, 1977.

2. R. C. Rice, ed., *SAE Fatigue Design Handbook,* 3rd ed., AE-22, SAE, Warrendale, PA, 1997.

3. L. Yang and A. Fatemi, "Cumulative Fatigue Damage Mechanisms and Quantifying Parameters: A Literature Review," *J. Test. Eval.,* Vol. 26, No. 2, 1998, p. 89.

4. A. Palmgren, "Durability of Ball Bearings," *ZDVDI,* Vol. 68, No. 14, 1924, p. 339 (in German).

5. M. A. Miner, "Cumulative Damage in Fatigue," *Trans. ASME, J. Appl. Mech.,* Vol. 67, 1945, p. A159.

6. S. S. Manson, J. C. Freche, and S. R. Ensign, "Application of a Double Linear Damage Rule to Cumulative Fatigue," *Fatigue Crack Propagation,* ASTM STP 415, ASTM, West Conshohocken, PA, 1967, p. 384.

7. A. Fatemi and L. Yang, "Cumulative Fatigue Damage and Life Prediction Theories: A Survey of the State of the Art for Homogeneous Materials," *Int. J. Fatigue,* Vol. 20, No. 1, 1998, p. 9.

8. S. M. Marco and W. L. Starkey, "A Concept of Fatigue Damage," *Trans. ASME,* Vol. 76, 1954, p. 627.

9. T. H. Topper, B. I. Sandor, and J. Morrow, "Cumulative Fatigue Damage Under Cyclic Strain Control," *J. Mater.,* Vol. 4, No. 1, 1969, p. 200.

10. J. H. Crews, Jr., "Crack Initiation at Stress Concentrations as Influenced by Prior Local Plasticity," *Achievement of High Fatigue Resistance in Metals and Alloys,* ASTM STP 467, ASTM, West Conshohocken, PA, 1970, p. 37.

11. R. I. Stephens, D. K. Chen, and B. W. Hom, "Fatigue Crack Growth with Negative Stress Ratio Following Single Overloads in 2024-T3 and 7075-T6 Aluminum Alloys," *Fatigue Crack Growth Under Spectrum Loads,* ASTM STP 595, ASTM, West Conshohocken, PA, 1976, p. 27.

12. D. V. Nelson and H. O. Fuchs, "Predictions of Cumulative Fatigue Damage Using Condensed Load Histories," *Fatigue Under Complex Loading: Analysis and Experiments,* AE-6, R. M. Wetzel, ed., SAE, Warrendale, PA, 1977, p. 163.

13. H. O. Fuchs, "Discussion: Nominal Stress or Local Strain Approaches to Cumulative Damage," *Fatigue Under Complex Loading: Analysis and Experiments,* AE-6, R. M. Wetzel, ed., SAE, Warrendale, PA, 1977, p. 203.

14. D. F. Socie, "Fatigue Life Prediction Using Local Stress-Strain Concepts," *Exp. Mech.,* Vol. 17, No. 2, 1977, p. 50.

15. J. M. Potter, "Spectrum Fatigue Life Predictions for Typical Automotive Load Histories," *Fatigue Under Complex Loading: Analysis and Experiments,* AE-6, R. M. Wetzel, ed., SAE, Warrendale, PA, 1977, p. 107.

16. H. O. Fuchs, D. V. Nelson, M. A. Burke, and T. L. Toomay, "Shortcuts in Cumulative Damage Analysis," *Fatigue Under Complex Loading: Analysis and Experiments,* AE-6, R. M. Wetzel, ed., SAE, Warrendale, PA, 1977, p. 145.

17. M. Matsuishi and T. Endo, "Fatigue of Metals Subjected to Varying Stress," paper presented to the Japan Society of Mechanical Engineers, Fukuoka, Japan, March 1968.

18. N. E. Dowling, W. R. Brose, and W. K. Wilson, "Notched Member Fatigue Life Predictions by the Local Strain Approach," *Fatigue Under Complex Loading: Analysis and Experiments,* AE-6, R. M. Wetzel, ed., SAE, Warrendale, PA, 1977, p. 55.

19. "Cycle Counting for Fatigue Analysis," ASTM Standard E1049, *Annual Book of ASTM Standards,* Vol. 03.01, ASTM, West Conshohocken, PA, 1998, p. 693.

20. A. Teichmann, "The Strain Range Counter," Vickers-Armstrong Aircraft Ltd., Technology Office VTO/M/46, April 1955.

21. J. E. Hayes, "Fatigue Analysis and Fail-Safe Design," *Analysis and Design of Flight Vehicle Structures,* E. F. Bruhn, ed., Tri-State Offset Co., Cincinnati, OH, 1965, p. C 13-1.

22. S. L. Bussa, N. J. Sheth, and S. R. Swanson, "Development of a Random Life Prediction Model," Ford Engineering Technology Office Report No. 702204, May 1970.

23. J. M. Barsom, "Fatigue Crack Growth Under Variable-Amplitude Loading in Various Bridge Steels," *Fatigue Crack Growth Under Spectrum Loads,* ASTM STP 595, ASTM, West Conshohocken, PA, 1976, p. 217.

24. N. E. Dowling, *Mechanical Behavior of Materials,* 2nd ed., Prentice-Hall, Upper Saddle River, NJ, 1998.

25. O. E. Wheeler, "Spectrum Loading and Crack Growth," *Trans. ASME, J. Basic Eng.,* Vol. 94, 1972, p. 181.

26. J. Willenborg, R. M. Engle, and H. A. Wood, "A Crack Growth Retardation Model Using an Effective Stress Concept," TM-71-1-FBR, Wright-Paterson Airforce Base, OH, 1971.

27. W. Elber, "The Significance of Fatigue Crack Closure," *Damage Tolerance in Aircraft Structures,* ASTM STP 486, ASTM, West Conshohocken, PA, 1971, p. 230.

28. S. Suresh, *Fatigue of Materials,* 2nd ed., Cambridge University Press, Cambridge, 1999.

29. R. G. Forman, V. Shivakumar, S. R. Mettu, and J. C. Newman, "Fatigue Crack Growth Computer Program 'NASGRO' Version 3.0," NASA JSC-22267B, NASA, 1996.

30. J. C. Newman, Jr., "FASTRAN-II-A Fatigue Crack Growth Structural Analysis Program," NASA TM 104159, NASA, 1992.

31. J. A. Harter, "MODGRO-User's Manual," AFWAL-TM-88-157-FIBE, Dayton, OH, 1988.

32. R. G. Forman, V. Shivakumar, J. C. Newman, S. Piotrowski, and L. Williams, "Development of the NASA FLAGRO Computer Program," *Fracture Mechanics: 18th Symp.,* ASTM STP 945, ASTM, West Conshohocken, PA, 1988, p. 781.

33. D. F. Socie and P. Artwahl, "Effects of Spectrum Editing on Fatigue Crack Initiation and Propagation in a Notched Member," University of Illinois, College of Engineering Report FCP No. 31, December 1978.

34. SAE Fatigue Design and Evaluation Committee meeting minutes, 1998.

PROBLEMS

1. A particular type of ball bearing operating at 3000 RPM has a rated life of 50 000 hours, 6500 hours, and 1000 hours when subjected to constant amplitude loads of 1 kN, 2 kN, and 4 kN, respectively. During each hour of operation, the load is 4 kN for 5 minutes, 2 kN for 15 minutes, and 1 kN for 40 minutes. (*a*) How many hours do you expect the bearings to last? (*b*) What percentage of the damage is caused by each of the load levels? (*c*) Are the load sequence effects expected to be important in this problem?

2. A rod is made of a steel with S_u = 450 MPa and a fatigue limit of S_f = 175 MPa, defined at 10^6 cycles. The rod is subjected to five fully reversed blocks of nominal stress cycling, as indicated below. The blocks are then repeated. (*a*) Predict the expected fatigue life of the part if the part is smooth. (*b*) Repeat part (*a*), assuming that the part had a circular groove with K_t = 1.5 and a groove root radius of 5 mm. For the notched rod, assume the given nominal stress blocks and K_t to be based on net stress.

S_a (MPa)	350	300	250	200	125
Applied cycles	1	5	50	500	10 000

3. Repeat Problem 2 for the following:

   (*a*) RQC-100 steel.
   (*b*) 4340 steel with S_u = 1240 MPa.
   (*c*) 4340 steel with S_u = 1470 MPa.
   (*d*) 2024-T3 aluminum.
   (*e*) 7075-T6 aluminum.

4. Assume that the ordinate in Fig. 9.1*d* is twice the nominal stress in MPa (2:1 scale). Construct a peak and valley histogram of occurrences versus stress. Choose a small stress increment (i.e., about 50 MPa) for the abscissa because of the small number of reversals. Then construct a cumulative distribution of peaks and valleys. Can you do the above for stress ranges?

5. Obtain a cycle count of the stress history in Problem 4 using

   (*a*) the range-pair counting method.
   (*b*) the level-crossing counting method.
   (*c*) the peak counting method.

6. Rainflow count the history of Fig. 9.1*d* and plot the range histogram of occurrences versus stress. Then construct a cumulative distribution of ranges.

7. An axially loaded member made of 2024-T3 aluminum is repeatedly subjected to the block of stress history shown. Using the *S–N* approach,

(*a*) estimate the expected life if the member is smooth, and (*b*) estimate the expected life if the member has a notch with $K_t = 2$ and the notch root radius is 1 mm. For the notched member, assume that the given nominal stress block and K_t are based on net stress.

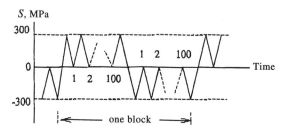

8. Repeat Problem 7 using the strain–life approach.

9. How many blocks of Fig. 9.1*d* loading can be applied to a small, smooth uniaxial rod of RQC-100 steel if the given axis has a scale of 200 units = 100 MPa, according to the *S–N* approach? Repeat the calculations if a notch of $K_t = 1.5$ and a notch root radius equal to 5 mm had existed, where K_t and nominal stress history are based on the net cross-sectional area.

10. Repeat Problem 9 for the notched rod, using the strain–life approach. Include the notch stress–strain hysteresis loops in your solution.

11. Repeat Problem 9 using the following:

   (*a*) 1090 steel with $S_u = 1090$ MPa.
   (*b*) 4340 steel with $S_u = 1470$ MPa.

12. A typical ground–air–ground (GAG) spectrum is shown in Fig. 4.1*a*. Substantial comparative testing of several prototype solutions is to be done. Simplify the given spectrum in order to make the tests. Provide five optional simplified spectra with the advantages and limitations of each. Do any of the spectra have some commonality?

13. Determine when and explain why a tensile overload such as that in Fig. 9.6*a* can (*a*) be beneficial, (*b*) be detrimental, and (*c*) have a negligible effect on the fatigue life of parts. Consider smooth, notched, and cracked parts for both axial and bending loads.

14. The same as Problem 13, but consider the overloads of Figs. 9.6*b* and 9.6*c*.

15. Using the two equivalent *K* methods described in Section 9.8, how many blocks of Fig. 9.1*d* loading can be applied to a wide center-cracked panel with an initial crack length $2a = 2$ mm? The given axis has a scale of 200 units = 50 MPa. Compare your results for the two methods used. Use the following materials and the crack growth rate properties given in Table 6.3:

(*a*) Hot-rolled 1020 steel.

(*b*) 4340 steel with $S_u = 1250$ MPa.

(*c*) 7075-T6 aluminum.

(*d*) Ti-6A1-4V mill annealed.

16. In Problem 13 from Chapter 6, reverse the order of loading and then solve parts *a* to *f* without considering interaction effects.

17. Repeat Problem 16, but use retardation influence. Assume that $m \approx 1.5$.

CHAPTER 10

MULTIAXIAL STRESSES

Multiaxial states of stress are very common, and multiaxial strain is difficult to avoid. The strains are triaxial, for example, in a tensile bar. With longitudinal strain ε we have two transverse strains $-\nu\varepsilon$, where ν is Poisson's ratio, which changes from about 0.3 in the elastic range to 0.5 in the plastic range. In a shaft that transmits torque, there are two principal stresses σ, equal in magnitude but opposite in sign. There is neither stress nor strain in the third principal direction. In a thin-walled pressure vessel subjected to cyclic pressure, the longitudinal and hoop stresses are also principal stresses whose directions remain fixed during pressure cycling. In a crankshaft we have torsion and bending. On points of its surface, two principal stresses exist that vary in magnitude and direction. The frequencies of the bending cycles and torsion cycles are not the same. Most points on the surface have triaxial strain. The state of stress in notches is usually multiaxial and is not the same as the state of stress in the main body. For example, at the root of a thread the state of stress is biaxial, although it may be uniaxial in the main body of a bolt. In addition, the stress concentration factor changes with the state of stress, and stress and strain concentration factors are not equal.

Can we somehow manage to apply our knowledge and data from uniaxial behavior and tests to multiaxial situations? This is the question with which this chapter is concerned. We first briefly review states of stress and strain and classify constant amplitude multiaxial loading into proportional and non-proportional loading. This is followed by a brief discussion of yielding and plasticity under multiaxial stress states. Multiaxial fatigue life estimation methods are then classified and described as stress-based approaches, strain-based and energy-based approaches including critical plane models, and the fracture mechanics approach for crack growth. In these discussions we assume isotropic material, although it is known that the fatigue properties of some materials may not be isotropic; for example, the fatigue strength in the direction of rolling or extruding may be substantially greater than that in the transverse direction. Also, unnotched behavior and constant amplitude loading are emphasized, while brief discussions of notched effects and variable amplitude loading are included at the end of the chapter.

10.1 STATES OF STRESS AND STRAIN AND PROPORTIONAL VERSUS NONPROPORTIONAL LOADING

An understanding of the state of stress and strain in a component or structure is essential for multiaxial fatigue analysis. The state of stress and strain at a point in the body can be described by six stress components (σ_x, σ_y, σ_z, τ_{xy}, τ_{xz}, τ_{yz}) and six strain components (ε_x, ε_y, ε_z, γ_{xy}, γ_{xz}, γ_{yz}) acting on orthogonal planes x, y, and z. Stresses and strains acting in any other direction or plane can be found by using transformation equations or graphically by using Mohr's circle.

Of special interest to fatigue analysis are the magnitudes and directions of the maximum normal principal stress, σ_1, the maximum shearing stress, τ_{max}, the maximum normal principal strain, ε_1, and the maximum shearing strain, γ_{max}, acting at a critical location in the component or structure. It is important to realize that even though only a few planes experience the maximum principal normal stress (or strain) and the maximum shearing stress (or strain), many other planes can experience a very large percentage of these quantities. For example, with simple uniaxial tension, even though only the loading plane experiences the maximum normal stress σ_1, all planes oriented between $\pm 13°$ from the loading plane experience at least 95 percent of σ_1. Also, a shearing stress is present on every stressed plane, except for the loading plane. In addition to the maximum principal stress or strain plane and the maximum shear planes, the octahedral planes are of importance in yielding prediction and fatigue analysis. There are eight octahedral planes making equal angles with the three principal stress directions. The shearing stress on these planes is given by

$$\tau_{oct} = \frac{1}{3} \sqrt{(\sigma_1 - \sigma_2)^2 + (\sigma_2 - \sigma_3)^2 + (\sigma_3 - \sigma_1)^2} \qquad (10.1)$$

The normal stress on an octahedral plane is the hydrostatic stress (also called the "average normal stress") given by

$$\sigma_{oct} = \sigma_h = \sigma_{ave} = \frac{1}{3}(\sigma_1 + \sigma_2 + \sigma_3) \qquad (10.2)$$

The shear strain acting on an octahedral plane is given by

$$\gamma_{oct} = \frac{2}{3} \sqrt{(\varepsilon_1 - \varepsilon_2)^2 + (\varepsilon_2 - \varepsilon_3)^2 + (\varepsilon_3 - \varepsilon_1)^2} \qquad (10.3)$$

During constant amplitude cyclic loading, as the magnitudes of the applied stresses vary with time, the size of Mohr's circle of stress also varies with time. In some cases, even though the size of Mohr's circle varies during cyclic loading, the orientation of the principal axes with respect to the loading axes remains fixed. This is called "proportional loading." In many cases, however, the principal directions of the alternating stresses are not fixed in the stressed part but change orientation. Crankshafts are a typical example. Shafts subjected to out-of-phase torsion and bending are another. This type of loading is called "nonproportional loading." Proportional and nonproportioinal loading are illustrated by Fig. 10.1 for combined axial-torsion loading of a shaft shown in Fig. 10.1a. The loads in Fig. 10.1b are applied in-phase, so that the maximum and minimum axial and torsion stresses occur simultaneously. The ratio of axial stress, σ_y, and torsion stress, τ_{xy}, remains constant during cycling, as shown by the linear relationship in Fig. 10.1c. This is therefore called "proportional loading." If the loads are applied 90° out-of-phase (Fig. 10.1d), the stress path, σ_y-τ_{xy}, follows an ellipse, as shown in Fig. 10.1c. The ratio of axial stress, σ_y, and torsion stress, τ_{xy}, varies continuously during the cycle. This is therefore an example of nonproportional loading. Mohr's circles of stress at times 2 and 3 during the in-phase loading cycle are shown in Fig. 10.1e. It can be seen that the orientation of the principal normal stress directions remains fixed (i.e., angle 2α remains constant), even though the size of the circle changes as the magnitudes of the loads vary with time. Mohr's circles of stress at three times (1, 2, and 3) during the out-of-phase loading cycle are shown in Fig. 10.1f. For this case, the orientations of the principal normal stress axes rotate continuously with respect to the loading axes (i.e., x-y axes). The maximum principal normal stress axis orientation starts out at $\alpha = 45°$ at time 1, decreases to a smaller α at time 2, and rotates to $\alpha = 0°$ at time 3. Then it continues to rotate in a clockwise direction until point 9, where it returns to $\alpha = 45°$.

10.2 YIELDING AND PLASTICITY IN MULTIAXIAL FATIGUE

As discussed in Chapter 3, cyclic plastic deformation is an essential component of the fatigue damage process. Therefore, an understanding of multiaxial cyclic plastic deformation is often necessary, particularly in situations where significant plasticity exists, such as at notches and in low-cycle fatigue. The basic elements of plasticity theory consist of a yield function to determine when plastic flow begins, a flow rule which relates the applied stress increments

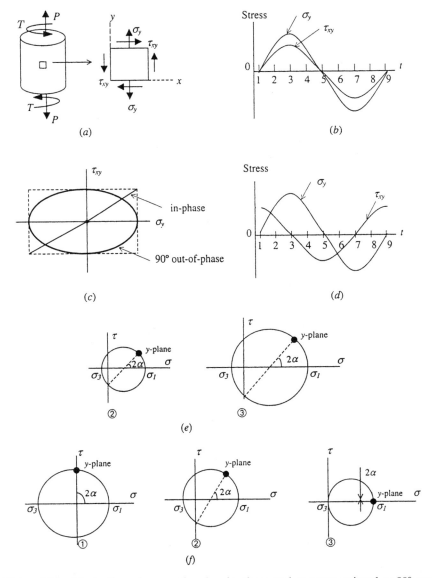

Figure 10.1 Illustration of proportional or in-phase and nonproportional or 90° out-of-phase loading. (*a*) Stress element in axial-torsion loading. (*b*) Applied in-phase axial and shear stress histories. (*c*) Stress path for in-phase and 90° out-of-phase loading. (*d*) Applied 90° out-of-phase axial and shear stress histories. (*e*) Mohr's circle of stress at times 2 and 3 in the cycle for in-phase loading. (*f*) Mohr's circle of stress at times 1, 2, and 3 in the cycle for 90° out-of-phase loading.

to the resulting plastic strain increments once plastic flow has begun, and a hardening rule that describes the change in the yield criterion as a function of plastic strains. Time effects such as creep and viscoelasticity are neglected here, so that yielding depends only on instantaneous increments of stress or strain and on the previous history of the material.

The yield criterion can be visualized in a "stress space" in which each of the coordinate axes represents one principal stress. A commonly used yield criterion for metals is the von Mises yield criterion, which can be visualized as a circular cylinder in the stress space. For unyielded material the axis of the cylinder passes through the origin of the coordinates. It is inclined equal amounts to the three coordinate axes and represents pure hydrostatic stress (i.e., $\sigma_1 = \sigma_2 = \sigma_3$). The von Mises yield criterion is given by

$$\sqrt{(\sigma_1 - \sigma_2)^2 + (\sigma_2 - \sigma_3)^2 + (\sigma_3 - \sigma_1)^2} = 2S_y^2 \qquad (10.4)$$

where S_y is the material yield strength. For biaxial or plane states of stress ($\sigma_3 = 0$), the yield condition is the intersection of the cylinder with the σ_1–σ_2 plane, which is a yield ellipse.

It is often convenient to convert the multiaxial stress state to an "equivalent" stress, σ_e, which is the uniaxial stress that is equally distant from (or located on) the yield surface. During initial loading, the von Mises equivalent stress is given by

$$\sigma_e = \frac{1}{\sqrt{2}} \sqrt{(\sigma_1 - \sigma_2)^2 + (\sigma_2 - \sigma_3)^2 + (\sigma_3 - \sigma_1)^2} \qquad (10.5)$$

or

$$\sigma_e = \frac{1}{\sqrt{2}} \sqrt{(\sigma_x - \sigma_y)^2 + (\sigma_y - \sigma_z)^2 + (\sigma_z - \sigma_x)^2 + 6(\tau_{xy}^2 + \tau_{yz}^2 + \tau_{zx}^2)} \qquad (10.6)$$

where σ_x, τ_{xy}, and so on are stresses in an arbitrary orthogonal coordinate system. Yielding occurs when $\sigma_e = S_y$. These expressions for the equivalent stress, σ_e, are also related to the octahedral shear stress and the distortion energy.

Once plastic deformation has begun, we need a flow rule to relate stresses and plastic strains. Equations relating stresses and plastic strains are also called "constitutive equations" and are typically based on the normality condition. This condition states that the increment of plastic strain caused by an increment of stress is such that the vector representing the plastic strain increment is normal to the yield surface during plastic deformation.

A hardening rule is needed to describe the behavior of the material once it is plastically deformed or yielded. One possible hardening rule is the isotropic rule, which assumes that strain hardening corresponds to an enlargement of the yield surface (i.e., an increase in S_y) without a change of shape or position in the stress space. Another is the kinematic rule, which assumes that strain hardening shifts the yield surface without changing its size or shape. A version of the kinematic hardening rule commonly used in cyclic plasticity is the Mroz kinematic hardening rule.

Nonproportional multiaxial cyclic loading often produces additional strain hardening which is not observed under proportional loading conditions. Therefore, the cyclic stress–strain curve for nonproportional loading is above that for proportional loading. The reason for the additional hardening is the interaction of slip planes, since many more slip planes are active during nonproportional loading due to the rotation of the principal axes, as discussed in Section 10.1. The amount of this additional hardening depends on the degree of load nonproportionality as well as on the material. Models to incorporate this additional hardening are discussed in [1].

10.3 STRESS-BASED CRITERIA

10.3.1 Equivalent Stress Approaches

Equivalent stress approaches are extensions of static yield criteria to fatigue. The most commonly used equivalent stress approaches for fatigue are the maximum principal stress theory, the maximum shear stress theory (also called Tresca theory), and the octahedral shear stress theory (or von Mises theory). An "equivalent" nominal stress amplitude, S_{qa}, can be computed according to each criterion as

Maximum principal stress theory: $\qquad S_{qa} = S_{a1}$ $\qquad$ (10.7)

Maximum shear stress theory: $\qquad S_{qa} = S_{a1} - S_{a3}$ $\qquad$ (10.8)

Octahedral shear stress theory:

$$S_{qa} = \frac{1}{\sqrt{2}} \sqrt{(S_{a1} - S_{a2})^2 + (S_{a2} - S_{a3})^2 + (S_{a3} - S_{a1})^2} \quad (10.9)$$

Here S_{a1}, S_{a2}, and S_{a3} are principal alternating nominal stresses with $S_{a1} > S_{a2} > S_{a3}$. Note that Eq. 10.9 is identical to Eq. 10.5 for von Mises equivalent stress for static loading, except that alternating stresses are used. This criterion can also be written in terms of the x, y, z stress components, with the resulting equation identical to Eq. 10.6 but with alternating stresses. Once the equivalent nominal stress amplitude, S_{qa}, is calculated, the multiaxial stress state is reduced to an equivalent uniaxial stress state. Therefore, the $S–N$ approach discussed in Chapter 4 can then be used for fatigue life calculations by setting S_{qa} equal to the appropriate fatigue strength S_{Nf} or S_f.

The octahedral shear stress criterion (von Mises) is the most widely used equivalent stress criterion for multiaxial fatigue of materials having ductile behavior. The maximum principal stress criterion is usually better for multi-axial fatigue of materials having brittle behavior. The three aforementioned criteria are shown in Fig. 10.2 for biaxial stress states ($S_3 = 0$) [2]. The coordinate axes in this figure are the principal alternating stresses (S_{a1} and S_{a2}), normalized by the uniaxial fatigue strength, S_f. Multiaxial in-phase fatigue data at long lives (10^7 cycles) for different combinations of fully reversed alternating stresses for a mild steel, a chromium–vanadium (Cr–V) steel, and a cast iron are also plotted in Fig. 10.2. It can be seen that data for the mild steel and Cr–V steel, which behave in a ductile manner, agree well with the octahedral shear stress criterion. The data for cast iron, which behaves in a brittle manner, agree better with the maximum principal stress criterion.

If mean or residual stresses are present, an equivalent mean nominal stress, S_{qm}, can be calculated based on the von Mises effective stress:

$$S_{qm} = \frac{1}{\sqrt{2}} \sqrt{(S_{m1} - S_{m2})^2 + (S_{m2} - S_{m3})^2 + (S_{m3} - S_{m1})^2} \quad (10.10)$$

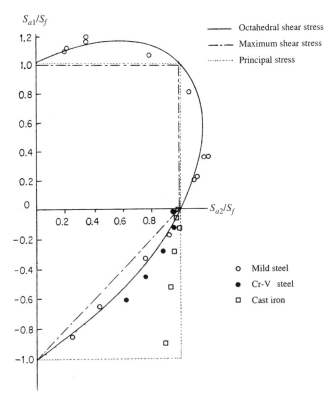

Figure 10.2 Comparison of biaxial fatigue data with equivalent stress failure criteria [2].

where S_{m1}, S_{m2}, and S_{m3} are principal mean nominal stresses. Another commonly used equivalent mean stress is the sum of mean normal stresses:

$$S_{qm} = S_{m1} + S_{m2} + S_{m3} = S_{mx} + S_{my} + S_{mz} \tag{10.11}$$

Note that the second equality in Eq. 10.11 exists because the sum of normal stresses represents a stress invariant (i.e., a stress quantity independent of the coordinate axes used). Both the Tresca and von Mises equivalent stresses are insensitive to a hydrostatic stress (i.e., $S_{m1} = S_{m2} = S_{m3}$). Therefore, if the mean stress is hydrostatic, Eq. 10.10 results in $S_{qm} = 0$. Since fatigue life has been observed to be sensitive to hydrostatic stress, the use of Eq. 10.11 is preferred for this case. In addition, Eq. 10.10 always results in a positive equivalent mean stress, whereas Eq. 10.11 can be either positive or negative. Equation 10.11, therefore, better represents the beneficial effect of compressive mean stress and the detrimental effect of tensile mean stress on fatigue life. Also, according to Eq. 10.11, mean torsion stress has no direct influence on fatigue (i.e., $S_{qm} = 0$ for mean torsion). This agrees with experimental evidence, as long as the maximum shear stress remains below yielding for the material. If the maximum shear stress is high enough to produce yielding, it may change residual stresses and indirectly affect fatigue resistance. This equation, however, suggests that the effect of a tensile mean stress acting in one direction can be nullified by a compressive mean stress acting in another direction (e.g., in plane stress $S_{qm} = 0$ if $S_{mx} = -S_{my}$), which does not necessarily agree with experimental evidence.

Stresses S_{qa} and S_{qm} are those equivalent alternating and mean stresses that can be expected to give the same life in uniaxial loading as in multiaxial loading. After S_{qa} and S_{qm} are calculated, the expected fatigue life is found from the formulas for uniaxial fatigue, such as the modified Goodman equation (Eq. 4.8). The formulas are used with the magnitude of S_{qa} in place of S_a and with the magnitude of S_{qm} in place of S_m.

Equivalent stress approaches have been commonly used because of their simplicity, but their success in correlating multiaxial fatigue data has been limited to a few materials and loading conditions. In addition, they should be used only for proportional loading conditions, in which the principal axes directions remain fixed during the loading cycle.

10.3.2 Sines Method

Sines method [2] uses the alternating octahedral shear stress for cyclic stresses (i.e., Eq. 10.1 in terms of principal alternating nominal stresses) and the hydrostatic stress for mean stresses (i.e., Eq. 10.2 in terms of principal mean or residual nominal stresses). It can be represented by

$$\sqrt{(S_{a1} - S_{a2})^2 + (S_{a2} - S_{a3})^2 + (S_{a3} - S_{a1})^2}$$
$$+ m (S_{mx} + S_{my} + S_{mz}) = \sqrt{2} S_{Nf} \tag{10.12}$$

where m is the coefficient of mean stress influence and S_{Nf} is the uniaxial, fully reversed fatigue strength that is expected to give the same fatigue life on uniaxial smooth specimens as the multiaxial stress state. The coefficient m can be determined experimentally by obtaining a fatigue strength with a nonzero mean stress level (e.g., uniaxial fatigue strength for $R = 0$ condition where $S_m = S_a$). The value of m is on the order of 0.5. If plotted in terms of principal alternating stresses for biaxial stress states, the Sines method is similar to the octahedral shear stress theory shown in Fig. 10.2, except that the ellipse becomes smaller for a positive mean stress term and larger for a negative mean stress term. In terms of x, y, z stress components, the Sines method is expressed by the equation

$$\sqrt{(S_{ax} - S_{ay})^2 + (S_{ay} - S_{az})^2 + (S_{az} - S_{ax})^2 + 6(\tau_{axy}^2 + \tau_{ayz}^2 + \tau_{azx}^2)}$$
$$+ m(S_{mx} + S_{my} + S_{mz}) = \sqrt{2}\, S_{Nf} \quad (10.13)$$

The multiaxial stress state is then reduced to an equivalent uniaxial stress, S_{Nf}. The Sines method should also be limited to those cases in which the principal alternating stresses do not change their directions relative to the stressed part (i.e., proportional loading). For this type of loading, this theory fits most observations for long life fatigue and can be extended for application to strain-controlled, low-cycle fatigue.

10.3.3 Examples Using the Stress–Life Approach

We use two examples to illustrate applications of the S–N approach. Both examples involve proportional biaxial loading. As discussed in other chapters, it should be recognized that stress-based approaches are, in general, suitable for long life fatigue situations, in which strains are mainly elastic.

Example 1 A closed-end, thin-walled tube made of 1020 sheet steel, with inside diameter $d = 100$ mm (4 in.) and wall thickness $t = 3$ mm (0.12 in.), is subject to internal pressure, p, which fluctuates from 0 to 15 MPa (2.18 ksi). What is the expected fatigue life?

Stress analysis shows a longitudinal stress varying from 0 minimum to $pd/4t = (15 \times 100)/12 = 125$ MPa maximum and a circumferential stress varying from 0 to $pd/2t = 250$ MPa maximum, which is in-phase with, or proportional to, the longitudinal stress. These stresses are also the principal stresses such that $S_1 = 250$ MPa and $S_2 = 125$ MPa. The radial stress in a thin-walled tube is small compared to the longitudinal and circumferential stresses such that $S_3 \approx 0$.

For fatigue analysis we separate the stresses into alternating and mean components:

$$S_{a1} = S_{m1} = 125 \text{ MPa} \quad \text{and} \quad S_{a2} = S_{m2} = 62.5 \text{ MPa}$$

We then form "equivalent" alternating and mean stresses. They are equivalent because we expect their joint effect to give the same life in uniaxial tests that we expect from the multiaxial situation. The equivalent alternating stress is calculated from Eq. 10.9:

$$S_{qa} = \frac{1}{\sqrt{2}} \sqrt{(125 - 62.5)^2 + (62.5 - 0)^2 + (0 - 125)^2} = 108 \text{ MPa}$$

The equivalent mean stress from Eq. 10.11 is simply the sum of the mean normal stresses in three mutually perpendicular directions, $S_{qm} = 125 + 62.5 = 187.5$ MPa. With S_{qa} and S_{qm} values known, we can use the modified Goodman equation (Eq. 4.8) to obtain the uniaxial, fully reversed fatigue strength, S_{Nf}. From Table A.2 for 1020 HR sheet steel, $S_u = 441$ MPa. Then

$$\frac{S_{qa}}{S_{Nf}} + \frac{S_{qm}}{S_u} = \frac{108}{S_{Nf}} + \frac{187.5}{441} = 1 \quad \text{or} \quad S_{Nf} = 188 \text{ MPa}$$

Now the fatigue life can be calculated using Basquin's S–N equation with cyclic properties of the material from Table A.2

$$S_{Nf} = \sigma_f'(2N_f)^b = 1384\,(2N_f)^{-0.156}$$

Substituting $S_{Nf} = 188$ MPa results in $N_f = 180\,000$ cycles. If we use the Sines method with $m = 0.5$, Eq. 10.12 results in

$$\sqrt{(125 - 62.5)^2 + (62.5 - 0)^2 + (0 - 125)^2}$$
$$+ 0.5\,(125 + 62.5 + 0) = \sqrt{2}\,S_{Nf}$$

from which we obtain $S_{Nf} = 175$ MPa, resulting in $N_f = 290\,000$ cycles. The difference between the two results in less than a factor of 2.

Example 2 Consider a shaft with a diameter of 20 mm (0.8 in.) carrying a gear with a pitch diameter of 110 mm (4.33 in.) on its end. The tangential force on the gear is 8000 N (1800 lb), and the radial force is neglected. The distance from the center of the gear to a bearing is 25 mm (1 in.). How do we compare this stress situation to test data from uniaxial stresses?

The bending moment is $M = 8000 \times 0.025 = 200$ N.m and the torque is $T = 8000 \times 0.055 = 440$ N.m. Both bending stress, S, and torsion stress, τ, are found on the surface of the shaft at the bearing as follows:

$$S = 200\,\frac{32 \times 10^6}{\pi \times 2^3}\,\text{Pa} = 255 \text{ MPa} \quad \text{and} \quad \tau = 440\,\frac{16 \times 10^6}{\pi \times 2^3}\,\text{Pa} = 280 \text{ MPa}$$

The bending stress is fully reversed since the shaft rotates with $S_a = 255$ MPa, while the torsion stress is a steady stress with $\tau_m = 280$ MPa. We calculate S_{qa} from Eq. 10.9 with $S_{a1} = S_a = 255$ MPa and $S_{a2} = S_{a3} = 0$.

$$S_{qa} = \frac{1}{\sqrt{2}} \sqrt{(255 - 0)^2 + 0^2 + (0 - 255)^2} = 255 \text{ MPa}$$

According to Eq. 10.11, $S_{qm} = 280 - 280 = 0$, since principal stresses in torsion are equal to shear stress, $S_{m1} = \tau_m$, $S_{m2} = 0$, and $S_{m3} = -\tau_m$. The fatigue life of this shaft is, therefore, expected to be about that for uniaxial stressing (i.e., rotating bending) of the same shaft, with $S_a = S_{qa} = 255$ MPa and without the mean torque (i.e., $S_m = S_{qm} = 0$). The Sines method predicts the same result as the octahedral shear stress criterion, since the mean stress term in Eq. 10.12 is zero. In this example, the principal axes do not remain fixed on the shaft. However, the principal axes for the alternating stress components do not rotate with time. As a result, this loading can also be classified as proportional cyclic loading, and no additional strain hardening, as discussed in Section 10.2, is associated with this loading.

10.4 STRAIN-BASED, ENERGY-BASED, AND CRITICAL PLANE APPROACHES

10.4.1 Strain-Based and Energy-Based Approaches

Strain-based approaches are used with the strain–life curve (Eq. 5.14) in situations in which significant plastic deformation can exist, such as in low-cycle fatigue or at notches. Analogous to equivalent stress approaches, equivalent strain approaches have been used as strain-based multiaxial fatigue criteria. The most commonly used equivalent strain approaches are strain versions of the equivalent stress models discussed in Section 10.3.1 as follows:

Maximum principal strain theory: $\varepsilon_{qa} = \varepsilon_{a1}$ (10.14)

Maximum shear strain theory: $\varepsilon_{qa} = \dfrac{\varepsilon_{a1} - \varepsilon_{a3}}{1 + \nu}$ (10.15)

Octahedral shear strain theory:

$$\varepsilon_{qa} = \frac{\sqrt{(\varepsilon_{a1} - \varepsilon_{a2})^2 + (\varepsilon_{a2} - \varepsilon_{a3})^2 + (\varepsilon_{a3} - \varepsilon_{a1})^2}}{\sqrt{2}\,(1 + \nu)}$$ (10.16)

where ε_{a1}, ε_{a2}, and ε_{a3} are principal alternating strains with $\varepsilon_{a1} > \varepsilon_{a2} > \varepsilon_{a3}$. Once an equivalent alternating strain, ε_{qa}, has been calculated from the multiaxial stress state, the strain–life equation (Eq. 5.14 with ε_a replaced by ε_{qa}) is used for fatigue life prediction. Similar to the equivalent stress approaches, equivalent strain approaches are also not suitable for nonproportional multi-axial loading situations, in which the principal alternating strain axes rotate during cycling.

Energy-based approaches use products of stress and strain to quantify fatigue damage. Several energy quantities have been proposed for multiaxial fatigue, such as the plastic work per cycle as the parameter for life to crack nucleation. Plastic work is calculated by integrating the product of the stress and the plastic strain increment (the area of the hysteresis loop) for each of the six components of stress. The sum of the six integrals is the plastic work per cycle. The determination of the hysteresis loops σ_x–ε_x, τ_{xy}–γ_{xy}, and so on requires careful consideration of hardening rules and flow rules, particularly during nonproportional loading, since the shape of the hysteresis loops depends on the loading path. Application of this method to high-cycle fatigue situations is difficult since plastic strains are small. Models based on total strain energy density per cycle, consisting of both elastic and plastic energy density terms, have also been developed by Ellyin [3] and co-workers and by Park and Nelson [4]. Energy-based approaches can be used for nonproportional loading. Energy, however, is a scalar quantity and, therefore, does not reflect fatigue damage nucleation and growth observed on specific planes.

10.4.2 Critical Plane Approaches and the Fatemi-Socie Model

Experimental observations indicate that cracks nucleate and grow on specific planes (also called "critical planes"). Depending on the material and loading conditions, these planes are either maximum shear planes or maximum tensile stress planes. Multiaxial fatigue models relating fatigue damage to stresses and/or strains on these planes are called "critical plane models." These models, therefore, can predict not only the fatigue life, but also the orientation of the crack or failure plane. Different damage parameters using stress, strain, or energy quantities have been used to evaluate damage on the critical plane.

A critical plane model based on stress quantities was developed by Findley for high-cycle fatigue in the 1950s [5]. Findley identified the cyclic shear stress and the normal stress on the shear stress plane as the parameters governing fatigue damage. Findley's model can be described as

$$\frac{\Delta\tau}{2} + k\sigma_n = C \tag{10.17}$$

where $\Delta\tau/2$ is the shear stress amplitude, σ_n is the normal stress acting on the $\Delta\tau$ plane, and k and C are constants. According to this model, failure (as defined by crack nucleation and growth of small cracks) occurs on the plane with the largest value of $(\Delta\tau/2 + k\sigma_n)$. Later, Brown and Miller [6] formulated a similar approach in terms of strains, where the cyclic shear strain and the normal strain on the maximum shear plane are the governing parameters. A simple formulation of this approach is given by

$$\frac{\Delta\gamma_{max}}{2} + s\,\Delta\varepsilon_n = C \tag{10.18}$$

where $\Delta\gamma_{max}/2$ is the maximum shear strain amplitude, $\Delta\varepsilon_n$ is the normal strain range on the $\Delta\gamma_{max}$ plane, and s is a material dependent constant that weights the importance of normal strain for different materials.

Critical plane approaches attempt to reflect the physical nature of fatigue damage (i.e., damage mechanisms) in their formulation. A common model based on the physical interpretation of fatigue damage is the Fatemi-Socie model [7]. In this model, the parameters governing fatigue damage are the maximum shear strain amplitude, $\Delta\gamma_{max}/2$, and the maximum normal stress, $\sigma_{n,max}$, acting on the maximum shear strain amplitude plane. The physical basis for this model is illustrated in Fig. 10.3. Cracks are usually irregularly shaped at the microscopic level, as the crack grows through the material grain structure. This results in interlocking and friction forces between crack surfaces during cyclic shear loading (i.e., crack closure), as shown in Fig. 10.3a. Consequently, the crack tip driving force is reduced and the fatigue life is increased. A tensile stress perpendicular to the crack plane tends to separate crack surfaces, and therefore, reduce interlocking and frictional forces, as shown in Fig. 10.3b. This increases the crack tip driving force, and the fatigue life is reduced. Fractographic evidence of this behavior has been reported [8,9]. The Fatemi-Socie model is expressed as

$$\frac{\Delta\gamma_{max}}{2}\left(1 + k\,\frac{\sigma_{n,max}}{S_y}\right) = C \tag{10.19}$$

where S_y is the material monotonic yield strength and k is a material constant. The maximum normal stress is normalized by the yield strength to preserve

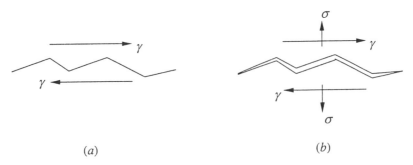

(a)   (b)

Figure 10.3 Physical basis of the Fatemi-Socie model. (a) Shear loading of a crack. (b) Effect of tensile stress on the shear crack.

the unitless feature of strain. The value of k can be found by fitting fatigue data from simple uniaxial tests to fatigue data from simple torsion tests (note that for the simple torsion test, $\sigma_{n,max} = 0$, and the left side of Eq. 10.19 reduces to $\Delta\gamma_{max}/2$). As a first approximation or if test data are not available, $k \approx 1$.

Mean or residual stress effects on fatigue life in this model are accounted for by the maximum normal stress term, since

$$\sigma_{n,max} = \sigma_{n,a} + \sigma_{n,m} \tag{10.20}$$

where $\sigma_{n,a}$ and $\sigma_{n,m}$ are the alternating normal and mean or residual normal stresses, respectively. Additional hardening resulting from nonproportional loading, as discussed in Section 10.2, is also incorporated by the maximum normal stress term since additional hardening results in an increase in the alternating normal stress, $\sigma_{n,a}$.

Equation 10.19 can be written in terms of shear strain–life properties obtained from fully reversed torsion tests (usually using thin-walled tube specimens) as

$$\frac{\Delta\gamma_{max}}{2}\left(1 + k\frac{\sigma_{n,max}}{S_y}\right) = \frac{\tau_f'}{G}(2N_f)^{b_0} + \gamma_f'\,(2N_f)^{c_0} \tag{10.21}$$

where G is the shear modulus, τ_f' is the shear fatigue strength coefficient, γ_f' is the shear fatigue ductility coefficient, and b_0 and c_0 are shear fatigue strength and shear fatigue ductility exponents, respectively. If these properties are not available, they can be estimated from uniaxial strain–life properties as $\tau_f' \approx \sigma_f'/\sqrt{3}$, $b_0 \approx b$, $\gamma_f' \approx \sqrt{3}\,\varepsilon_f'$, and $c_0 \approx c$. Equation 10.19 can also be expressed in terms of the uniaxial strain–life properties in Eq. 5.14 by equating the left side of Eq. 10.19 to C for fully reversed, uniaxial straining:

$$\frac{\Delta\gamma_{max}}{2}\left(1 + k\frac{\sigma_{n,max}}{S_y}\right) = \left[(1 + \nu_e)\frac{\sigma_f'}{E}(2N_f)^b + (1 + \nu_p)\,\varepsilon_f'\,(2N_f)^c\right]$$

$$\times\left[1 + k\frac{\sigma_f'}{2S_y}(2N_f)^b\right] \tag{10.22}$$

where ν_e and ν_p are elastic and plastic Poisson ratios, respectively. Therefore, once the left side of Eq. 10.21 or Eq. 10.22 has been calculated for the multiaxial loading condition, fatigue life can be obtained. Equations 10.21 and 10.22 usually result in close agreement if shear fatigue properties and the value of k are available for the material. If these properties are estimated using the above approximations, discrepancies between lives predicted from the two equations can exist, as illustrated by the example problem in Section 10.4.3.

Figure 10.4 illustrates the correlation of multiaxial fatigue data for Inconel 718 using thin-walled tube specimens, based on the Fatemi-Socie parameter [10]. The data shown represent a wide variety of constant amplitude loading conditions obtained from in-phase (proportional) axial-torsion tests with or without mean stress, out-of-phase (nonproportional) axial-torsion tests, and biaxial tension tests (using internal/external pressurization) with or without mean stress. Park and Nelson [4] have evaluated this model for a variety of metallic alloys including mild steel, high-strength and high-temperature steel alloys, stainless steels, and several aluminum alloys under a wide variety of proportional and nonproportional constant amplitude loading conditions. The value of k for these materials ranged between 0 and 2. Good correlations of multiaxial fatigue lives are reported for the Fatemi-Socie model, as well as for the Park-Nelson energy-based model mentioned in Section 10.4.1.

According to the Fatemi-Socie model, cyclic shear strain must be present for fatigue damage to occur (Eq. 10.19 indicates no damage if $\Delta\gamma_{max}/2 = 0$). Therefore, this model is suitable for materials where the majority of the fatigue life is spent in crack nucleation and small crack growth along the maximum shear planes. This situation is representative of many metals and alloys. For some materials such as cast iron, however, crack nucleation and/or crack growth is along the maximum tensile stress or strain planes. In this case, the Smith-Watson-Topper parameter discussed in Section 5.6 can be used as the damage model, where the governing parameters are the maximum principal strain amplitude, ε_{a1}, and the maximum normal stress on the maximum principal strain amplitude plane, $\sigma_{n,max}$. The critical plane is then the plane with the largest value of $(\varepsilon_{a1}\sigma_{n,max})$. Similar to the Fatemi-Socie model, mean and residual stress effects and additional hardening due to nonproportional loading are incorporated through the maximum normal stress term.

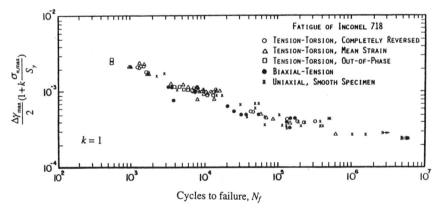

Figure 10.4 Correlation of Inconel 718 multiaxial fatigue data using the Fatemi-Socie parameter [10] (reprinted by permission of D. L. Morrow).

10.4.3 Example of Nonproportional Loading

Consider a smooth, thin-walled tube made of 1045 hot-rolled steel and sub-jected to cyclic 90° out-of-phase axial-torsion straining. Both axial and shear strains are fully reversed, with axial strain amplitude of $\varepsilon_{ax} = 0.0026$ and shear strain amplitude of $\gamma_{axy} = 0.0057$. We want to predict the fatigue life using the maximum shear strain theory and the Fatemi-Socie critical plane model.

The material has a yield strength of $S_y = 380$ MPa and the following strain–life properties [11]:

$$\text{Torsion: } G = 80 \text{ GPa}, \ \tau_f' = 505 \text{ MPa}, b_0 = -0.097, \gamma_f' = 0.413, c_0 = -0.445$$
$$\text{Uniaxial: } E = 205 \text{ GPa}, \sigma_f' = 948 \text{ MPa}, b = -0.092, \ \varepsilon_f' = 0.26, \ c = -0.445$$

The constant k in the Fatemi-Socie parameter for this material is 0.6, as found by fitting fully reversed uniaxial and torsional fatigue data.

The applied strain histories are given by the following equations:

$$\varepsilon_x = \varepsilon_{ax} \sin \omega t \quad \text{and} \quad \gamma_{xy} = \gamma_{axy} \sin (\omega t + 90°)$$

where the x axis is the tube's longitudinal axis. The applied strain path (γ_{xy}-ε_x plot), the resulting measured axial and torsional hysteresis loops, and the stress path (τ_{xy}-σ_x plot) are shown in Fig. 10.5 [12]. Variation of the applied strains (ε_x and γ_{xy}) and stable measured stresses (σ_x and τ_{xy}) during a cycle are shown in Fig. 10.6a. Note that due to the nonproportional nature of the loading, the hysteresis loops shown in Figs. 10.5b and 10.5c are rounded. As a result, there is a phase lag between the maximum values of axial strain and axial stress, and between the maximum values of shear strain and shear stress during a cycle, which can be observed in Fig. 10.6a.

To evaluate fatigue damage, it is important to realize that the maximum damage does not necessarily occur on the axial loading plane (i.e., x-plane). Instead, damage on different planes must be evaluated in order to determine the critical plane(s) with the maximum value of the damage parameter. Stress and strain transformation equations can be used to identify the critical plane(s). For example, shear strain on an arbitrary plane, γ_θ, is given by

$$\gamma_\theta = (\varepsilon_x - \varepsilon_y) \sin 2\theta - \gamma_{xy} \cos 2\theta$$

where θ is the angle between this arbitrary plane and the x-plane, ε_x and γ_{xy} are the applied strains, and ε_y is the transverse normal strain due to Poisson's effect calculated from

$$\varepsilon_y = -v\varepsilon_x = -\left[v_e \, (\varepsilon_x)_{\text{elastic}} + v_p \, (\varepsilon_x)_{\text{plastic}}\right] = -[0.3 \, \sigma_x/E + 0.5(\varepsilon_x - \sigma_x/E)]$$
$$= 0.2 \, \sigma_x/E - 0.5 \, \varepsilon_x$$

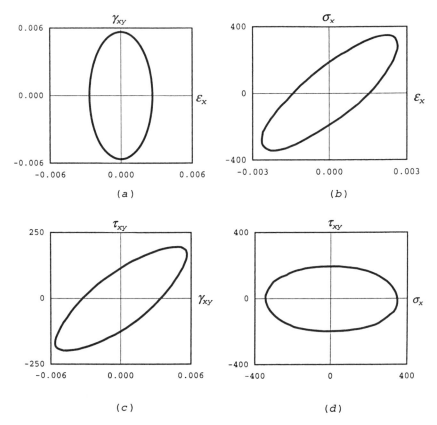

Figure 10.5 Deformation response from 90° out-of-phase axial-torsion straining [12]. (*a*) Applied strain path. (*b*) Stable axial hysteresis loop. (*c*) Stable torsion hysteresis loop. (*d*) Stress path (reprinted with permission of the Society of Automotive Engineers).

where elastic and plastic Poisson ratios are 0.3 and 0.5, respectively. Variation of the shear strain amplitude, $\Delta\gamma/2$, as a function of plane orientation angle, θ, is shown in Fig. 10.6*b*. As can be seen, the planes oriented at 0° and 90° (i.e., planes oriented perpendicular and parallel to the tube axis) are the planes experiencing the maximum shear strain range $\Delta\gamma_{max} = 0.0113$. The maximum shear strain amplitude can be related to the fatigue life in an analogous manner to the uniaxial strain–life relationship (Eq. 5.14), but with torsional strain–life properties as follows:

$$\frac{\Delta\gamma_{max}}{2} = \frac{\tau_f'}{G}(2N_f)^{b_0} + \gamma_f'(2N_f)^{c_0} = \frac{505}{80\,000}(2N_f)^{-0.097} + 0.413\,(2N_f)^{-0.445}$$

Substituting $\Delta\gamma_{max}/2 = 0.0057$ in this equation results in $N_f = 23\,000$ cycles.

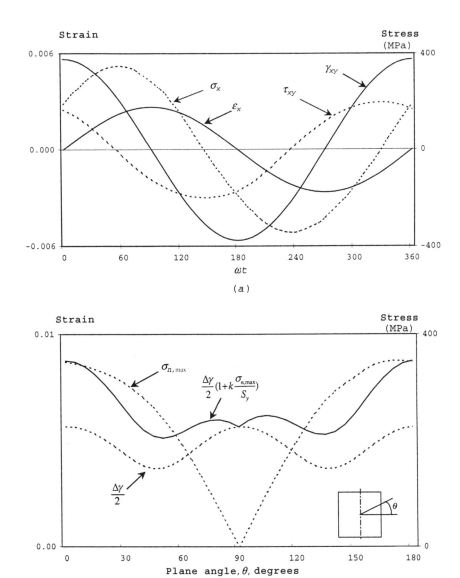

Figure 10.6 Stress and strain variations in a cycle and with plane orientation for 90° out-of-phase axial-torsion straining. (*a*) Stress and strain variations during one cycle. (*b*) Variation of shear strain range, maximum normal stress, and the Fatemi-Socie parameter with plane orientation.

To evaluate the Fatemi-Socie parameter, the normal stress on the maximum shear strain amplitude plane, $\sigma_{n,max}$, must also be obtained. Variation of the normal stress as a function of plane orientation angle can be obtained by using the stress transformation equation and is shown in Fig. 10.6b. Spreadsheet calculations using digital computers are usually used to calculate different stress or strain quantities on different planes and at different times in the loading cycle. It can be seen that for this example, the maximum normal stress, $\sigma_{n,max}$, occurs on the x-plane ($\theta = 0°$) with $\sigma_{n,max} = 350$ MPa. Therefore, the critical plane is predicted to be the x-plane (perpendicular to the tube axis), since this plane experiences the largest values of both $\Delta\gamma$ and σ_n. Experimental results (i.e., cracking observations) for the material and loading in this example agree with this prediction [8]. Variation of the Fatemi-Socie parameter as a function of plane orientation angle is shown in Fig. 10.6b. To calculate fatigue life, Eq. 10.21 with the given values of k and S_y can be used:

$$\frac{\Delta\gamma_{max}}{2}\left[1 + 0.6\,\frac{\sigma_{n,max}}{380}\right] = \frac{505}{80\,000}\,(2N_f)^{-0.097} + 0.413\,(2N_f)^{-0.445}$$

Substituting $\Delta\gamma_{max}/2 = 0.0057$ and $\sigma_{n,max} = 350$ MPa from Fig. 10.6b in this equation results in $N_f = 6000$ cycles. Experimental life for the material and the loading condition described in this example problem was 5300 cycles [7], which is in close agreement with the prediction. If the fatigue properties from torsion tests and the value of k were not available, Eq. 10.22 with uniaxial fatigue properties could be used and the value of k can be approximated to be unity. In this case, the predicted life will be 9500 cycles, which is still reasonably close to the experimental life. The maximum shear strain theory overestimates the fatigue life (i.e., $N_f = 23\,000$ according to this theory), as it ignores the contribution of the normal stress to the damage process.

Average experimental fatigue life for in-phase (proportional) axial-torsion loading of the material in this example at the same strain amplitudes of $\varepsilon_{ax} = 0.0026$ and $\gamma_{axy} = 0.0057$ was observed to be 18 500 cycles [13]. This indicates a substantial reduction in fatigue life, down to 5300 cycles, as a result of applying the loads 90° out-of-phase. In general, fatigue life for in-phase versus out-of-phase loading can be compared based either on the magnitudes of the applied stress/strain amplitudes or on the magnitude of the maximum equivalent stress/strain amplitude. In other words, one method of comparing fatigue life can be based on the amplitudes of individual components of applied stress or strain, as was done in this example problem. A second method is to first combine the applied stress or strain components into an equivalent stress or strain and then use the amplitude of this equivalent stress or strain as the basis for comparison. If comparison is based on equal maximum equivalent stress/strain amplitude, out-of-phase loading is always equally or more damaging than in-phase loading. If comparison is based on equal applied stress/strain amplitudes, out-of-phase loading can be either less or more damaging than in-phase loading, depending on the material and its degree of additional cyclic hardening from out-of-phase loading. At high loads or in the low-cycle regime, where strains are mainly plastic, the additional cyclic hardening is higher that at lower loads or in the high-cycle regime, where strains are mainly elastic.

10.5 FRACTURE MECHANICS MODELS FOR FATIGUE CRACK GROWTH

Fracture mechanics is used to characterize the growth of cracks where multi-axial stress states can result in mixed-mode crack growth. A characteristic of mixed-mode fatigue cracks is that they can grow in a non-self similar manner, i.e., the crack changes its direction of growth. Therefore, under mixed-mode loading conditions, both crack growth direction and crack growth rate are important.

Different combinations of mixed-mode loading can exist from multiaxial loads. For example, in simple torsion of smooth shafts, surface cracks can form and grow in longitudinal and/or transverse directions along the maximum shear planes, where mixed-mode II and III exists along the crack front. Cracks can also form and grow along the ±45° angle to the axis of the shaft along planes of maximum principal stress where they grow in mode I. Plate components with edge or central cracks under in-plane biaxial tension or under three- or four-point bending and shear loading often produce mixed-mode I and II crack growth. In plate components, however, the mode I contribution often becomes dominant after a short period of crack growth.

Experimental observations suggest that the growth of cracks with small plastic zone size relative to the crack length depends mainly on the alternating stresses that pull the crack faces open (i.e., stresses normal to the crack plane). The other alternating stresses parallel to the crack do not often significantly affect the crack growth rate. However, if crack tip plasticity is large, stress components parallel to the crack can also significantly affect the crack growth rate.

Several parameters have been used to correlate fatigue crack growth rates under mixed-mode conditions. These include equivalent stress intensity factors, equivalent strain intensity factors, strain energy density, and the J-integral. For example, an equivalent stress intensity factor range, ΔK_q, based on crack tip displacements was proposed by Tanaka [14] and is given by

$$\Delta K_q = \left(\Delta K_I^4 + 8\, \Delta K_{II}^4 + \frac{8\, \Delta K_{III}^4}{1-v} \right)^{0.25} \tag{10.23}$$

Another form of equivalent stress intensity factor is based on the equivalence of energy release rate, G, and the stress intensity factor, K, for nominal elastic loading, as described in Chapter 6. Adding the individual energy release rates for a planar crack under plane stress conditions results in

$$G = G_I + G_{II} + G_{III} = \frac{K_I^2}{E} + \frac{K_{II}^2}{E} + \frac{(1 + v) \, K_{III}^2}{E} \tag{10.24}$$

which leads to the following equivalent stress intensity factor range:

$$\Delta K_q = [\Delta K_I^2 + \Delta K_{II}^2 + (1 + v) \, \Delta K_{III}^2]^{0.5} \tag{10.25}$$

ΔK_q has been used for the mixed-mode loading condition in a Paris-type equation (Eq. 6.19) to obtain the crack growth rate, da/dN, or cycles to failure through integration.

In situations in which the plastic zone size is large compared to the crack length, such as in low-cycle fatigue or in growth of small cracks, equivalent strain intensity factors or the J-integral may be more suitable parameters to use. An equivalent strain intensity factor based on the Fatemi-Socie model (Eq. 10.19) has been shown to be successful in correlating small crack growth data from a variety of proportional and nonproportional axial-torsion tests with thin-walled tube specimens made of Inconel 718 and 1045 steels [15]. Detailed discussion of mixed-mode crack growth behavior can be found in [1,16].

10.6   NOTCH EFFECTS AND VARIABLE AMPLITUDE LOADING

Notches cannot be avoided in many components and structures. The severity of a notch is often characterized by the stress concentration factor, K_t. However, as discussed in Chapter 7, the stress concentration factor depends not only on the notch geometry, but also on the type of loading. It should also be recognized that the stress state at the root of the notch is often multiaxial even under uniaxial loading conditions. For example, in axial loading of a circumferentially notched bar, both axial and tangential stresses exist at the root of the notch. The tangential stress component results from the notch constraint in the transverse direction. In the S–N approach, fatigue strength can be divided by the fatigue notch factor, K_f. If the theoretical stress concentration factors, K_t's, for multiaxial loading differ too greatly for different principal directions, each nominal alternating stress component can be multiplied by its corresponding fatigue notch factor, K_f. This can be done because the S–N approach is used mainly for elastic behavior and, therefore, superposition can be used to estimate notch stresses from combined multiaxial loads. For example, for combined bending and torsion of a notched shaft and in the absence of mean stresses, an equivalent stress, based on the octahedral shear stress, can be computed as

$$S_{qa} = \sqrt{(K_{fB} S_B)^2 + 3 (K_{fT} S_T)^2} \qquad (10.26)$$

where S_B and S_T are nominal bending and torsion stresses, respectively, and K_{fB} and K_{fT} are fatigue notch factors in bending and torsion, respectively. Notch effects in the ε–N approach for multiaxial loading when notch root plastic deformation exists are more complex and often require the use of cyclic plasticity models. For this case, Neuber's rule is often generalized to the multiaxial loading situation by using equivalent stresses and strains. Detailed discussion of notch effects in the ε–N approach for multiaxial proportional and nonproportional stress states is provided in [1].

Multiaxial fatigue analysis for variable amplitude loading is quite complex, particularly when the applied loads are out-of-phase or nonproportional. Nevertheless, excellent axles and crankshafts subject to complex multiaxial fatigue are produced based on experience. Similar to uniaxial loading, the two main steps in cumulative damage analysis from multiaxial variable amplitude loading are identification or definition of a cycle and evaluation of damage for each identified or counted cycle. Several methods for predicting cumulative damage in multiaxial fatigue have been developed, such as those of Bannantine and Socie [17] and Wang and Brown [18]. Bannantine and Socie's method is based on the critical plane approach, in which cycles on various planes are counted and a search routine is then used to identify the plane experiencing the most damage (i.e., the critical plane). It should be noted, however, that the definition of the damage plane for the critical plane model used may not be unique during variable amplitude cyclic loading. For example, for the Fatemi-Socie critical plane model discussed in Section 10.4.2, the critical plane can be defined as either the maximum shear plane with the largest normal stress or the plane which experiences the maximum value of the damage parameter (i.e., the left side of Eq. 10.19). For constant amplitude proportional and nonproportional loading, the two definitions result in the same plane. This is illustrated in Fig. 10.6b, where the $0°$ plane is shown to have the maximum shear strain range, $\Delta\gamma_{max}$, with the largest normal stress, $\sigma_{n,max}$, as well as the maximum value of the damage parameter (shown by the solid curve). For complex variable amplitude loading, the two planes may be different. In this case, the plane having the largest value of the damage parameter should be chosen as the critical plane. Wang and Brown's method is based on counting cycles using an equivalent strain. Once cycles have been counted using one of the aforementioned methods, damage is accumulated for each cycle according to the Palmgren-Miner linear damage rule discussed in Chapter 9. Additional discussion of multiaxial variable ampltitude loading can be found in [1].

10.7 SUMMARY

Multiaxial states of stress are very common in engineering components and structures. Multiaxial cyclic loading can be categorized as proportional where the orientation of the principal axes remains fixed or as nonproportional where the principal directions of the alternating stresses change orientation during cycling. Nonproportional cyclic loading involves greater difficulties and often produces additional cyclic hardening compared to proportional loading. Multiaxial fatigue analysis for situations in which significant plasticity exists often requires the use of a cyclic plasticity model consisting of a yield function, a flow rule, and a hardening rule.

Stress-based, strain-based, energy-based, and critical plane approaches have been used for life prediction under multiaxial stress states. The most common stress-based methods are those based on equivalent octahedral shear stress and the Sines method. Equivalent nominal stress amplitude based on the octahedral shear stress criterion and equivalent mean nominal stress based on hydrostatic stress are obtained from

$$S_{qa} = \frac{1}{\sqrt{2}}\sqrt{(S_{a1} - S_{a2})^2 + (S_{a2} - S_{a3})^2 + (S_{a3} - S_{a1})^2} \qquad (10.9)$$

$$S_{qm} = S_{m1} + S_{m2} + S_{m3} \qquad (10.11)$$

The S–N approach is limited primarily to elastic and proportional loading situations. Strain-based methods are based on equivalent strain. They may be suitable for inelastic loading, but their use is still limited to proportional loading conditions. Energy-based and critical plane approaches are more general approaches and are suitable for both proportional and nonproportional multiaxial loading. Critical plane approaches reflect the physical nature of fatigue damage and can predict both fatigue life and the orientation of the failure plane. A common critical plane approach is the Fatemi-Socie model:

$$\frac{\Delta\gamma_{max}}{2}\left(1 + k\frac{\sigma_{n,max}}{S_y}\right) = \frac{\tau_f'}{G}(2N_f)^{b_0} + \gamma_f'(2N_f)^{c_0} \qquad (10.21)$$

Fracture mechanics models are used to characterize the growth of macrocracks in multiaxial fatigue, where mixed-mode crack growth often exists. Equivalent stress intensity factors have been used to relate mixed-mode crack growth data to mode I data from uniaxial loading. Notches and variable amplitude loading introduce additional complexities, particularly for nonproportional loading. Several methods to incorporate these effects have been developed.

10.8   DOS AND DON'TS IN DESIGN

1. Don't ignore the presence of multiaxial stress states, as they can significantly affect fatigue behavior. The state of stress at the root of a notch is usually multiaxial even under uniaxial loading conditions.

2. Do determine whether the alternating stresses or strains have fixed principal directions. If so, the loading is proportional and fairly simple methods for life estimation can be used.

3. Don't ignore the effects of nonproportional cyclic loading, since it can produce additional cyclic hardening and often results in a shorter fatigue life compared to proportional loading.

4. Do recognize that application of equivalent stress and equivalent strain approaches to multiaxial fatigue is limited to simple proportional or in-phase loading. Other methods such as the critical plane approach are more suitable for more complex nonproportional or out-of-phase loading.

5. Do determine whether the fatigue damage mechanism for the given material and loading is dominated by shear or by tensile cracking. Different fatigue damage models apply to each case.

REFERENCES

1. D. F. Socie and G. B. Marquis, *Multiaxial Fatigue,* Society of Automotive Engineers, Warrendale, PA, 1999.

2. G. Sines, "Behavior of Metals Under Complex Static and Alternating Stresses," *Metal Fatigue,* G. Sines and J. L. Waisman, eds., McGraw-Hill Book Co., New York, 1959, p. 145.

3. F. Ellyin, *Fatigue Damage, Crack Growth and Life Prediction,* Chapman and Hall, London, 1997.

4. J. Park and D. V. Nelson, "Evaluation of an Energy-Based Approach and a Critical Plane Approach for Predicting Multiaxial Fatigue Life," *Int. J. Fatigue,* Vol. 22, No. 1, 2000, p. 23.

5. W. N. Findley, "A Theory for the Effect for Mean Stress on Fatigue of Metals Under Combined Torsion and Axial Load or Bending," *Trans. ASME, J. Eng. Industry,* Vol. 81, 1959, p. 301.

6. M. W. Brown and K. J. Miller, "A Theory for Fatigue Under Multiaxial Stress–Strain Conditions," *Proceedings of the Institute of Mechanical Engineers,* Vol. 187, 1973, p. 745.

7. A. Fatemi and D. F. Socie, "A Critical Plane Approach to Multiaxial Fatigue Damage Including Out-of-Phase Loading," *Fatigue Fract. Eng. Mater. Struct.,* Vol. 11, No. 3, 1988, p. 149.

8. D. F. Socie, "Critical Plane Approaches for Multiaxial Fatigue Damage Assessment," *Advances in Multiaxial Fatigue,* ASTM STP 1191, D. L. McDowell and R. Ellis, eds., ASTM, West Conshohocken, PA, 1993, p. 7.

9. P. Kurath and A. Fatemi, "Cracking Mechanisms for Mean Stress/Strain Low-Cycle Multiaxial Loadings," *Quantitative Methods in Fractography,* ASTM STP 1085, B. M. Strauss and S. K. Putatunda, eds., ASTM, West Conshohocken, PA, 1990, p. 123.

10. D. L. Morrow, *Biaxial-Tension Fatigue of Inconel 718,* Ph.D. diss., University of Illinois, Urbana, 1988.

11. A. Fatemi and P. Kurath, "Multiaxial Fatigue Life Predictions Under the Influence of Mean Stresses," *Trans. ASME, J. Eng. Mater. Tech.,* Vol. 110, 1988, p. 380.

12. A. Fatemi and R. I. Stephens, "Cyclic Deformation of 1045 Steel Under In-Phase and 90 Deg Out-of-Phase Axial-Torsional Loading Conditions," *Multiaxial Fatigue: Analysis and Experiments,* SAE AE-14, G. E. Leese and D. F. Socie, eds., SAE, Warrendale, PA, 1989, p. 139.

13. A. Fatemi and R. I. Stephens, "Biaxial Fatigue of 1045 Steel Under In-Phase and 90 Deg Out-of-Phase Loading Conditions," *Multiaxial Fatigue: Analysis and Experiments,* SAE AE-14, G. E. Leese and D. F. Socie, eds., SAE, Warrendale, PA, 1989, p. 121.

14. K. Tanaka, "Fatigue Propagation from a Crack Inclined to the Cyclic Tensile Axis," *Eng. Fract. Mech.,* Vol. 6, 1974, p. 493.

15. S. C. Reddy and A. Fatemi, "Small Crack Growth in Multiaxial Fatigue," *Advances in Fatigue Lifetime Predictive Techniques,* ASTM STP 1122, M. R. Mitchell and R. W. Landgraf, eds., ASTM, West Conshohocken, PA, 1992, p. 276.

16. J. Qian and A. Fatemi, "Mixed Mode Fatigue Crack Growth: A Literature Survey," *Eng. Fract. Mech.,* Vol. 55, No. 6, 1996, p. 969.

17. J. A. Bannantine and D. F. Socie, "A Variable Amplitude Multiaxial Fatigue Life Prediction Method," *Fatigue Under Biaxial and Multiaxial Loading,* ESIS 10, Mechanical Engineering Publications, London, 1991, p. 35.

18. C. H. Wang and M. W. Brown, "Life Prediction Techniques for Variable Amplitude Multiaxial Fatigue—Part I: Theories," *Trans. ASME, J. Eng. Mater. Tech.* Vol. 118, 1996, p. 367.

PROBLEMS

1. Three common stress states are simple uniaxial tension, simple torsion, and equibiaxial tension. (*a*) Show the principal stress directions and the planes of maximum stress or strain and maximum shear stress or strain. (*b*) Plot the variation of the normal and shearing stresses as a function of plane orientation for the case of simple uniaxial tension.

2. An AISI 8640 steel shaft is Q&T to $S_u = 1400$ MPa (200 ksi) and $S_y = 1250$ MPa (180 ksi) with $RA = 10$ percent. A step with a fillet radius of 5 mm exists in the shaft that causes $K_t = 2.1$ in torsion. If the smaller shaft diameter is 50 mm, what maximum pulsating torque ($R = 0$) can be applied over a period of several years and last for at least 1 million cycles?

3. Repeat Problem 2 if the shaft is shot-peened, causing compressive residual stresses at the step equal to 700 MPa in both the longitudinal and transverse directions.

4. In Problems 2 and 3, what maximum torque can be applied for 50 000 cycles?

5. A 25 mm diameter 2024-T3 aluminum shaft is subject to constant amplitude in-phase bending and torsion. The bending moment varies from 0 to 144 N.m, and the torque varies from -100 to $+200$ N.m. (*a*) What is the safety factor against static yielding? (*b*) Will the shaft withstand 10 million cycles of this combined loading?

6. For Problem 5, how much could the bending moment be increased and still have a life of 50 000 cycles of combined loading?

7. Prove that the stress path for the 90° out-of-phase stressing in Fig. 10.1*d* is elliptical.

8. A smooth circular shaft is made of 1038 normalized steel and subject to completely reversed, constant amplitude torque. The torsion shear strain on the surface of the shaft is measured to be ± 0.005. Determine the expected life according to (*a*) the maximum shear strain theory and (*b*) the octahedral shear strain theory.

9. Constant amplitude axial and torsion strain histories applied to a thin-walled tube are trapezoidal, as shown in part (*a*) of the following figure. (*a*) Plot the strain path (γ_{xy} versus ε_x) for this load history. (*b*) Is this proportional or nonproportional loading? (*c*) The tube is made of 1045 hot-rolled steel with properties given in Section 10.4.3 and the resulting stress response shown in part (*b*) of the figure. What is the expected fatigue life according to the maximum shear strain theory? (*d*) Repeat part (*c*) using the Fatemi-Socie critical plane model.

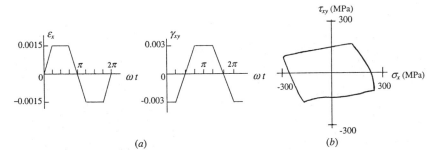

(*a*)   (*b*)

CHAPTER 11

ENVIRONMENTAL EFFECTS

Environmental effects on fatigue of metals may be more severe than sharp stress concentrations or almost harmless. Quantitative fatigue life predictions are often not possible because of the many interacting factors that influence environmental fatigue behavior and because of lack of significant data. For example, in corrosion fatigue, frequency or hold time effects can be quite substantial, while in noncorrosive environments these are usually second order concerns. Fretting fatigue is often missed in full-scale testing due to accelerated testing. At elevated temperatures, mean stress and thermomechanical effects are complex because of the interaction between creep, fatigue, and environment. The stress intensity factor, K, also has more limitations at elevated temperature because of appreciable plasticity; thus, additional elastic-plastic parameters have been introduced. Substantial reduction in fracture toughness can occur at low temperatures, which reduces critical crack sizes at fracture. Irradiation can reduce both fatigue resistance and fracture toughness. Despite these severe difficulties, fatigue design with environmental considerations must be accomplished. This places great emphasis on better understanding of material/environment/stress interaction, real-life product testing, inspection, service history analysis, experience, and accelerated acquisition of experience. The above environmental topics of corrosion, fretting, and low and high temperatures, as well as the effects of irradiation on fatigue behavior, are discussed in this chapter. These topics are of great importance to many engineering fields including aerospace, marine, ground vehicles, bioimplants, and nuclear systems.

11.1   CORROSION FATIGUE

There are many environments that affect fatigue behavior; however, most engineering components and structures interact with air, moisture or humidity, water, and salt water. Thus, most of the following discussion involves these environments. The principles, however, are also applicable to other corrosive environments.

"Corrosion fatigue" refers to the joint interaction of corrosive environment and repeated stressing. The synergistic combination of both acting together is more detrimental than that of either one acting separately. That is, repeated stressing accelerates the corrosive action, and the corrosive action accelerates the mechanical fatigue mechanisms. Under static loads, corrosive environments may also be detrimental, particularly in higher-strength alloys. Environmental-assisted fracture under static loading has been called "stress corrosion cracking" (SCC). More recently, the term "environment-assisted cracking" (EAC) has been adopted by ASTM because it is a more general term that includes SCC and other forms of environmentally influenced cracking. We briefly consider SCC (EAC) before discussing corrosion fatigue.

11.1.1   Stress Corrosion Cracking/Environment-Assisted Cracking

Let us consider a number of identical precracked test specimens (e.g., Fig. 4.5g–j) or components made from the same medium or high-strength alloy. Subject each specimen to a different initial force that causes a different initial stress intensity factor. Keep those initial forces applied to each specimen until fracture or test termination. Simultaneously expose the specimens or components to a corrosive environment. Depending on the initial stress intensity factor, the cracks will grow at different rates. The crack growth mechanisms can be transcrystalline cleavage or microvoid coalescense, intercrystalline, or combinations. The time to fracture can be monitored, and results similar to those shown in Fig. 11.1 for plane strain conditions are obtained. Here, at a very short life, the initial stress intensity factor, K_{Ii}, is essentially the same as the fracture toughness, K_{Ic}, obtained without the corrosive environment. As K_{Ii} is reduced, the time to fracture increases. A limiting threshold value, K_{ISCC} (K_{IEAC}), is finally obtained. This is the plane strain stress intensity factor below which a crack is not observed to grow in the specific environment. It should be noted that fracture would not occur at K_{Ii} values less than K_{Ic} if the specimens or components were not exposed to the corrosive environment. The subscripts refer to plane strain conditions (usually mode I) and to SCC or EAC. The subscript I is omitted if plane stress conditions exist. Values of 10^3 to 10^4 hours have been recommended in ASTM Standard E1681 [1] as minimum termination times for determining threshold values K_{ISCC} (K_{IEAC}) or K_{SCC} (K_{EAC}) using constant load tests. With constant crack opening displacement tests, the time of testing can be shorter. K_{ISCC} (K_{IEAC}) or K_{SCC}

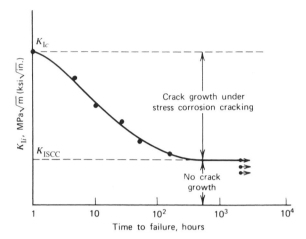

Figure 11.1 Determination of K_{ISCC} (K_{IEAC}) with precracked constant load specimens.

(K_{EAC}) have become common design properties, and substantial data have been tabulated in the *Damage Tolerant Design Handbook* [2] and the *Atlas of Stress-Corrosion and Corrosion Fatigue Curves* [3]. Values of K_{ISCC} (K_{IEAC}) range from approximately 10 to 100 percent of K_{Ic}, indicating a wide range of sensitivity to corrosive environments. Under plane stress conditions, threshold values, K_{SCC} (K_{EAC}), will depend upon thickness and have higher values than plane strain values.

Figure 11.2 shows typical values of K_{ISCC} (K_{IEAC}) as a function of yield strength for 4340 steel immersed in a salt water environment [4]. As the

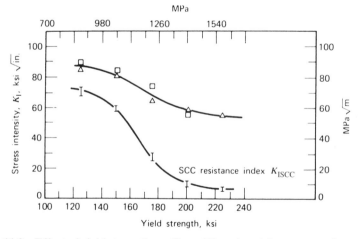

Figure 11.2 Effect of yield strength on K_{ISCC} (K_{IEAC}) and fracture toughness in 4340 steel tested in flowing seawater (Key West, FL) [4] (reprinted with permission of the National Association of Corrosion Engineers). (□) K_{IX} (dry brake), (△) $K_{I\delta}$.

yield strength increased, K_{ISCC} (K_{IEAC}) decreased even more rapidly than an approximate fracture toughness index labeled K_{Ix} or $K_{I\delta}$ for the noncorrosive environment. Thus, in general, as the yield strength increases, the susceptibility to stress corrosion cracking (environment assisted cracking) becomes more pronounced. Lower ratios of K_{ISCC}/K_{Ic} (K_{IEAC}/K_{Ic}) usually occur with the higher yield strength materials.

The above discussion on stress corrosion cracking is very much related to corrosion fatigue. It implies that repeated loads are not needed for cracks to extend if applied stress intensity factors are above the applicable K_{SCC} (K_{EAC}) or K_{ISCC} (K_{IEAC}). Thus a complex interaction exists between static and repeated loads in the presence of corrosive environments. Corrosion fatigue cracks can grow at stress intensity factors below the applicable K_{SCC} (K_{EAC}) or K_{ISCC} (K_{IEAC}).

11.1.2 Stress–Life (*S*–*N*) Behavior

Since most fatigue data are obtained in laboratory air, we have primarily only corrosion fatigue data at our disposal because the air environment can certainly be corrosive, with relative humidity or moisture content and oxygen being the two major influencing factors. However, laboratory air fatigue data are often used as a reference for comparing other environmental effects on fatigue.

Figure 11.3 schematically shows typical constant amplitude *S*–*N* diagrams obtained under four key room temperature environmental conditions. The relative fatigue behavior shown is very realistic. If air tests are taken as the

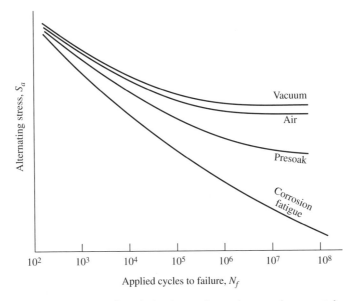

Figure 11.3   Relative *S*–*N* fatigue behavior under various environmental conditions.

reference, it is seen that tests in a vacuum environment can have a small beneficial effect primarily at long fatigue life. This benefit, however, depends on the material and the reference air environment. High-humidity air can be quite detrimental to fatigue behavior at long life. This is particularly true with aluminum alloys. Presoaking fatigue specimens in a corrosive liquid followed by testing in air often causes a detrimental effect primarily at long life. However, the combination of simultaneous environment and repeated stressing causes the most drastic decrease in long-life fatigue strength, as shown in Fig. 11.3. The relatively flat long-life S–N behavior that can occur in air or vacuum is eliminated under corrosion fatigue conditions. The long-life $R = -1$ corrosion fatigue strengths in various environments can vary from approximately 10 to 100 percent of air fatigue strengths, which is approximately 3 to 40 percent of the ultimate tensile strength. For some materials, air fatigue properties can even be improved by some environments as a result of surface protection from oxygen and water vapor. At short lives, all four test conditions shown in Fig. 11.3 tend to converge. This is primarily due to insufficient time for corrosion to be influential.

The mechanisms of corrosion fatigue are very complex. Corrosion fatigue is an electrochemical process dependent on the environment/material/stressing interaction. Pits with dimensions of 1 mm or less which act as micro or mini stress raisers for cracks to nucleate more readily can form very early in life. The preferential sites for these pits to form are along slip bands, grain boundaries, and inclusions and at ruptured oxide films. A greater number of surface fatigue cracks usually occur under corrosive fatigue conditions, and these can grow transcrystalline, intercrystalline, or both. As the cracks get larger they also tend to grow in a mode I manner, perpendicular to the maximum tensile stress. Fracture surfaces are often discolored in the fatigue crack growth region. Very little loss of material due to corrosion occurs, and thus loss of material is not a major contributing factor. Both crack nucleation life and crack growth life are reduced under corrosion fatigue conditions. Small cracks, as discussed in Sections 6.8 and 7.4, have a dominant role in corrosion fatigue. Frequency effects are important and alter fatigue behavior as a result of the time-dependent nature of corrosion fatigue. Lower frequency and longer hold times at peak loads cause decreased corrosion fatigue resistance. Only at very short lives or at high crack growth rates are frequency and hold time effects small. However, under these conditions, air and corrosion fatigue behavior in general are similar. Corrosion fatigue also depends on how the corrosive environment is applied. For example, specimens or components submerged in fresh or salt water have better corrosion fatigue resistance than those subjected to a water spray, drip, or wick, or those submerged in continuously aerated water. This is due to the great importance of oxygen and the formation of oxide films in the corrosion fatigue process.

The above description only partially explains the results shown in Fig. 11.3. The vacuum environment is beneficial because of the elimination of water vapor and oxygen. Presoaking can cause pits to form, and these pits then act as stress raisers. Thus, presoaking can cause fatigue behavior similar to that for notched specimens. The low corrosion fatigue curve is the result of the synergistic interaction among environment, material, and stressing. Protective oxide films that can form at pits and crack tips only under corrosive conditions can be continually broken by repeated stresses such that new fresh surfaces are continually exposed to the corrosive environment. Deep cracks in large, thick components may not be exposed to this environmental interaction and may grow in a manner similar to the growth of cracks in the air environment.

McAdam [5] in the late 1920s carried out a comprehensive air and corrosion fatigue test program with carbon, low-alloy, and chromium steels using rotating bending specimens subjected to fresh water spray. The results of these tests are summarized in Fig. 11.4 [6], where fatigue strengths in air and water at 2×10^7 cycles are plotted against ultimate tensile strength. The results show the tendency of unnotched fatigue strengths in air to increase with ultimate tensile strength. The fatigue strengths in water for the carbon and low-alloy steels, however, are almost independent of the ultimate tensile strength. These quantitative data substantiate the qualitative schematic behavior shown in Fig. 4.9. All the corrosion fatigue strengths for these annealed, quenched and tempered carbon and low-alloy steels were between 85 and 210 MPa (12 and 30 ksi), yet their ultimate tensile strengths varied from 275 to 1720 MPa (40 to 250 ksi). These very low corrosion fatigue strengths are one of the most often overlooked aspects in fatigue design. The corrosion fatigue strengths for steels containing 5 percent or more chromium were much better, but they were still less than the air fatigue strengths. The above results were obtained at a test frequency of 24 Hz, which is very high. At lower frequencies, where more time is available, these values would be even lower.

Mean stresses and residual stresses under corrosion fatigue conditions behave in a similar manner as in air [3]. That is, tensile mean/residual stresses are

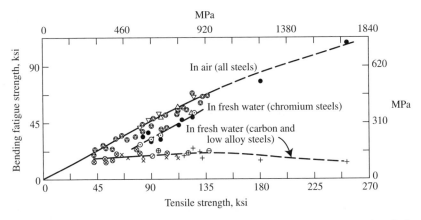

Figure 11.4 Influence of ultimate tensile strength on fresh water corrosion fatigue limits [5] (reprinted with permission of Pergamon Press Ltd.).

often detrimental and compressive mean/residual stresses are often beneficial. Corrosion fatigue behavior can be similar to that shown in the Haigh diagram of Fig. 7.10. However, significantly less mean stress corrosion fatigue data are available for the many material/environment conditions compared to inert or air conditions.

Table A.5 includes corrosion fatigue strengths for selected engineering alloys subjected to water or salt water environments [6]. These data only serve to illustrate corrosion fatigue effects and should not be used as design values. In general, salt water is more detrimental than fresh water, but variations in water chemistry can affect fatigue behavior. Fresh water and salt water were detrimental to fatigue strengths in all alloys listed in Table A.5 except for copper and phosphor bronze. Detrimental effects are shown for steels, brass, aluminum, magnesium, and nickel. Additional data are available [3,6]. In general, those materials with the best corrosion resistance in a specific environment also had the best corrosion fatigue resistance. Thus, a key to determining good fatigue design is to look for high corrosion resistance for a given material and environment.

11.1.3 Strain–Life (ε–N) Behavior

ε–N testing in a corrosive environment, such as fresh water or salt water, is much more difficult than S–N corrosion fatigue testing because a transducer such as a strain gage or extensometer is usually needed to control and measure strains. The transducer must be isolated electrically and mechanically from the corrosive environment, which is somewhat difficult to achieve. Because of experimental difficulties, much less ε–N corrosion fatigue information exists compared to that for S–N corrosion fatigue. In addition, a significant portion of ε–N curves involve low-cycle fatigue where less corrosion fatigue influence is expected, depending upon the frequency.

Bernstein and Loeby [7] investigated 3.5 percent salt water corrosion fatigue behavior under strain control for 2024-T4 aluminum, 1045 normalized steel, and 304 stainless steel. They used two levels of strain range, 1.2 and 1.7 percent, with low-cycle fatigue lives between 500 and 5 000 cycles. Smooth, cylindrical axial test specimens were tested in aerated salt water. Both elastic and plastic strains occurred with the two strain ranges. Tests were performed in air and salt water at 20 cpm (0.33 Hz) and in salt water at a lower frequency between 0.1 and 1 cpm (0.00167 and 0.0167 Hz). Thus, both corrosive environment and corrosive frequency effects were evaluated. Under the low-cycle fatigue strain-controlled tests at 20 cpm, the 3.5 percent salt water fatigue life for the three metals was reduced by factors between 1.6 and 2.4. At the lower frequencies, the fatigue life reduction factors were between 1.6 and 4. The 1045 steel and the 2024-T4 aluminum were affected by both the corrosive environment and the frequency, while the 304 stainless steel was affected only by the corrosive environment. The 1045 steel and 2024-T4 aluminum had continuous surface corrosion, pitting, and multiple fatigue cracking, while the

304 stainless steel had little surface corrosion, but showed pitting and multiple fatigue cracking. The cyclic stress–strain response at half-life for a given material was similar in both air and 3.5 percent salt water. This similarity is due to the cyclic stress–strain response being more dependent upon material bulk properties rather than the localized surface corrosion fatigue influence.

Stephens et al. [8] evaluated the ε–N behavior of AZ91E-T6 cast magnesium alloy in air and aerated 3.5 percent salt water using smooth, cylindrical axial specimens. They performed strain-controlled fatigue tests with $R = \varepsilon_{min}/\varepsilon_{max} = -1, 0,$ and -2. Corrosion fatigue tests were run between 0.5 and 2 Hz. Specimens were presoaked for 12 hours prior to testing. The $R = -1$ results are shown in Fig. 11.5, where it is seen that salt water significantly reduced the fatigue life by factors of 5 to 100, depending upon the strain range. Under $R = 0$ and -2 air and corrosive conditions, mean stress relaxation occurred at higher strain ranges where plastic strains were significant; thus, little mean stress effects occurred here. At lower strain ranges, where little plasticity existed, the mean stresses did not relax, and compressive mean stresses were beneficial and tensile mean stresses were detrimental. The tensile mean stress with the salt water environment had the largest effect on decreasing fatigue life. The cyclic stress–strain response was similar for both air and salt water environments, and the ε–N fatigue model (Eq. 5.14) was a good fit to both the air and salt water environment fatigue data.

11.1.4 Fatigue Crack Growth (*da/dN–ΔK*) Behavior

Fatigue crack growth plays an important role in corrosion fatigue. However, complete sigmoidal *da/dN–ΔK* behavior is difficult to determine due to the

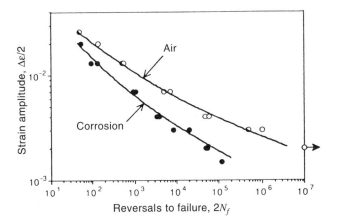

Figure 11.5 Strain–life fatigue behavior for a cast magnesium alloy, $R = -1$, in air and a 3.5 percent salt water environment [8] (reprinted with permission of the Society of Automotive Engineers, Inc.).

great expense and time needed to obtain useful low fatigue crack growth rates. Normal frequencies of 5 to 25 Hz, which are often used in noncorrosive environments, are too high for realistic corrosion fatigue crack growth behavior. Frequencies of 1 to 10 cpm (0.017 to 0.17 Hz) are required to obtain realistic corrosion fatigue crack growth rates. This implies that months of test time are needed that usually have not been taken. Thus, most corrosion fatigue crack growth rates available in the literature are for more than 10^{-8} m/cycle, which comprises regions II and III of the sigmoidal curve of Fig. 6.12. Region I corrosion fatigue crack growth data have usually been obtained at higher frequencies than in regions II and III.

Corrosive environments are usually detrimental to fatigue crack growth. However, at very high crack growth rates, where less time is available for the interaction between corrosion and repeated stresses, the differences between corrosion and noncorrosion rates are often small. In near-threshold and threshold regions, crack closure plays a dominant role and can provide anomalous results. Due to significant oxide film buildup on mating surfaces of the fatigue crack, region I corrosion fatigue crack growth resistance based on applied ΔK can sometimes appear to be better than that for noncorrosive environments. When oxide-induced crack closure loads are measured, and applied ΔK is replaced by $\Delta K_{\text{eff}} = \Delta K - K_{\text{op}}$, the anomalous effects are usually removed. This indicates the harmful effects of corrosion fatigue crack growth in region I. Fatigue crack growth da/dN–ΔK curve shapes are also influenced, depending on whether K_{max} is above or below the appropriate K_{SCC} (K_{EAC}). Above this value the corrosive environment can cause a plateau or near plateau in the growth rate that alters the sigmoidal shape, similar to that found in stress corrosion cracking. References 2, 3, and 9 provide comprehensive data on corrosion fatigue crack growth behavior.

Moisture or humidity in air can be detrimental to fatigue crack growth behavior, particularly in aluminum alloys. Bowles [10] showed just how important moisture effect is on fatigue crack growth of 2024-T3 and 7075-T6 aluminum sheet. An order of magnitude difference between laboratory moist air and dry air fatigue crack growth rates occurred at lower ΔK values for 2024-T3 aluminum, and a factor of 2 or less was found for the 7075-T6 aluminum, indicating varied sensitivity to moisture for the two alloys. These differences are quite significant, considering the very high frequency (20 Hz) used in these tests. At higher crack growth rates the laboratory air and dry air curves for the two alloys tended to converge. Given these various differences due to moisture alone, it is evident how complex corrosion fatigue crack growth can be.

The effect of fresh, natural flowing seawater and electrochemical potential on the fatigue crack growth of a common marine alloy HY-130 steel obtained by Crooker et al. [11] is shown in Fig. 11.6. The three seawater curves were obtained at 1 cpm (0.017 Hz) or 10 cpm (0.17 Hz), and the reference ambient air data were obtained at 0.5 Hz. In all three seawater test conditions, the fatigue crack growth rates were greater than those in ambient air. Higher negative electrochemical potential and lower frequency significantly increased

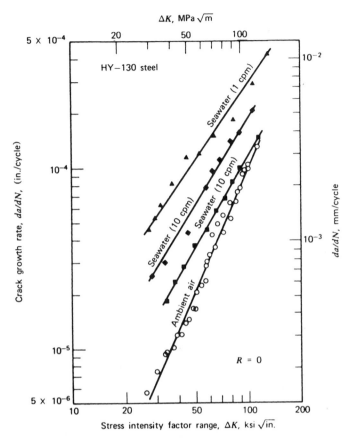

Figure 11.6 Corrosion fatigue crack growth rates for HY-130 steel in natural flowing seawater [11] (reprinted by permission of the American Society for Testing and Materials). (○) Ambient air (30 cpm), (■) seawater (−665 mV), (◆) seawater (−1050 mV), (▲) seawater (−1050 mV).

crack growth rates. Imhof and Barsom [12] found that increasing the yield strength of 4340 steel increased corrosion fatigue crack growth rates in 3 percent salt water, as shown in Fig. 11.7. Crack growth rates in air for the two steels were essentially the same. K_{ISCC} (K_{IEAC}) values for the two steels are shown in Fig. 11.7, and all corrosion fatigue crack growth data were obtained at stress intensity levels below these respective values. Mean stress effects on corrosion fatigue crack growth behavior indicate that for $R \geq 0$, increasing the R ratio can cause an increase in fatigue crack growth rates [3]. ΔK_{th} corrosion fatigue values are also reduced at higher R ratios.

11.1.5 Protection Against Corrosion Fatigue

The principal way corrosion fatigue problems can be reduced is to choose materials that resist corrosion in the expected environment. Increased corro-

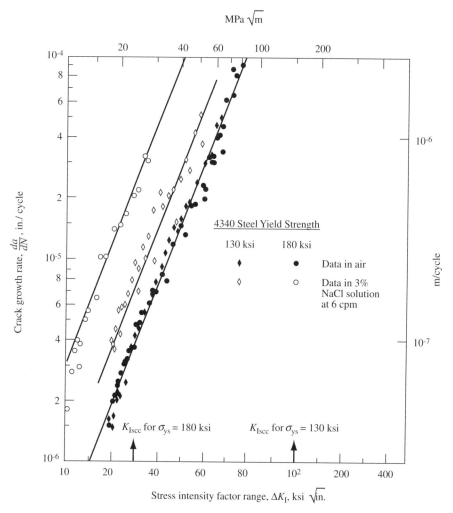

Figure 11.7 Corrosion fatigue crack growth rates for 4340 steel in 3 percent salt water solution [12] (reprinted by permission of the American Society for Testing and Materials).

sion fatigue resistance can also be achieved through various surface treatments discussed in Chapter 8, such as shot-peening, cold working, and nitriding, which induce desirable surface compressive residual stresses. Zinc and cadmium coatings have produced improved corrosion fatigue resistance. Zinc coating methods such as galvanizing provide better improvements. Chromium or nickel electrolytic plating of steel can increase corrosion fatigue resistance, but they also produce undesirable tensile residual stresses, hairline surface cracks, and possibly hydrogen embrittlement.

Surface coatings such as paint, oil, polymers, and ceramics can protect against corrosive air and liquid environments if they remain continuous. However, under service conditions for many components and structures it is difficult or maybe impossible for these coatings to retain complete continuity. Broken or disrupted coatings can eliminate their beneficial effects. Oxide coatings can increase corrosion fatigue resistance. Cladding of higher-strength aluminum alloys with a pure aluminum surface layer (alcladding) has caused substantial increases in corrosion fatigue resistance of the base alloy. However, it often decreases air fatigue resistance. Shot-peening in conjunction with oxide coatings produces even greater increases in corrosion fatigue resistance. In fact, shot-peening alone increases corrosion fatigue resistance, as well as air fatigue resistance. Corrosion inhibitors that form an adherent corrosion-resistant chemical film on the metal surface have been somewhat successful. Chromates and dichromates have been widely used.

11.1.6 Corrosion Fatigue Life Estimation

Corrosion fatigue life estimations for both constant and variable amplitude loading are far more complex than those for air or inert environments. The $S-N$, $\varepsilon-N$, and $da/dN-\Delta K$ models are applicable and have been used. However, lack of appropriate corrosion fatigue data, variability of the environment, electropotential, stress corrosion cracking (environment assisted cracking), frequency, and time-dependent aspects make these corrosion fatigue life estimates uncertain. A good example of corrosion fatigue life estimates is found in British Standard BS 7608 [13] on fatigue of weldments in seawater. Here, air $S-N$ design curves have been quantitatively reduced for seawater conditions. Damage-tolerant inspection periods incorporating corrosion and corrosion fatigue crack growth behavior have been applied to aircraft. Superposition of SCC (EAC) and fatigue crack growth models has been formulated, but quantitative use of these models has been minimal. With less confidence in corrosion fatigue life estimates, significant importance is placed on inspection and realistic testing in proper corrosive environments. However, much of the component, structure, and vehicle testing involves accelerated procedures discussed in Section 9.9.1, which can conceal intrinsic corrosion fatigue behavior due to frequency and time importance. This suggests that under corrosion fatigue conditions, realistic load–time histories should be used.

11.1.7 Summary

Corrosive environments usually reduce $S-N$, $\varepsilon-N$, or $da/dN-\Delta K$ fatigue resistance. The reduction is time dependent and is a function of environment, material, and stress level. At short lives the reduction may be small because of insufficient time, and at long lives the fatigue limit is normally nonexistent. Fatigue strengths at long lives can be a small fraction of the ultimate tensile strength. A synergistic interaction exists between the corrosive environment and stressing due to continuous breaking of protective oxide films during cycling. Very little loss of material is involved in corrosion fatigue that includes surface pitting, cracks nucleating from pits, and small and long fatigue crack growth to fracture. Better corrosion fatigue resistance exists in those environment/material conditions where corrosion resistance itself is better. Fatigue crack growth can occur with ΔK or K_{max} below the pertinent K_{SCC} (K_{EAC}). The interaction between stress corrosion cracking (environment assisted cracking) and corrosion fatigue crack growth can alter the usual sigmoidal da/dN–ΔK behavior. Constant and variable amplitude corrosion fatigue life estimates have been made using S–N, ε–N, and da/dN–ΔK models incorporating appropriate corrosion fatigue data. These estimates are less reliable than for air or inert environments. Thus, inspection along with component, structure, or vehicle testing in appropriate environments play a key role in corrosion fatigue design. However, accelerated testing may give nonconservative results due to time effects. Protection against corrosion fatigue includes choosing materials that resist corrosion in the specific environment and introducing surface compressive residual stresses, surface coatings, and inhibitors.

11.1.8   Dos and Don'ts in Design

1. Do consider that materials are susceptible to stress corrosion cracking under static loads when stress intensity factor values are greater than the appropriate K_{ISCC} (K_{IEAC}), which varies from about 10 to 100 percent of K_{Ic}.
2. Don't relate water or salt water corrosion fatigue resistance of steels to ultimate tensile strength. Many carbon and low-alloy steels have similar corrosion fatigue strengths in water and salt water. Thus, high-strength steels may not be advantageous unless surface compressive residual stresses and/or protective coatings are used.
3. Do obtain better corrosion fatigue resistance by choosing a material that exhibits low corrosion in the service environment.
4. Don't overlook the deleterious effects of moisture and humidity on fatigue resistance, particularly in aluminum alloys.
5. Do consider the many factors that can improve corrosion fatigue resistance, such as shot-peening, surface cold working, nitriding, anodic coatings, cadmium and zinc, cladding, paint, oil, ceramic and polymeric coatings, and chemical inhibitors.

11.2   FRETTING FATIGUE

The nature of fretting-induced fatigue failures is complex, and the terms used to describe this phenomenon are not universal. Terms such as "fretting," "fretting corrosion," "fretting fatigue," "fretting corrosion fatigue," and "fretting-initiated fatigue" are common. All of the terms, however, include the word "fretting," and all of them involve the behavior of two surfaces in contact subjected to small, repeated relative motion. We define these terms as follows and note that they are often interrelated and interchanged.

Fretting: A surface wear phenomenon occurring between two contacting surfaces having oscillating relative motion of small amplitude.

Fretting corrosion: A form of fretting in which chemical reaction predominates.

Fretting corrosion fatigue: The combined action of fretting, chemical reactions, and fatigue.

Fretting fatigue: The combined action of fretting and fatigue, which can also include chemical reactions.

Fretting-initiated fatigue: Can mean the same as fretting fatigue or can refer to fatigue at sites of previously formed fretting pits.

We primarily use the terms "fretting" and "fretting fatigue" and imply that corrosion products may also be involved. This is the usual case for metals operating in air.

Design against fretting fatigue is probably the least quantitative of all fatigue topics, yet it is involved with all assembled structures and components that experience repeated motion. Examples include riveted, bolted, pinned, and lug fasteners, shrink and press-fits, splines, keyways, clamps, universal joints, bearing/shaft/housing interfaces, gear/shaft interfaces, fittings, leaf springs, bio-implants, and wire ropes. Many of these assembled structures and components are intended to have no relative motion between mating surfaces. Fretting involves wear mechanisms and can occur with less than 10 μm of relative motion between mating surfaces. However, fretting damage has been observed with as little as a few nanometers of relative motion [14]. Fretting can result in seizure of mating parts, loss of fit, or fatigue failures. Seizure is caused by the buildup of fretting debris that do not escape, and loss of fit is due to fretting debris escaping from the contacting surfaces. Fretting fatigue failures result from cracks nucleating at the interface region and then growing under cyclic stresses until fracture occurs. The reduction in fatigue resistance due to fretting can be equal in importance to notch effects and corrosion fatigue. Mann [15] indicates that in extreme cases, fretting fatigue strengths may be as low as 5 to 10 percent of the base unnotched fatigue strengths, which implies that fatigue strength reduction factors of up to 10 to 20 may occur. However, Table 11.1 [16] shows that long-life fatigue strength reduction factors for various combinations of metals in contact range from almost 1 to 5, depending on the material combinations. This type of fatigue strength reduction certainly indicates the importance of considering fretting fatigue in design.

Fretting fatigue is of less importance in short-life components because of the insufficient number of repeated cycles necessary to form the fretting

TABLE 11.1 Fatigue Strength Reduction Factors Produced by the Fretting of Various Materials Against Steels and Aluminum Alloys [16]

Specimen	Hardness (HB)	Fatigue Strength (MPa)	Clamp	Fretting Fatigue Strength (MPa)	Strength Reduction Factor
Carbon steels					
0.1C steel	137	172	0.1C steel	122	1.41
			Brass	95	1.81
			Zinc	137	1.25
0.33C steel	165	372	0.33C steel	254	1.47
0.4C steel	420	550	0.2C steel	450	1.22
			0.4C steel	257	2.14
			70/30 brass	325	1.70
			Al–4.4Cu–0.5Mn–1.5Mg	500	1.10
0.7C steel, cold-drawn	365	525	0.7C steel, cold drawn	147	3.57
0.7C steel, normalized	270	371	0.7C steel, normalized	178	2.08
Alloy steels					
0.25Cr–0.25Ni–1.0Mn	285	372	0.1C steel	294	1.26
			18Cr–8Ni steel	264	1.41
1.3Cr–2.6Ni–0.4Mo	217	304	3 Si steel	241	1.51
1.1Cr–3.7Ni–0.4Mo	176	272	18Cr–8Ni steel	212	1.28
			Aluminum	238	1.14
0.6Cr–2.5Ni–0.5Mo	330	542	0.6Cr–2.5Ni–0.5Mo	124	4.37
1.4Cr–4.0Ni–0.3Mo	510	850	1.4Cr–4.0Ni–0.3Mo	240	3.55
Aluminum alloys					
Al–Cu–Mg	—	276	Al–Cu–Mg	99	2.79
Al–4.4Cu–0.5Mn–1.5Mg	140	159	Al–4.4Cu–0.5 Mn–1.5 Mg	32	1.94
Al–4.4Cu–0.8Mn–0.7Mg	160	134	Al–4.4Cu–0.8 Mn–0.7 Mg	49.5	2.72
			Mild steel	35.6	3.76
Al–4Cu	117	83.5	Al–4Cu	52.5	1.59
Copper alloys					
70/30 brass	175	139	70/30 brass	93	1.50

Reprinted with permission of R. B. Waterhouse.

damage, nucleate the cracks, and grow the cracks to fracture. That is, fretting fatigue is of greatest importance at long lives of more than 10^4 or 10^5 cycles. This most often involves small load amplitude and high frequency. Major factors affecting fretting action are the normal contact pressure between the mating surfaces, the amplitude of the relative motion, the frictional shear stresses between the mating surfaces, residual stresses, environment (corrosive and temperature), mating materials including their hardness and surface roughness, magnitude of applied alternating and mean stresses, frequency, and number of applied cycles. Many additional factors could be added, but they often influence those already stated. Combined factors often have an interactive synergistic effect on fretting and fretting fatigue that makes it difficult to determine the effect of a change in one factor. Fretting damage has been compared on the basis of both material weight loss and fatigue strength reduction. These two criteria can imply opposite effects, which can be confusing. We emphasize the reduction in fatigue resistance, since weight losses are generally small and are not the major concern in fatigue design.

11.2.1 Mechanisms of Fretting Fatigue

A fretting fatigue failure of a stainless steel strap subjected to alternating bending loads and rubbing against a cadmium-plated member is shown in Fig. 11.8 [17]. The fretting effect was greater than the combined effect of the reduced cross section at the hole and its accompanying stress concentration factor. Substantial fretting debris (dark regions) are found in the rubbing area. They consist primarily of oxides of the base metal. For steels the iron oxide is rust-colored, while for aluminum, titanium, and magnesium alloys the fretting oxide debris is black. Micro fatigue cracks nucleate in the fretting zone, often at the interface between fretted and nonfretted regions. Multiple microcracks can nucleate from both shallow dish-like pits and deeper pits or grooves that form under the fretting action. These microcracks are usually oblique to the surface, similar to the stage I cracks shown in Fig. 3.13. They can grow to perhaps 1/4 to 1/2 mm or so before significant angle changes occur. During this small crack growth, both mode I and II fatigue crack growth are involved, along with a multiaxial state of stress caused by the normal contact pressure,

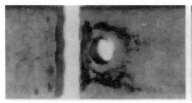

Figure 11.8 Failure caused by fretting fatigue [17] (reprinted with permission of R. B. Heywood).

the frictional shear stress, and the applied alternating stress. In addition, the small crack problems discussed in Section 6.8 are pertinent. Eventually the multiple oblique cracks become a single predominant mode I crack perpendicular to the surface or perpendicular to the maximum applied tensile stress. Once the crack tip extends beyond the vicinity of the fretting region, it is no longer controlled by the fretting process but rather by the local stress field near the crack tip caused by the applied alternating loads. LEFM principles, and hence the stress intensity factor range, ΔK, play the predominant roles in both small crack and long crack growth. Mode I, mixed-mode I and II, small crack, and long fatigue crack growth LEFM models for fretting fatigue have been reviewed by Rayaprolu and Cook [18]. These models involve integration of the appropriate da/dN–ΔK behavior to obtain the fretting fatigue crack growth life. If ΔK is greater than pertinent threshold levels, the crack grows under cyclic loading conditions. The stress intensity factor and the stress distribution, however, may be difficult to model for complex fretting conditions.

The total fretting fatigue life consists of the above-mentioned crack growth life plus the crack nucleation life and the cycle-dependent surface damage life preceding crack nucleation. The mechanism of fretting that eventually results in localized microcracks is conjectural [16,19–27]. On the one hand, we have the basic fatigue mechanisms similar to those described in Section 3.2. On the other hand, we have the ideas of adhesion, abrasion, and corrosion. Regardless, there are certain known phenomena that do occur under fretting conditions.

The contact of two mating surfaces occurs at local high asperities. An oscillatory rubbing action produces tangential cyclic shear stresses, which, along with high Hertzian stresses due to normal contact pressure, can cause local plastic deformation in these asperities. Microwelding and fracture of these asperities can occur and can be repeated under the small oscillatory relative motion, causing transfer of metal from one surface to another. Localized high temperatures can also occur, which can accelerate oxidation. The fretting debris that depend on the corrosive environment consist of oxides and metal. The debris are harder than the base metal, which can accelerate abrasion. The fretting debris are often platlets a few microns thick and 50 to 100 μm in diameter. The volume of the debris is usually larger than the volume of base particles removed due to oxidation. Hence, the trapped accumulated particles can become embedded in the base material and also cause seizure. Pits and grooves form at the contact surface and act as micro stress concentrations for microcracks to nucleate. The environment plays a major role in the fretting process. Both air and humidity contribute to reduced fretting fatigue resistance. Fretting fatigue resistance is greater in vacuum and inert atmospheres. Fretting debris still occur in vacuum and inert atmospheres, but the debris particles are the base metal, not oxides. Corrosive environments tend to decrease fretting fatigue resistance; thus, fretting fatigue mechanisms are both mechanical and chemical.

Fretting surface damage and crack nucleation life often represent only a small percentage of the total fretting fatigue life. Remember from Fig. 4.7

that for unnotched or smooth specimens, fatigue life at long lives is dominated by fatigue crack nucleation. However, under fretting fatigue conditions, this large crack nucleation life is significantly reduced or practically lost. Thus, the dominant fretting fatigue life is fatigue crack growth under mixed-mode I and II for small cracks and under mode I for longer cracks. Also, a fatigue limit is essentially nonexistent under fretting fatigue conditions.

Prefretting followed by fatigue testing without fretting can reduce fatigue strength as a result of the notch effect caused by pitting. This reduction, however, is substantially less than that under continuous fretting fatigue conditions. Thus, joint action involving fretting, cyclic stressing, and corrosion exists that is similar to corrosion fatigue, as discussed in Section 11.1.2.

Figure 11.9 shows a shrink or press-fitted shaft in a housing and with a hub, wheel, or gear. Both the unloaded and loaded conditions are shown. It is seen that the major contact area exists at the junction on the compression side of the shaft. It is this compression region where fretting fatigue starts. If the loading is reversed, fretting fatigue occurs on the top and bottom surfaces. If the shafts rotate, the entire shaft perimeter at the interface can be subjected to fretting fatigue.

11.2.2 Influence of Variables

Many variables that affect fretting fatigue behavior have been investigated using a variety of test procedures. Axial, rotating bending, torsion, and two

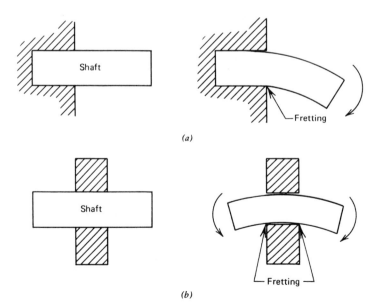

(a)

(b)

Figure 11.9 Deformation of a shaft/fixture indicating fretting locations. (a) Shrink or press fit shaft. (b) Hub, wheel, or gear on a shaft.

crossed cylinders have been used as test specimens. Both flat and spherical pressure pads have been used under either load control or displacement control (proving ring). These different test conditions can lead to different comparative results. For example, under displacement control of pressure pads normal pressure can change, depending upon the buildup and retention of debris or debris escape. In some situations, a particular variable change can either increase or decrease fretting fatigue resistance. This can be due to test procedures, scatter, and/or the interactive synergistic behavior of the many fretting fatigue variables. Most tests have been conducted with constant amplitude loading, but these tests can produce different comparative results than with variable amplitude loading. Vincent et al. [28] indicate that laboratory test conditions are often more severe than service conditions and could thus lead to overconservative designs.

The magnitude and distribution of the normal pressure between the mating surfaces appreciably affect fretting fatigue strengths. Waterhouse [16] compiled research by others and showed that increases in normal pressure can produce a substantial decrease in fretting fatigue strength as the pressure increases from 0 to about 50 MPa (7 ksi) for aluminum, titanium, nickel, and steel alloys. Pressure increases beyond this value, however, do not cause significant additional decreases in fretting fatigue strength. If the pressure increase eliminates the relative motion, fretting fatigue would be prevented. Fenner and Field [21] found that increasing the relative nominal slip amplitude from 0 to 2.5 μm and then to 22.5 μm decreased the fatigue strength at 2×10^7 cycles in an aluminum alloy by factors of about 1.7 and 5, respectively.

Lubrication to reduce the coefficient of friction ranges widely in effectiveness, depending on the characteristics of the lubricant and on whether constant load or constant displacement is used to apply the pressure load. Improvement in fretting fatigue strengths due to oils or greases ranges from zero to a few percent. Dry lubricants such as molybdenum disulfide have produced improvements in fretting fatigue strength of up to 20 percent. Repeated application of the lubricants is often necessary to retain the improvement.

Lower temperatures can produce greater fretting. This was first determined by noting that more damage in the form of fretting debris, pitting, or grooves occurred during the winter than in the summer. Absorbed moisture as a lubricant is important at low temperatures. Quantitative fretting fatigue resistance at low temperature is not clear. At elevated temperatures a higher oxidation rate may exist, and fretting fatigue resistance is usually lower than at room temperature for most materials.

Corrosive environments plus fretting fatigue can be extremely deleterious, and the S–N behavior can fall below the corrosion fatigue curve shown schematically in Fig. 11.3. Isolation from corrosive environments is therefore most desirable.

The influence of hardness on fretting fatigue resistance involves the interaction of both mating materials. In general, fretting fatigue resistance decreases with higher hardness, and hence with higher-strength materials. However, the opposite results have also been reported.

Compressive residual stresses at the surface can substantially increase fretting fatigue resistance. They can retard both fretting damage and fatigue crack growth. Shot-peening, surface rolling, tensile prestrain, nitriding, carburizing, and surface induction hardening can increase fretting fatigue strengths to almost the value of the nonfretted material. Introduction of surface residual compressive stresses appears to be a very consistent technique for improving fretting fatigue resistance.

Special soft pads or surface coatings between the mating surfaces, such as pure aluminum or magnesium metal shims and nylon, Teflon, and plasma spray coatings, have increased fretting fatigue resistance. These pads or coatings, however, can be fretted away and need to be replaced periodically.

Proper stress transfer through improved geometrical design details can also significantly improve fretting fatigue resistance. For example, the press-fits shown in Fig. 11.9 can have increased fretting fatigue resistance by machining stress relief grooves in the shaft adjacent to the junction, as shown in Fig. 7.6. A reduction in shaft diameter at the junction is also beneficial. The use of tapered plates (in both width and thickness) in lap joints also increases fretting fatigue resistance.

11.2.3 Summary

Fretting fatigue resistance depends on many interactive synergistic variables. Fretting fatigue mechanisms include surface damage, crack nucleation, multiple small mixed-mode oblique crack growth, mode I long crack growth, and final fracture. Under fretting conditions, cracks nucleate at a small percentage of the total life, with most of the fretting fatigue life involved in crack growth. LEFM thus plays an important role in fretting fatigue life calculations. In addition, S–N models with most fretting fatigue strength reduction factors ranging from almost 1 to 5 have been used in design. However, life calculations are less reliable than those for nonfretting conditions. Fretting will not often be eliminated, but it can be reduced by judicious alterations of the variables discussed in Section 11.2.2. Accelerated test programs, although a must, can often miss the fretting failures.

11.2.4 Dos and Don'ts in Design

1. Don't overlook fretting fatigue, since it can be quite deleterious and can occur under extremely small relative motion amplitudes (<10 μm) in most assembled components and structures, particularly at high frequency and long lives.
2. Do consider methods of reducing fretting fatigue through the use of compressive residual stresses such as shot-peening, surface rolling, or nitriding and through proper stress transfer.

3. Do consider using softer materials as inserts and coatings between two hard surfaces if possible. However, these may have to be replaced frequently.

4. Don't expect substantial improvements in fretting fatigue strength from lubricants such as oil and grease. Dry lubricants such as molybdenum disulfide are only slightly more beneficial

11.3   LOW-TEMPERATURE FATIGUE

Many fatigue designs in diverse fields of engineering must operate at temperatures below room temperature. These operating temperatures may be climatic temperatures as low as $-54°C$ ($-65°F$) for ground vehicles, civil structures, pipelines, and aircraft or cryogenic temperatures of $-163°C$ (110 K) for natural gas storage and transport, $-196°C$ (77 K) for liquid nitrogen storage and transport, $-253°C$ (20 K) for aerospace structures, and $-269°C$ (4 K) for superconducting electrical machinery. Fatigue behavior at these low temperatures has received much less attention than that at room and elevated temperatures. Most reports of low-temperature fatigue behavior have been based on constant amplitude tests, and few verifications of real-life fatigue results and predictions have been published for low temperatures. We consider low-temperature fatigue behavior first by reviewing the effect of low temperatures on monotonic material properties and then by considering *S–N, ε–N, da/dN–ΔK,* variable amplitude loading, and life predictions.

11.3.1   Monotonic Behavior at Low Temperatures

In general, unnotched ultimate tensile strength and yield strength increase at lower temperatures for metals, with the ratio of the ultimate strength to the yield strength tending toward a value of 1 at lower temperatures. Ductility, as measured by the percent elongation or reduction in area at fracture, usually decreases with lower temperatures, while the modulus of elasticity usually increases slightly. Total strain energy or toughness at fracture usually decreases at lower temperatures, as measured by the area under the stress–strain curve. Under notched conditions, toughness and ductility can decrease even further. This is true for both low and high strain rates. Impact energy absorbed, as measured from the Charpy V notch (CVN) impact test, the precracked Charpy (K_{Id}) test, or the dynamic tear (DT) test, shows substantial decreases. An upper and a lower shelf, characterized by a significant difference in energy-absorbing capacity and ductility, and a transition region usually exist for low- and medium-strength steels. Higher-strength steels and other metals usually have a more gradual impact energy versus temperature curve rather than an upper and a lower shelf. Both plane stress fracture toughness, K_c, and plane strain fracture toughness, K_{Ic}, often decrease with lower temperatures. The nil-ductility temperature (NDT), as measured from the drop weight test using a brittle weld bead with a machined notch, has varied from above room temperature to almost absolute zero Kelvin for steels. Thus, it is well known that the impact energy-absorbing capabilities of notched or cracked components can be drastically reduced at lower temperatures, depending on their composition, microstructure, and alloy system. This implies that greater notch and crack sensitivity exists at lower temperatures. Final fatigue crack lengths at fracture can then be drastically reduced at lower temperatures. The lower fracture toughness, lower ductility, and higher unnotched tensile strength do not, however, provide sufficient information indicating how cracks will nucleate and grow in components under real-life fatigue loadings at low temperatures.

11.3.2 Stress–Life (S–N) Behavior

Comprehensive summaries of S–N fatigue behavior at low temperatures were provided by Teed [29] in 1950, by Forrest [30] in 1962, and by Stephens et al. [31] in 1979. A tabular summary by Forrest for carbon steels, alloy steels, and cast steels is shown in Fig. 11.10. Here the averages of long-life, fully reversed fatigue strengths at low temperature divided by the fully reversed fatigue strengths at room temperature are shown for unnotched and notched specimens. No effort was made to correlate strength levels or stress concentration factors. The goal was to provide a general trend for long-life fatigue strengths at low temperatures compared to room temperature. The number of materials is given at the bottom of each column. The average ratios for smooth specimens ranged from essentially 1 to 2.5, with the higher ratios occurring at lower temperatures. For the notched specimens, the average ratios ranged from essentially 1 to 1.5, again with the higher ratios at lower temperatures. From a design standpoint, the most important aspect of Fig. 11.10 is the substantially smaller increases in fatigue strength in the notched specimens.

Figure 11.10 does not indicate the complete S–N behavior at low temperatures relative to room temperature. Spretnak et al. [32] determined the complete S–N behavior of unnotched and notched specimens between 10^3 and 10^7 cycles at low temperatures for many materials. Their results and others, e.g. [31], can be summarized as follows. At short and long lives, low temperatures are usually beneficial for constant amplitude, unnotched S–N fatigue behavior. At short lives (but more than 10^3 cycles) low temperatures do little good or harm to constant amplitude, notched S–N fatigue behavior. At long lives, notched fatigue strengths are usually slightly better than or similar to those at room temperature. However, repeated impact loadings, and thus high strain rates, at low temperatures can show quite different behavior from this in some cases. For example, Shul'ginov and Matreyev [33] tested two low-carbon/low-alloy steels at room temperature and $-60°C$ ($-76°F$) using smooth ($K_t = 1$), notched ($K_t = 2.5$), and sound butt welded joints ($K_t \approx 3.5$) with

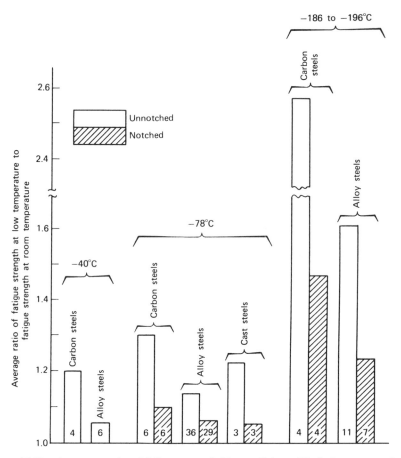

Figure 11.10 Average ratio of fully reversed ($R = -1$) long-life fatigue strengths at low temperatures and room temperature for unnotched and notched steels [30] (data used with permission of Pergamon Press Ltd.).

both continuous sine wave loading and repeated impact energy loading. At room temperature, the two loading patterns provided similar *S–N* behavior for a specific specimen configuration. At −60°C (−76°F) the fatigue resistance using the sine wave increased relative to that at room temperature. At −60°C (−76°F) the impact loadings caused a small decrease in *S–N* fatigue resistance with the smooth specimens, some additional decrease with the notched specimens, and a noticeable reduction in fatigue resistance with the butt welded joints, particularly at shorter lives compared to that at room temperature. Thus, repeated impact loadings with notches and welds were detrimental at this low temperature.

11.3.3 Strain–Life (ε–*N*) Behavior

Very little ε–*N* fatigue data at low temperatures exist. Under strain-controlled testing at low temperatures, metals can cyclic strain harden and/or soften, and their fatigue behavior generally fits the strain–life model of Eq. 5.14. Nachtigall [34] determined the ε–N behavior of 10 different materials using unnotched, smooth axial specimens at room temperature 27°C (300 K) and at two cryogenic temperatures: −195°C (78 K) liquid nitrogen, and −269°C (4 K) liquid helium. Comparative strain–life curves for three of the materials at three different temperatures from Nachtigall's report are shown in Fig. 11.11. In all 10 cases investigated by Nachtigall, at high cyclic fatigue lives, where the elastic strain range component is dominant, fatigue resistance increased at the cryogenic temperatures. Conversely, at low cyclic lives, where the plastic strain range component is dominant, fatigue resistance generally decreased with decreasing temperature. Only one nickel base alloy, Inconel 718, showed increased fatigue resistance over the entire life range at the cryogenic temperatures. A substantial decrease in fatigue resistance at short lives occurred for the 18Ni maraging steel at −269°C (4 K). This was accompanied by a drastic reduction in ductility, as measured by the percent reduction in area. This great

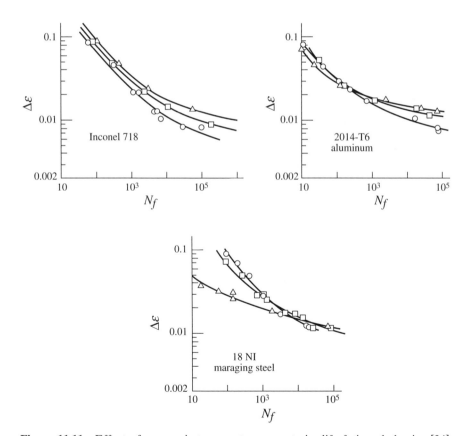

Figure 11.11 Effect of cryogenic temperatures on strain–life fatigue behavior [34]. (○) 20°C (300 K) ambient air, (□) −195°C (78 K) liquid nitrogen, (△) −296°C (4 K) liquid helium.

loss in ductility explains the substantial decrease in fatigue resistance at short lives, where the plastic strain range should be predominant. All 10 materials had an increase in ultimate tensile strength and a decrease in ductility at the cryogenic temperatures. Nachtigall used the Manson method of universal slopes to predict the strain–life fatigue behavior of the 10 materials at cryogenic temperatures with a degree of accuracy similar to that obtained for room temperature results. He concluded that low-cycle fatigue behavior of these materials at cryogenic temperatures can be predicted by using material tensile properties obtained at the same temperatures. Stephens et al. [35] reported ε–N fatigue behavior of five different cast steels using unnotched smooth axial specimens at room temperature and −45°C (−50°F). For all five cast steels, the −45°C (−50°F) fatigue resistance at longer lives was either similar to or slightly better than that at room temperature. However, at shorter lives, the −45°C (−50°F) fatigue resistance was either similar to or slightly lower than that at room temperature. Both monotonic and cyclic stress–strain curves at −45°C (−50°F) were higher than at room temperature for all five cast steels.

Polák and Klesnil [36] obtained strain–life curves for mild steel at room temperature, −60°C (213 K), and −125°C (148 K). Their data were obtained between about 200 and 10^5 cycles to failure. They found lower fatigue resistance at the lower temperatures for the shorter lives, which they attributed to very short fatigue cracks at fracture, along with brittle fracture. Kikukawa et al. [37] showed that the plastic strain range–life curves between about 5 and 10^3 cycles tend to be lower at lower temperatures. They showed this detrimental effect at low temperatures for both low- and medium-strength steels.

A summary of low-temperature strain–life fatigue behavior indicates that unnotched long-life fatigue resistance is unchanged or increased at lower temperatures, while short-life fatigue resistance may be decreased as a result of lower ductility and lower fracture toughness. At short lives, ductility is a controlling factor in strain-control behavior, while at longer lives strength is a more important controlling factor.

11.3.4   Fatigue Crack Growth (da/dN–ΔK) Behavior

To make a complete analysis of fatigue crack growth behavior under constant amplitude conditions, the complete sigmoidal da/dN–ΔK curve with three regions, I, II, and III, as shown in Fig. 6.12, must be considered. However, it is the fatigue crack growth rates associated with the lower part of region II, and also region I, that account for most of the fatigue crack growth life in many components and structures. Complete sigmoidal da/dN–ΔK curves at room temperature and at −160°C (113K) for a low-carbon steel are shown in Fig. 11.12a [38]. At lower crack growth rates (region I and the lower part of region II), the low temperature was quite beneficial, with approximately a 100 percent increase in ΔK_{th}. However, at higher crack growth rates the low temperature was detrimental. This crossover from beneficial to detrimental

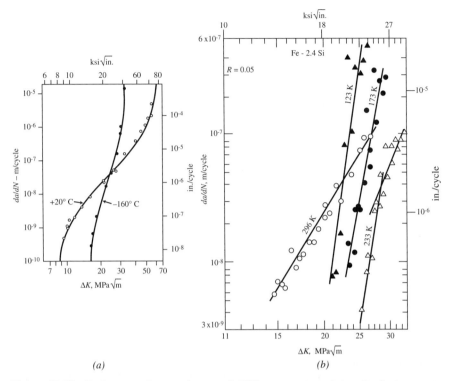

Figure 11.12 Fatigue crack growth rate, *da/dN,* versus stress intensity factor range, Δ*K*. (*a*) Low carbon steel at room temperature and −160°C (113 K) showing a crossover [38]. (*b*) Fe–2.4 Si steel for region II at room temperature and at low temperatures [40] (copyright ASTM; reprinted with permission). (Note: °C = K - 273).

at low temperatures has occurred in many constant amplitude fatigue crack growth tests. It may occur in region II, but it occurs most frequently in region III, where the crack growth rates are very high and dependent upon fracture toughness. Since fracture toughness often decreases with lower temperatures, as shown in Fig. 6.9, this is the principal reason for the crossover. The crossover is often accompanied by a transition from transgranular ductile fatigue crack growth mechanisms to combined ductile and cleavage or to significantly cleavage mechanisms.

A review of region I fatigue crack growth behavior at room and cryogenic temperatures involving *R* ratios of −1, 0, and up to 0.8 for 18 different metals including copper, steels, stainless steels, and nickel, aluminum, and titanium alloys was made by Liaw and Logsdon [39]. They found that region I *da/dN*–Δ*K* and ΔK_{th} behavior depended significantly on material, temperature, and *R* ratio. All region I *da/dN*–Δ*K* and ΔK_{th} data indicated that low temperatures were more beneficial than room temperature. As the temperature was decreased, region I *da/dN*–Δ*K* resistance increased for all *R* ratios.

Typical region I R ratio effects existed for both room and low temperatures when they were based upon ΔK, but were eliminated when ΔK_{eff} values based upon crack closure were used. However, $\Delta K_{th,eff}$ values at low temperatures were still greater than corresponding $\Delta K_{th,eff}$ values at room temperature. Thus, the low-temperature region I fatigue crack growth behavior improvements were not just manifestations of crack closure. Real intrinsic improvements occurred.

Gerberich and Moody [40] reviewed substantial fatigue crack growth behavior for a variety of materials at low temperatures. They indicated that many alternate fatigue crack growth fracture processes existed in metals at various low temperatures, and they emphasized the importance of microstructure on fatigue crack growth behavior. They showed that region II fatigue crack growth rates could be substantially altered at different low temperatures. For example, Fig. 11.12b shows region II constant amplitude results for an Fe–2.4 Si steel with $S_y = 200$ MPa (29 ksi) at four different temperatures using compact tension specimens. When the temperature was decreased from room temperature, 23°C (296 K), to -40°C (233 K), substantial decreases in da/dN occurred. As the temperature was further decreased to -100°C (173 K) and then to -150°C (123 K), the crack growth rates increased for a given stress intensity factor range, with some crack growth rates becoming higher than room temperature rates. Thus, a reversal in region II fatigue crack growth behavior occurred at about -40°C (233 K). Electron fractographs revealed that cleavage became the predominant mode of fatigue crack growth at temperatures below that at which the reversal in behavior occurred. Thus, a ductile-brittle fatigue transition temperature exists for these steels. Gerberich and Moody showed that region II slopes, n in the Paris equation, Eq. 6.19, could increase sharply for many steels at low temperatures except for some of the Fe–Ni steels. They suggested that similitude exists between a ductile-brittle fatigue transition temperature and the monotonic ductile-brittle transition temperature. Tobler and Reed [41] showed that Fe–Ni alloys provided similar or better fatigue crack growth resistance as long as the temperature remained in the "upper shelf" range, which was defined as the region where dimpled rupture or fibrous fractures occur during static fracture toughness tests. Cleavage cracking led to drastic acceleration of fatigue crack growth rates at temperatures below the transition region. Kawasaki et al. [42], and Stephens et al. [35], however, found that the fatigue crack growth transition temperature was substantially below the NDT or CVN temperature transitions. Stonesifer [43] also indicated that CVN ductile-brittle transition temperature mechanisms can be completely different from ductile-brittle transition temperature fatigue crack growth mechanisms. When large decreases in fracture toughness occur at low temperatures, crack nucleation and short crack growth may constitute almost the entire low-temperature fatigue life.

11.3.5   Variable Amplitude Behavior and Fatigue Life Estimation

Only a few variable amplitude low-temperature fatigue tests and fatigue life calculations have been reported. Cox et al. [44], using 7475-T761 aluminum, and Abelkis et al. [45], using 2024-T351 and 7475-T7651 aluminum, applied random flight spectra to center-cracked sheet specimens at $-54°C$ ($-65°F$) and at room temperature. Under these flight spectra, fatigue crack growth resistance at $-54°C$ ($-65°F$) was always equal to or better than that at room temperature. This was despite the lower fracture toughness for the 7475 alloy at low temperature and a crossover in the constant amplitude da/dN–ΔK curve. Cox et al. calculated the fatigue crack growth life for both temperatures using an interaction retardation model (the Willenborg model, described in Section 9.8) and a model neglecting retardation. The retardation model gave excellent results for both temperatures; neglecting retardation, the calculations were quite conservative. They also performed nine single tensile overload tests at each temperature with the 7475 alloy. Only three of the nine tests provided higher fatigue crack growth delay cycles at the low temperature compared to room temperature. At the highest overload value with the highest reference ΔK, fracture occurred during the overload at low temperature. Single tensile overloads with Ti-62222 titanium were also investigated at these two temperatures by Stephens et al. [46]. They found that 9 of 10 tests produced equivalent or greater fatigue crack growth delay at the low temperature. Additional single overload/underload studies with both mill annealed and solution treated Ti-62222 showed that for 25 of 32 tests, low-temperature fatigue crack growth delay was equivalent to or greater than that at room temperature. However, at low temperature the highest overload with the highest reference ΔK caused fracture in the mill annealed metal during the overload. Repeated single tensile overloads applied every 2500 cycles for six different overload ratios also indicated similar or better fatigue crack growth behavior at $-54°C$ ($-65°F$) than at room temperature. Fatigue crack growth using the standardized miniTwist variable amplitude load spectrum with β-21S titanium and Ti-62222 also indicated similar or better fatigue crack growth behavior at the low temperature. Fatigue crack growth delay and miniTwist life calculations for both temperatures gave similar variations with respect to test data.

Stephens et al. [47] tested five cast steels at four temperatures—room, $-34°C$ ($-29°F$), $-45°C$ ($-50°F$), and $-60°C$ ($-76°F$)—using a keyhole specimen with a stress concentration factor $K_t = 4$, and the SAE transmission history shown in Fig. 4.1b. They found that as the temperature was lowered, fatigue crack nucleation life (defined as a 0.25 mm crack) and total fatigue life (fracture) increased or remained similar to those at room temperature. However, fatigue crack growth life increased at $-34°C$ ($-29°F$) and then decreased at the two lower temperatures, except for one steel, where it only decreased, as the temperature was lowered. Calculated fatigue lives at room temperature and $-50°C$ using models described in Chapter 9 were within the same scatter bands for both temperatures with respect to test data. Liu and Duan [48], using a simulated random sea ice variable amplitude load spectrum at $21°C$ ($70°F$) and $-25°C$ ($-13°F$) with ASTM A131 steel, showed that fatigue crack growth resistance at the low temperature was lower than at ambient temperature.

These few examples indicate that variable amplitude fatigue behavior at low temperatures may be equivalent to or better than that at room temperature; however, exceptions occur. They also indicate that fatigue life prediction models developed for room temperature can be used at low temperatures using low-temperature reference data.

11.3.6   Summary

Long-life constant amplitude smooth and notched fatigue resistance generally increases at lower temperatures. At shorter lives, low temperature can be similar, beneficial, or detrimental to fatigue life. Repeated impact loading with sharp notches can be detrimental at low temperatures. Region I fatigue crack growth resistance is usually increased at low temperatures, and crossover may occur in region II or III. The crossover is caused by lower ductility and fracture toughness at low temperatures and by a change from ductile fatigue crack growth mechanisms to cleavage. Under variable amplitude loading, low temperature is beneficial, detrimental, or has little effect on fatigue life. It appears that a rather positive attitude toward low temperature increased fatigue resistance may be justified with many materials and components. However, sufficient data exist to raise concern about whether room temperature fatigue designs are satisfactory at low temperatures, particularly for short lives, for sharply notched or cracked parts, and for reduced low-temperature ductility and fracture toughness. This concern is important, since low-temperature load interaction or sequence effects are less well known than those at room temperature for both crack nucleation and crack growth. Composition and microstructure are much more influential at low temperatures. It appears reasonable that fatigue life predictions for low-temperature real-life load histories can be made using low-temperature fatigue properties in a manner similar to that used at room temperature.

11.3.7   Dos and Don'ts in Design

1. Don't adopt the design philosophy that room temperature fatigue life predictions and test programs will always be satisfactory for all low-temperature conditions. Each material/loading situation should be considered independently, since low temperatures can be beneficial or detrimental or have little influence on total fatigue life.

2. Don't believe that because monotonic tensile strengths increase with lower temperatures, fatigue properties also increase.

3. Do consider that large reductions in fracture toughness and ductility can occur at low temperatures and result in short crack sizes at fracture that might otherwise require a longer time for growth and better inspection.

4. Do note that NDT and CVN transition temperatures are generally different from fatigue transition temperatures, which may be similar or substantially lower. Lower NDT and CVN transition temperatures, however, appear to be accompanied by lower fatigue transition temperatures.

11.4 HIGH-TEMPERATURE FATIGUE

Many components in most fields of engineering, e.g., automobile engines, aircraft engines, power, chemical, and nuclear plants, are subjected to fatigue at elevated temperatures. High-temperature fatigue is mainly a concern at temperatures above 30 or 40 percent of the absolute melting temperature. Steam turbine components made of low-alloy ferritic steels such as turbine shafts and disks encounter temperatures in excess of 565°C (1050°F), while some gas turbine engine components, typically made of nickel-base superalloys, are exposed to temperatures in excess of 980°C (1800°F). Since some of these components are costly and safety-critical, it is understandable that there is significant interest in proper characterization of fatigue behavior at high temperatures.

Fatigue behavior and life predictions are more complicated at high temperatures than at room temperature. A complex interaction between thermally activated, time-dependent processes is involved. These include environmental oxidation, creep/relaxation, and metallurgical aspects acting jointly with mechanical fatigue mechanisms. Factors such as frequency, wave shape, and creep/relaxation, which are usually of secondary importance at room temperature, have appreciable importance at high temperatures. Time-dependent fatigue may thus be a better description at high temperatures. Extrapolation of short-term test results to long-term product requirements is common. The mode of both static and fatigue crack nucleation, crack growth, and fracture tends to shift from transcrystalline to intercrystalline as the temperature is raised. This mode change occurs at higher temperatures for fatigue conditions compared to creep conditions. In general, fatigue resistance for a given metal in an air environment decreases as the temperature increases.

Components that operate at elevated temperatures are often subjected to transient temperature gradients due to start-up and shut-down. During the start-up to shut-down cycle, thermally induced cyclic stresses can occur. Components subjected to this behavior often operate at temperatures such that both fatigue and creep damage occur, and each must be taken into account. Gas-turbine engine blades are a prime example of components subjected to creep/fatigue.

Unequal heating of parts of a component can produce thermal stresses that can lead to fatigue failure, a condition called "thermal fatigue." Operating environments in which simultaneous changes in mechanical loads (or strains) and temperature are common, i.e., pressure vessels and piping and turbine disks and blades, produce a condition commonly referred to as "thermomechanical fatigue." This is in contrast to isothermal fatigue, in which the temperature remains constant during cyclic loading.

Oxidation plays a key role in high-temperature fatigue and creep. Protective oxide formation is a major factor in the fatigue resistance of a given material. These protective oxide films, however, can be broken down by reversed slip (local plasticity), causing a much shorter high-temperature crack nucleation life. Crack growth rates are also accelerated by high-temperature environmental oxidation. Freshly exposed surfaces produced by local plasticity can oxidize rapidly. Grain boundaries are selectively attacked by oxygen. Tests at high temperature in a vacuum or inert atmosphere have shown substantial increases in fatigue/creep resistance compared to high-temperature air tests. Thus, local oxidation is one of the primary factors in degradation of fatigue/creep resistance at high temperatures. Frequency and wave shape effects are also substantially reduced at high temperatures in a vacuum or inert atmosphere. High-temperature fatigue cracks in a vacuum are more frequently transcrystalline, which indicates that oxygen is mainly responsible for the intercrystalline cracks in an air environment. The mechanisms of damage at elevated temperature are numerous, and only the most common ones are discussed here. A more detailed description of these mechanisms can be found in [49–51].

11.4.1 Creep Deformation

Materials can deform slowly and continuously over time under constant load or stress until failure occurs. This thermally assisted, time-dependent deformation is known as "creep." A common example of this type of failure is the filament creep that occurs in constantly burning light bulbs. The temperature regime for which creep plays a dominant role in material deformation is typically $T > 0.5T_m$, where T_m is the melting point of the material in degrees Kelvin. Degradation of materials due to creep is often categorized into two groups, mechanical and environmental. Mechanical degradation implies that materials undergo inelastic deformation due to an applied load, and their dimensions change with time. Environmental degradation is due to the reaction of the material with the environment or the diffusion of external elements, i.e., cavitation and environmental attack at grain boundaries.

Creep curves, which are a graphical representation of strain versus time under constant stress (or load) and temperature, are schematically shown in Fig. 11.13. Using the curve associated with σ_1 in Fig. 11.13 as an example, after the initial strain, ε_0, develops upon loading, the period of time between ε_0 and ε_1 is called "primary creep" (stage I). Between ε_1 and ε_2 a steady creep rate exists (stage II); this portion of the curve is termed "secondary creep." At creep strains above ε_2 the creep rate increases quickly; and this portion of the curve is termed "tertiary creep" (stage III). As stress or temperature increases, the strain rate for a given time increases, while the time to fracture, t_r, decreases, as shown in Fig. 11.13. The shape of the creep curve is a function of several competing mechanisms, such as strain hardening or softening, recrys-

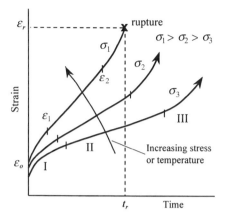

Figure 11.13 Schematic of creep curves for different stress levels or temperatures.

tallization, cavitation, grain boundary sliding, oxidation, void nucleation and growth, and necking. Various models have been developed that describe stage I (primary creep) and stage II (secondary creep). Creep tests are usually conducted to determine either a "creep limit", i.e., the stress necessary to obtain a predetermined strain for a specified period of time, or the creep rupture strength. The creep rupture strength, S_R, is the value of stress that causes rupture (fracture) of a specimen or component at a given temperature after a specified time, t_r. Failure due to creep can result from widespread or localized damage. Components subjected to uniform loading and temperature such as boiler tubes and pressure vessels are candidates for widespread creep damage, typically failing by creep rupture. Components subjected to stress (or strain) and temperature gradients such as turbine blades often experience creep crack growth as a result of a crack developing at a critical location such as a stress concentration. The later failure mode is addressed using fracture mechanics and is briefly discussed in Section 11.4.5.

11.4.2 Stress–Strain Behavior under Cyclic Loading and Hold Times

High-temperature load histories often contain hold times at a given stress or strain. Gas turbine engine disks or blades can fall into this temperature-load history condition. Under constant stress conditions, creep or creep crack extension may occur, which results in a change in component shape. Under constant strain conditions relaxation may occur, which results in a reduction of the applied stress. Coffin [52] summarized the basic stress–strain hysteresis loops for various cyclic loads and hold-time histories, as shown in Fig. 11.14. Here it is seen that the hysteresis loops are quite complex and discontinuous, even under isothermal conditions (Fig. 11.14a–d). Figure 11.14e shows thermomechanical fatigue with simultaneous changes in mechanical loading and

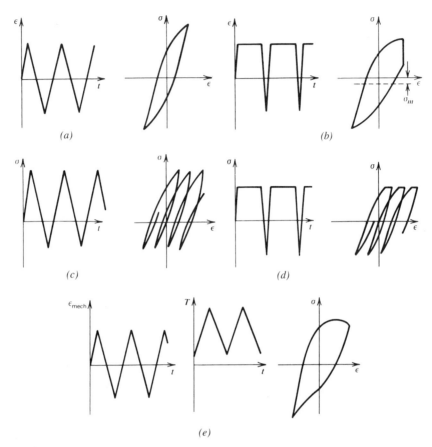

Figure 11.14 Stress–strain hysteresis loops for various cyclic and hold-time histories [52] (reprinted with permission from *Fracture, 1977: Advances in Research on the Strength and Fracture of Materials,* edited by D. M. R. Taplan, Pergamon Press). (*a*) Continuous strain cycling. (*b*) Strain hold/cycling. (*c*) Continuous stress cycling. (*d*) Stress hold/cycling. (*e*) Combined mechanical-thermal (thermomechanical) cycling.

temperature. From Fig. 11.14, we can appreciate the difficulty of predicting the fatigue life of parts subjected to real-life load histories at high temperature, whether isothermal or thermomechanical.

11.4.3 Stress–Life (*S–N*)/Creep Behavior

Metals at high temperatures do not usually show a fatigue limit. The fatigue strength continuously decreases with cycles to failure. Thus, 10^8 cycles may be a reasonable value for obtaining a long-life fatigue strength. Figure 11.15, obtained by Forrest [6], provides a very comprehensive view of how the long-life, fully reversed fatigue strengths of many materials are influenced by high

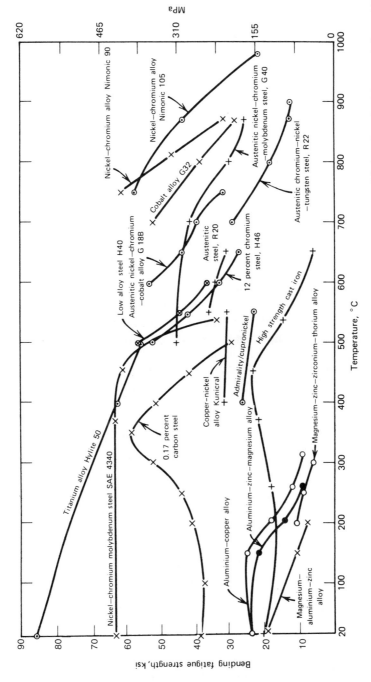

Figure 11.15 Temperature influence on fully reversed fatigue strengths of metals at long life [6] (reprinted with permission of Pergamon Press).

temperature. The range of temperature is from 20 to almost 1000°C (68 to 1830°F). Data for a given material do not include the entire temperature range because each material has a specific working temperature range over which it is economically and structurally feasible. This range is usually a function of the melting temperature. The aluminum and magnesium alloys are feasible only at temperatures up to about 200 to 300°C (390 to 570°F), while the nickel–chromium and cobalt alloys are predominant at higher temperatures between 600 and 900°C (1110 and 1650°F). In all cases except for mild steel and cast iron, fatigue strengths decrease with increasing temperature. This anomaly for mild steel and cast iron is due to cyclic strain aging and is accompanied by a decrease in ductility. Fatigue strengths for the temperatures shown in Fig. 11.15 vary by a factor of about 2.5 or less for a given material. If all the curves shown were extended and compared to room temperature fatigue strengths, even larger reductions would be seen. Thus, degradation of fatigue strengths at reasonable high working temperatures can be quite substantial.

Notches at high temperatures are detrimental under predominantly fatigue conditions. However, under predominantly creep conditions, notches can either decrease or increase the strength based on net section stresses. Interaction between creep and fatigue can thus provide different notch effects. For example, Vitovec and Lazan [53] determined that net section creep rupture strengths at 900°C (1650°F) in S-816 alloy using notched specimens with $K_t = 3.4$ were higher than those for unnotched specimens, as shown in Fig. 11.16a, while under fully reversed fatigue conditions the unnotched fatigue strengths were superior, as shown in Fig. 11.16c. With an alternating stress ratio $A = S_a/S_m$ = 0.67 or $R = 0.2$ (Fig. 11.16b), mixed results occurred. At shorter lives the notch strength was less, while at longer lives it was greater than the unnotched strength. In general, metals are less notch sensitive at high temperatures because of localized plastic and creep flow at notches and the general oxidation of the unnotched or notched surfaces. Residual stresses also have less effect at high temperatures as a result of stress relaxation from lower-yield strengths and cyclic plastic and/or creep plastic flow. However, they can still be beneficial in many situations.

A general effect of tensile mean stress, notches, and creep at high temperature on fatigue, as obtained with S-816 alloy by Vitovec and Lazan [53], is shown in Fig. 11.17. These results were obtained under load control at a constant frequency. The curves of Fig. 11.17 are for lives of 2.16×10^7 cycles or 100 hours. Since the tests were performed at constant frequency, a direct relationship between cycles and hours exists. The solid curves represent unnotched behavior and the dashed curves represent notched behavior for $K_t = 3.4$. Four different test temperatures in air ranging from room temperature to 900°C (1650°F) are shown. The vertical axis represents fully reversed $(R = -1)$ fatigue conditions, and the horizontal axis represents creep rupture strengths at high temperature and ultimate tensile strength at room temperature. As the temperature is increased, both creep rupture strengths and fully

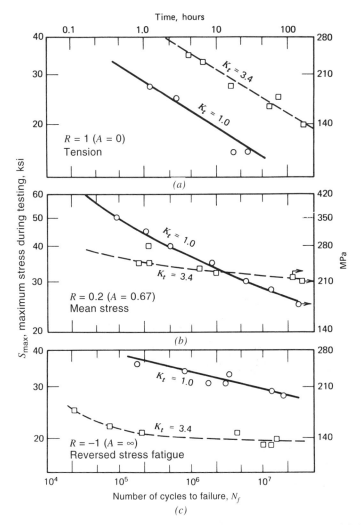

Figure 11.16 Stress rupture and S–N diagrams for unnotched and notched S-816 alloy specimens at 900°C (1950°F) [53]. (*a*) Stress rupture. (*b*) Mean stress. (*c*) Fully reversed.

reversed fatigue strengths decrease. The curves tend to approximate ellipses or circles as the temperature increases. Thus, a first approximation for mean stress effects when both unnotched fatigue and creep are involved is

$$\left(\frac{S_a}{S_f}\right)^2 + \left(\frac{S_m}{S_R}\right)^2 = 1 \tag{11.1}$$

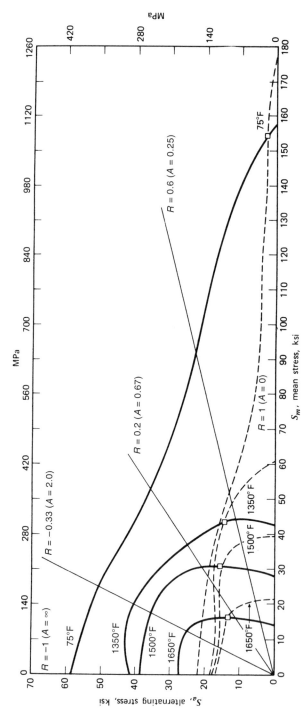

Figure 11.17 Tensile mean stress effects for unnotched and notched S-816 alloy specimens for 100 hours of life or 2.16×10^7 cycles [53]. (——) Unnotched specimens, $K_t = 1$. (---) notched specimens, $K_t = 3.4$.

where

S_a = alternating stress

S_m = mean stress

S_f = fully reversed fatigue strength

S_R = creep rupture strength

Note that S_R is used in place of S_u in the modified Goodman equation (Eq. 4.5a) or the Gerber equation (Eq. 4.5b). The modified Goodman curve or the Gerber curve, Eqs. 4.5a and 4.5b, would generally be conservative for high-temperature unnotched mean stress effects. However, they may be nonconservative for some notched behavior, especially under predominately fatigue conditions (high A ratio). Substantial increases in S_m can occur before S_a decreases. The open squares represent the intersection of the unnotched and notched behavior for a given temperature. For conditions with high mean stress and thus high creep involvement, notch strengthening is shown. Similar trends occur for other materials and elevated temperatures.

11.4.4 Strain–Life (ε–N) Behavior

The gas turbine, steam turbine, and nuclear power fields have created the principal motivation for strain–life high-temperature fatigue design information. These are fields where cyclic loads are periodically superimposed on long-term static creep loads. Typical strain–life fatigue behavior obtained in air by Berling and Slot [54] using 304 stainless steel is shown in Fig. 11.18 for

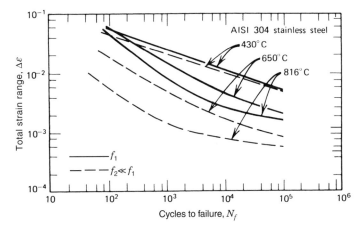

Figure 11.18 The effect of temperature and frequency on strain–life fatigue of 304 stainless steel [54] (reprinted by permission of the American Society for Testing and Materials).

three temperatures and two cyclic frequencies. For a given high temperature, the lower frequency has less fatigue resistance, and as the temperature increases the fatigue resistance decreases. Coffin [52,55] indicated that this high-temperature strain–life fatigue behavior is typical for air environments, and he and others have attributed the frequency and temperature effects to environmental aspects, primarily oxidation.

Figure 11.19, from Berling and Conway [56], shows the influence of tensile hold time on strain–life fatigue behavior for 304 stainless steel at 650°C (1200°F) in air. The hold time periods for each cycle are indicated and range from 1 minute at low strain range to 3 hours (180 minutes) at high strain range. As the hold time increased, the life decreased drastically at both test strain ranges. Tensile hold time in most materials is generally found to be more detrimental than equal hold times in tension/compression or compression only [52].

The application of strain–life fatigue behavior to the design of notched components at high temperature has been used with the local notch strain approach, as is described in earlier chapters for life to a small crack. Shortcomings exist, however, as a result of creep/fatigue/environment interaction.

Thermomechanical Fatigue   Many components, such as gas turbine blades subjected to low-cycle, high-temperature fatigue conditions, involve both thermal and mechanical loading. This is referred to as "thermomechanical fatigue" (TMF) and implies that the component is simultaneously subjected to both cyclic stress and cyclic temperature, as shown schematically in Fig. 11.14*e*. The strain-temperature-time relationship can be classified into two categories, in-phase and out-of-phase, according to the phase relationship between strain and temperature. In-phase (0°) TMF implies that the maximum normal strain and maximum temperature occur simultaneously, while out-of-phase (180°)

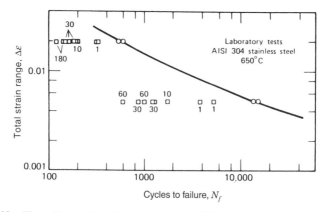

Figure 11.19   The effect of hold time on strain–life fatigue of 304 stainless steel at 650°C (1200°F) [56]. (○) No hold time, (□) tensile hold time in minutes as indicated.

TMF implies that the maximum normal strain coincides with the minimum temperature. Only 0° and 180° phasing results in proportional variation of strain and temperature. Other degrees of out-of-phase cycling (such as 90° or 270°) can produce other TMF strain variations, referred to as "nonproportional phasing." Each in-phase and out-of-phase TMF cycle has a unique hysteresis loop that gives different responses when in tension or compression.

Because of the complexity of the analysis of TMF, simplifications have been made. In some cases, thermomechanical fatigue life has been approximated by isothermal fatigue (IF) at the maximum temperature of the TMF cycle using the same mechanical strain range. This has proven satisfactory, for example, in analyzing out-of-phase TMF conditions [57]. Halford et al. [58] showed that bithermal tests conducted on B-1900+Hf alloy provided a link between isothermal and thermomechanical testing. In bithermal testing, the tensile and compressive halves of the fatigue cycle are conducted isothermally at two significantly different temperatures. Many studies, however, have shown that the strain–life (ε–N) behavior for TMF tests can be significantly different than that for isothermal fatigue tests. Thus, it is evident that the TMF lives cannot be predicted accurately based on isothermal fatigue for all conditions. Kuwabara et el. [59] compiled information for a broad range of materials subjected to TMF and categorized them into four groups based on their in-phase (IP) and out-of-phase (OP) cyclic behavior. The four material types, where N is cycles to failure, are categorized as follows:

Type I: $N_{IP} < N_{OP}$ at lower strain ranges

Type O: $N_{OP} < N_{IP}$ at lower strain ranges

Type E: $N_{IP} \approx N_{OP}$

Type E′: $N_{IP} < N_{OP}$ at higher strain ranges, but $N_{IP} \approx N_{OP}$ at lower strain ranges

In general, type I behavior is characteristic of creep damage, type O is characteristic of environmentally enhanced damage, and types E and E′ occur when neither creep nor environment makes a significant contribution to the damage, i.e., time-independent or cycle-dependent fatigue damage. Low-strength, high-ductility materials are typically types O, E, and E′, while high-strength, low-ductility materials tend to exhibit type I behavior. Kuwabara [59] concluded that the life estimation techniques previously discussed, in combination with isothermal strain–life data, were satisfactory for type O and type E materials, while for type I materials, fatigue life calculations were nonconservative.

Life Prediction Models Coffin [60,61] modified Eq. 5.14 by introducing empirical frequency factors to allow for different frequency effects including cycle shape, hold time, and the extent of tension/compression time in the cycle. The strain range partitioning (SRP) method, first presented by Manson et al. [62], has also been used successfully in several high-temperature strain–

life fatigue situations. For isothermal fatigue, this approach involves partitioning of the total inelastic strain range ($\Delta\varepsilon_{in}$) into four possible components associated with the direction of straining (tension or compression) and the type of inelastic strain (time-independent or time-dependent). Figure 11.20a shows a schematic of the four basic types of inelastic strain derived from the hysteresis loop from a creep-fatigue test. $\Delta\varepsilon_{pp}$ and $\Delta\varepsilon_{cc}$ represent reversed plastic or time-independent (pp) and reversed creep or time-dependent (cc) strain ranges, respectively, while $\Delta\varepsilon_{pc}$ and $\Delta\varepsilon_{cp}$ represent combined plastic and creep (pc and cp) strain ranges. $\Delta\varepsilon_{pc}$ and $\Delta\varepsilon_{cp}$ are mutually exclusive in a cycle. The Coffin-Manson relation relating plastic strain and life, Eq. 5.16, can be applied for each of the four types of strain range where

$$\Delta\varepsilon_{ij} = A'(N_{ij})^c \tag{11.2}$$

A' and c are material constants, and the subscripts ij denote the various combinations of c and p. By adding up the fractional damage for each type of strain, the damage caused by each cycle, d, is estimated by the expression

$$d = \frac{1}{N_f} = \Sigma\frac{f_{ij}}{N_{ij}} = \frac{f_{pp}}{N_{pp}} + \frac{f_{cc}}{N_{cc}} + \frac{f_{pc}}{N_{pc}}\left(\text{or }\frac{f_{cp}}{N_{cp}}\right) \tag{11.3a}$$

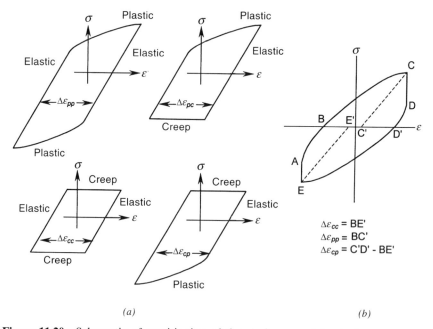

(a)

(b)

Figure 11.20   Schematic of partitioning of the strain range into strain components. (a) Four types of inelastic strain range. (b) Hysteresis loop containing $\Delta\varepsilon_{pp}$, $\Delta\varepsilon_{cc}$, $\Delta\varepsilon_{cp}$.

and at failure,

$$\sum_{1}^{N_f} d = D = 1 \tag{11.3b}$$

where D is the cumulative damage. The fractional strain, f_{ij}, for each type of strain is written as

$$f_{ij} = \frac{\Delta\varepsilon_{ij}}{\Delta\varepsilon_{in}} \tag{11.3c}$$

where $\Delta\varepsilon_{in}$ is the total inelastic strain range

$$\Delta\varepsilon_{in} = \Delta\varepsilon_{pp} + \Delta\varepsilon_{cc} + \Delta\varepsilon_{pc} \text{ (or } \Delta\varepsilon_{cp}) \tag{11.3d}$$

The number of cycles to failure is estimated by $N_f = 1/d$. To use the strain range partitioning method, a stable hysteresis loop of the stress–strain cycle (Fig. 11.20b) is required from which the partitioned strain-range components $\Delta\varepsilon_{pp}$, $\Delta\varepsilon_{cc}$, and $\Delta\varepsilon_{pc}$ (or $\Delta\varepsilon_{cp}$) are obtained, as well as the total inelastic strain, $\Delta\varepsilon_{in}$. The relationship given in Eq. 11.2 is required for each partitioned strain range and is obtained by experiments. While the inelastic strain range consists of the plastic (time-independent) strain and the creep (time-dependent) strain, partitioning of the two strains is not as straightforward for thermomechanical fatigue as it is for isothermal fatigue. However, plastic and creep strains can be determined experimentally or based on constitutive models involving plasticity and creep.

Damage effects under combined creep/fatigue conditions also have been calculated by linearly summing the fractions of damage due to creep with that of fatigue such that

$$\sum_{1}^{N_f} \left(\frac{N}{N_f}\right)_{\text{fatigue}} + \sum_{1}^{N_f} \left(\frac{t}{t_r}\right)_{\text{creep}} = D \text{ or } 1 \tag{11.4}$$

where N is the number of cycles at a given strain range, N_f is the time-independent fatigue life at the given strain range, t is the time at a given applied stress and t_r is the rupture time at that stress. This is a simple approximation since it treats fatigue and creep damage separately, whereas they are often interactive processes. Extensions and modifications of these models, as well as several other life prediction models, exist. These models range from strain- or stress-based to oxidation- and environment-based. A review of many high-temperature, strain-based life estimation models can be found in [57,63,64].

11.4.5 Fatigue Crack Growth (da/dN–ΔK) Behavior

The stress intensity factor range, ΔK, and the stress ratio, R, can be used to describe high-temperature fatigue crack growth behavior. However, due to the complex interaction of thermally activated processes and fatigue, fatigue crack growth behavior is much more difficult to predict. The effects of temperature, environment, frequency, waveform, and stress ratio have been investigated and have given mixed results. James [65] showed that region II fatigue crack growth rates increased for a given ΔK as temperature was increased for Hastelloy X-280 (Fig. 11.21). Here it is seen that da/dN for a given ΔK is more than an order of magnitude higher at 649°C (1200°F) than at room

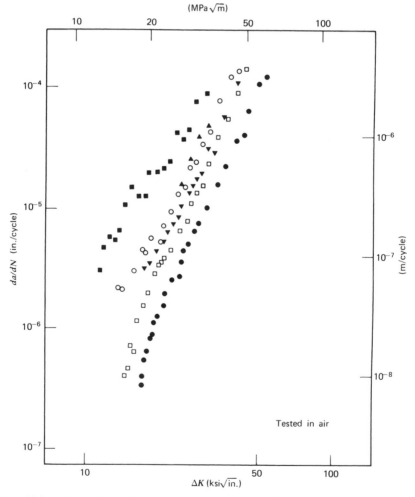

Figure 11.21 The effect of temperature on fatigue crack growth behavior, Hastelloy X-280, $R = 0.05$ [65]. Temperature in °C: (●) 24, (□) 316, (▲▼) 427, (○) 538, (■) 649.

temperature. All tests were performed at a given frequency and in an air environment. Similar behavior has been observed for nickel-base superalloys and a CrMoV steel in which region II fatigue crack growth rates increased as temperatures increased [66–68]. In most cases, for region II fatigue crack growth, as temperatures increase the fatigue crack growth rates increase for a given ΔK.

The effects of temperature on threshold and near-threshold fatigue crack growth have produced mixed results. Based on applied ΔK, tests on mild steel have shown that temperature has little effect on threshold values at low stress ratios, while threshold values for stainless steel have been shown to increase with an increase in temperature in air. Hicks and King [67] showed that for a fine-grain powder nickel-base superalloy, threshold values were insensitive to temperature, but at higher stress intensity factor ranges, as the temperature was increased from 20 to 600°C (68 to 1110°F), crack growth rates increased (Fig. 11.22a). For a coarse-grained series of the same material, threshold values were reduced as the temperature was increased (Fig. 11.22b). The decrease in ΔK_{th} as the temperature was increased for the coarse-grain material was attributed to a change in fracture morphology, as the fracture surface was much flatter. For the fine-grain material, fracture surface roughness was found to be similar at 20 and 600°C (68 and 1110°F), thus leading to similar threshold behavior. Fine-grain materials tend to give higher fatigue resistance, while coarse-grain materials give higher creep resistance. Therefore, at moderate high temperatures, fine grain gives better fatigue resistance where fatigue dominates, while coarse grain is more desirable at very high operating temperatures where creep and stress rupture dominate. Stephens et al. [69] showed that threshold and near-threshold fatigue crack growth rates for Ti-62222 titanium were lower at 175°C (350°F) than at 25°C (77°F) at $R = 0.1$ by nearly an order of magnitude. This was attributed to the formation of an oxide layer at 175°C (350°F). Liaw et al. [68] showed that for a CrMoV steel, near-threshold crack growth rates increased as the temperature was raised from 25 to 260°C (75 to 500°F), but as the temperature was raised from 260 to 425°C (500 to 800°F), a significant decrease in crack growth rate and an increase in ΔK_{th} occurred. They attributed a decrease in surface roughness to the decrease in ΔK_{th} with increasing temperature from 25 to 150°C (75 to 300°F), while an increase in oxide-induced crack closure with increasing temperature from 150 to 425°C (300 to 800°F) was correlated with increasing ΔK_{th}. Near-threshold fatigue crack growth rates were found to increase in Al-Li 8090 as the temperature was increased from 25 to 150°C (75 to 300°F) [70]. However, at higher ΔK, slower fatigue crack growth rates were observed as the temperature was increased from 25 to 150°C (75 to 300°F). Thus, a cross-over within the region I/region II fatigue crack growth curves occurred.

The fatigue crack growth results at elevated temperatures presented above have most often been attributed to changes in the crack growth fracture processes as a result of an interaction between microstructure, creep damage, and environment, i.e., oxidation. As exposure temperatures reach approxi-

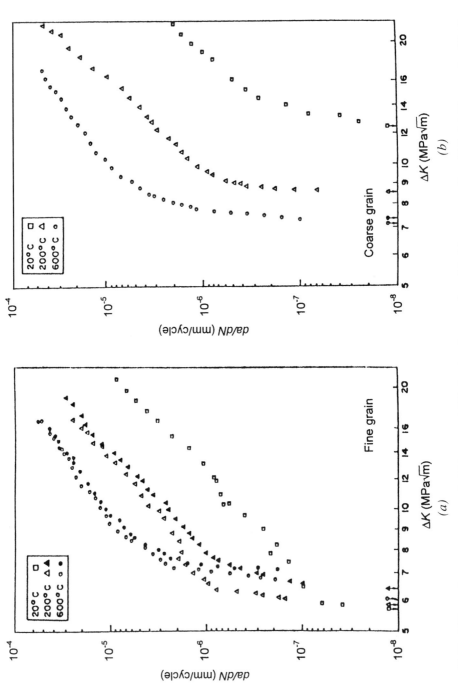

Figure 11.22 Effect of temperature on *da/dN* versus Δ*K* behavior for powder nickel-base superalloy [67]. (*a*) Fine grain. (*b*) Coarse grain (reprinted with permission from Elsevier Science).

mately one-half of the melting temperature, $0.5T_m$, the slip characteristics for many materials change from planar to wavy. Additionally, dislocation climb and cross slip are enhanced, providing less resistance to crack growth. While these results were mixed based on nominal threshold behavior for room and elevated temperatures, elevated temperatures were mostly found to be detrimental based on ΔK_{eff}. This was mostly attributed to accelerated environmental action such as oxidation, in which oxide products formed in the crack, causing oxide-induced crack closure. Thus, in cases where the nominal ΔK_{th} was found to be greater at elevated temperature in comparison to room temperature, these were most often a manifestation of crack closure. Liaw et al. [68] and Stephens et al. [69] showed that increasing the stress ratio, R, thus decreasing the crack closure contribution to crack growth, decreased the influence of temperature on fatigue crack growth rates.

James and Knecht [71] showed that the growth of fatigue cracks at high temperature in a vacuum or inert atmosphere can be significantly less than in air and very similar to that at room temperature. They attributed the greater fatigue crack growth rate at high temperature to environmental (oxidation) effects. Typical frequency effects at high temperature on fatigue crack growth rates obtained by James [72] for 304 austenitic stainless steel at 540°C (1000°F) are shown in Fig. 11.23. Frequencies varying from 0.083 to 4000 cpm are shown, and more than an order of magnitude difference exists for these data between the two extreme frequencies. Again, this was attributed primarily to environmental effects where, at lower frequencies, more environmental interaction in the presence of air occurs.

Mean stress effects on fatigue crack growth at elevated temperatures are similar to those presented for room temperature in Section 6.5. At higher R ratios the fatigue crack growth rates are higher for a given ΔK value. The largest differences appear at threshold levels based on nominal ΔK. Accounting for crack closure, thus based on ΔK_{eff}, tends to cause convergence of the various R ratio fatigue crack growth curves for a given material.

Using techniques similar to those presented in Chapter 6 and Section 9.8, fatigue crack growth life estimations at high temperatures can also be made. However, care must be taken in recognizing the limitations of LEFM at elevated temperatures, as yield strengths decrease significantly and plasticity takes place early on in the fatigue process. At high temperatures and low frequencies, nonlinear deformation can dominate. While ΔJ has been used to characterize fatigue crack growth under high-plasticity conditions, it cannot be extended to include time-dependent deformation. Thus, ΔJ is limited if low frequencies or hold times are involved. Other approaches that have been proposed [73,74] are based on creep parameters, such as C^* and $C(t)$, and include small-scale and transitional creep crack growth due to hold times of a fatigue cycle.

A linear superposition model that accounts for both cycle-dependent (fatigue) and time-dependent (creep) crack growth in a cycle has been used. Based on this model, the total crack growth rate can be written as

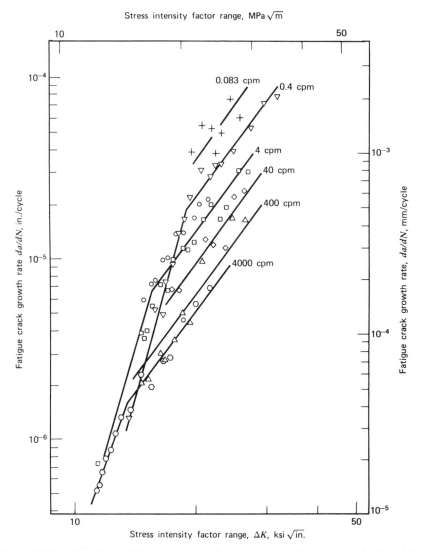

Figure 11.23   The effect of frequency on fatigue crack growth behavior of 304 stainless steel at 538°C (1000°F), $R = 0.05$ [72] (reprinted by permission of the American Society for Testing and Materials).

$$\frac{da}{dN} = \left(\frac{da}{dN}\right)_{\text{fatigue}} + \left(\frac{da}{dt}\right)_{\text{creep}} \qquad (11.5)$$

Other approaches and modifications of Eq. 11.5 have also been proposed that consider both cycle- and time-dependent crack growth.

11.4.6 Summary

Fatigue behavior and life predictions are more complicated at high temperatures because of the complex interaction between time-dependent (creep) and cycle-dependent (fatigue) processes. Oxidation, cavitation, grain boundary sliding, and other environmental conditions/processes contribute heavily to high-temperature fatigue and creep behavior. Most materials exhibit a decrease in fatigue resistance with increasing temperature. Notches at high temperatures are detrimental under predominantly fatigue conditions, while under predominantly creep conditions notches can either increase or decrease the strength. Notch strain approaches, strain range partitioning, and the linear superposition method for creep and fatigue have proven to be successful fatigue estimation models in some applications. However, a universal model for all materials and load-temperature conditions does not exist. Fatigue crack growth behavior is highly influenced by increased temperatures, and oxidation plays a key role in threshold and near-threshold behavior. Lower frequencies typically allow more environmental interaction to occur, resulting in higher crack growth rates. LEFM life estimation techniques can be used to evaluate high-temperature fatigue crack growth conditions if material behavior remains predominately elastic. However, care must be taken due to the potential for nonlinear deformation, as the yield strength reduces with increased temperature and plastic deformation dominates. Time-dependent parameters have been developed for these conditions.

11.4.7 Dos and Don'ts in Design

1. Do consider ways of protecting the surface from air at high temperatures, since oxidation is one of the principal causes of fatigue resistance degradation at high temperatures.
2. Don't assume that notch sensitivity is substantially reduced at high temperatures. Reduce stress concentrations if possible.
3. Do consider residual stresses for increased high-temperature fatigue resistance, but remember that they can have less influence at high temperature than at room temperature because of possible high-temperature stress relaxation from plastic strains.
4. Do consider both crack nucleation life and crack growth life in considering the entire high-temperature fatigue/creep interaction life.
5. Don't neglect thermal stresses if temperatures are not uniform in a part or component.

11.5 NEUTRON IRRADIATION

The nuclear reactor field has the greatest interest in neutron irradiation. Here long-term, steady-state creep loadings at elevated temperatures plus repeated loadings are superimposed on irradiation. Many variables affect the mechanical properties of reactor materials, namely, neutron fluence, neutron flux, time, irradiation temperature, operating temperature, prior thermomechanical treatment, and basic microstructure. The most common parameter for measuring irradiation effects is the neutron fluence, which is the neutron flux integrated over the exposure time and given in units of neutrons/cm^2 for energies greater than 0.1 MeV.

The general effect of increasing neutron irradiation fluence on monotonic properties is to increase the ultimate tensile strength and yield strength while decreasing elongation and reduction in area at fracture. This increases the susceptibility to hydrogen-assisted cracking, which has been an important factor with bcc steels in high-temperature reactor water [75]. Fracture toughness, K_{Ic}, and Charpy V notch (CVN) energy usually decrease with increased neutron irradiation fluence. The nil-ductility temperature (NDT) for ferritic steels is increased along with an accompanying decrease in upper shelf CVN energy. At these higher strain rates, one finds even greater undesirable embrittlement.

The effect of neutron irradiation on fatigue properties is not sufficiently known. This is due to the greater emphasis on creep resistance and the very complex nature and expense of fatigue tests of irradiated parts and specimens. Beesten and Brinkman [76] and Brinkman et al. [77] compared low-cycle, high-temperature [400 to 700°C (750 to 1290°F)] strain–life fatigue behavior for unirradiated and irradiated 304 and 316 austenitic stainless steels. For the strain levels investigated, fatigue lives ranged between 400 and 20 000 cycles. They found that irradiation reduced fatigue life by factors between 1.5 and 2.5.

James [78,79] has made comprehensive reviews of fatigue crack growth behavior for metals under various neutron fluences, irradiation temperatures, test temperatures, and frequency. The materials included ferritic steels (ASTM alloys A302B, A533B, A508, and A543), austenitic stainless steels (304, 308, and 316), and nickel-base alloys (Inconel 625, Inconel 718, and Nimonic alloys). These materials have received the most irradiation fatigue attention. Most fatigue crack growth tests had growth rates higher than 10^{-7} m/cycle (4×10^{-6} in./cycle), with only a few tests with data slightly less than 2.5×10^{-8} m/cycle (10^{-6} in./cycle). Thus, all the data reviewed by James were in regions II and III of the sigmoidal da/dN–ΔK curve. No threshold values were reported or have been reported since his reviews. Conclusions, therefore, cannot be made for the entire fatigue crack growth region.

Figure 11.24 [79] shows fatigue crack growth rate results for irradiated and unirradiated A533B steel tested at 288°C (550°F) for various frequencies (10 to 600 cpm) from three different irradiation experiments. Much of the scatter can be attributed to frequency effects in air at high temperature, as shown in Section 11.4. For a given frequency, the irradiated crack growth resistance was both slightly better and slightly worse than for the unirradiated material. This type of small difference was found for most of the ferritic steels. Both irradiated and unirradiated fatigue crack growth resistance decreased at higher

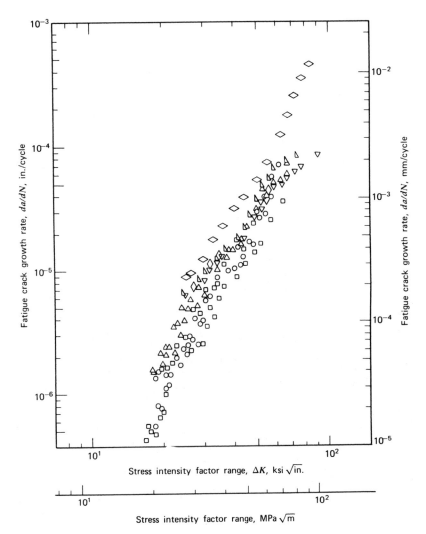

Figure 11.24 Fatigue crack growth behavior of irradiated and unirradiated ASTM A533 Grade B Class 1 steel tested in air at 288°C (550°F) [79]. (▽), unirradiated; others, irradiated.

temperatures. For the ferritic steels in regions II and III, James concluded that constant amplitude fatigue crack growth resistance after irradiation was neither significantly less than nor significantly greater than in air environment alone. For irradiated austenitic steels James [79] found mixed results, with some studies indicating no difference, some studies showing higher growth rates in irradiated material, and others showing just the opposite effect. Cullen et al. [80] and Cullen [81] showed that da/dN–ΔK for neutron irradiated steel in 288°C (550°F) water was no different than unirradiated steel in the same

288°C (550°F) water environment. *da/dN–ΔK* was higher in water than in air, but neutron irradiation had no additional synergistic effect in water. Thus, no final generalization concerning total fatigue behavior under neutron irradiation can be made. It appears, however, that any constant amplitude fatigue degradation is not as severe as the embrittlement caused by the irradiation. Data on fatigue effects with irradiation under spectrum loading are not available.

REFERENCES

1. *Standard Test Method for Determining a Threshold Stress Intensity Factor for Environment-Assisted Cracking of Metallic Materials,* ASTM E1681, Vol. 3.01, ASTM, West Conshohocken, PA, 2000, p. 984.

2. *Damage Tolerant Design Handbook, A Compilation of Fracture and Crack Growth Data for High Strength Alloys,* CINDAS/Purdue University, Lafeyette, IN, 1994.

3. A. J. McEvily, ed., *Atlas of Stress-Corrosion and Corrosion Fatigue Curves,* ASM International, Materials Park, OH, 1990.

4. M. H. Peterson, B. F. Brown, R. L. Newbegin, and R. E. Grover, "Stress Corrosion Cracking of High Strength Steels and Titanium Alloys in Chloride Solutions at Ambient Temperature," *Corrosion,* Vol. 23, 1967, p. 142.

5. D. J. McAdam, "Corrosion Fatigue of Metals," *Trans., Am. Soc. Steel Treating,* Vol. 11, 1927, p. 355.

6. P. G. Forrest, *Fatigue of Metals,* Pergamon Press, London, 1962.

7. H. Bernstein and C. Loeby, "Low Cycle Corrosion Fatigue of Three Engineering Alloys in Salt Water," *Trans. ASME, J. Eng. Mater. Tech.,* Vol. 110, 1988, p. 234.

8. R. I. Stephens, C. D. Schrader, D. L. Goodenberger, K. B. Lease, V. V. Ogarevic, and S. N. Perov, "Corrosion Fatigue and Stress Corrosion Cracking of AZ91E-T6 Cast Magnesium Alloy in 3.5% NaCl Solution" *Magnesium Properties and Applications for Automobiles* (SP-962), Paper No. 930752, SAE, Warrendale, PA, 1993, p. 85.

9. R. N. King, A. Stacey, and J. V. Sharp, "Review of Fatigue Crack Growth Rates for Offshore Steels in Air and Seawater Environments," *Proceedings of the International Conference on Offshore Mechanics and Arctic Engineering- OMAE,* ASME, Vol. 3, 1996, p. 341.

10. C. Q. Bowles, "The Role of Environment, Frequency and Wave Shape During Fatigue Crack Growth in Aluminum Alloys," Delft University of Technology, Department of Aerospace Engineering, Report LR-270, May 1978.

11. T. W. Crooker, F. D. Bogar, and W. R. Cares, "Effects of Flowing Natural Seawater and Electrochemical Potential on Fatigue-Crack Growth in Several High-Strength Marine Alloys," *Corrosion-Fatigue Technology,* ASTM STP 642, ASTM, West Conshohocken, PA, 1978, p. 189.

12. E. J. Imhof and J. M. Barsom, "Fatigue and Corrosion-Fatigue Crack Growth of 4340 Steel at Various Yield Strengths," *Progress in Flaw Growth and Fracture Toughness Testing,* ASTM STP 536, ASTM, West Conshohocken, PA, 1973, p. 182.

13. "Code of Practice for Fatigue Design and Assessment of Steel Structures," BS 7608: 1993, British Standards Institution, London, 1993.

14. S. J. Shaffer and W. A. Glaeser, "Fretting Fatigue," *Fatigue and Fracture,* ASM Handbook, Vol. 19, ASM International, Materials Park, OH, 1996, p. 221.

15. J. Y. Mann, *Fatigue of Materials,* Melbourne University Press, Melbourne, Australia, 1967.

16. R. B. Waterhouse, *Fretting Corrosion,* Pergamon Press, London, 1972.

17. R. B. Heywood, *Designing Against Fatigue of Metals,* Reinhold Publishing Corp., New York, 1962.

18. D. B. Rayaprolu and R. Cook, "A Critical Review of Fretting Fatigue Investigations at the Royal Aerospace Establishment," *Standardization of Fretting Fatigue Test Methods and Equipment,* H. M. Attia and R. B. Waterhouse, eds., ASTM STP 1159, ASTM, West Conshohocken, PA, 1992, p. 129.

19. *Fretting in Aircraft Systems,* AGARD Conference Proceedings No. 161, Agard, Munich, October 1974.

20. D. W. Hoeppner, V. Chandrasekaran, and T. L. Elliot, eds. *Fretting Fatigue Current Technology and Practices,* ASTM STP 1367, ASTM, West Conshohocken, PA, 2000.

21. A. J. Fenner and J. E. Field, "Fatigue Under Fretting Conditions," *Rev. Metall.,* Vol. 55, 1958, p. 475.

22. R. B. Waterhouse, ed., *Fretting Fatigue,* Applied Science Publishers, London, 1981.

23. H. M. Attia and R. B. Waterhouse, eds., *Standardization of Fretting Fatigue Test Methods and Equipment,* ASTM STP 1159, ASTM, West Conshohocken, PA, 1992.

24. D. A. Hills and D. Nowell, *Mechanics of Fretting Fatigue,* Kluwer Academic Publishers, Dordrecht, the Netherlands, 1994.

25. T. C. Lindley, "Fretting Fatigue Characteristics and Integrity Assessment," *The Theoretical Concepts and Numerical Analysis of Fatigue,* A. F. Blom and J. C. Beevers, eds., Engineering Materials Advisory Service, Cradle Heath Warley, West Midland, UK, 1992, p. 73.

26. R. B. Waterhouse, "Fretting Fatigue," *Int. Mater. Rev.,* Vol. 37, No. 2, 1992, p. 77.

27. R. B. Waterhouse and T. L. Lindley, eds., *Fretting Fatigue,* ESIS Publication 18, Mechanical Engineering Publications, London, 1994.

28. L. Vincent, Y. Berthier, and M. Godet, "Testing Methods in Fretting Fatigue: A Critical Appraisal," *Standardization of Fretting Fatigue Test Methods and Equipment,* H. M. Attia and R. B. Waterhouse, eds., ASTM STP 1159, ASTM, West Conshohocken, PA, 1992, p. 33.

29. P. L. Teed, *The Properties of Metallic Materials at Low Temperatures,* Chapman and Hall, London, 1950.

30. P. G. Forrest, *Fatigue of Metals,* Pergamon Press, Oxford, 1962.

31. R. I. Stephens, J. H, Chung, and G. Glinka, "Low Temperature Fatigue Behavior of Steels—A Review," Paper No. 790517, SAE, Warrendale, PA, April 1979.

32. J. W. Spretnak, M. G. Fontana, and H. E. Brooks, "Notched and Unnotched Tensile and Fatigue Properties of Ten Alloys at 25 and −196°C," *Tran. ASM,* Vol. 43, 1951, p. 547.

33. B. S. Shul'ginov and V. V. Matreyev, "Impact Fatigue of Low-Alloy Steels and Their Welded Joints at Low Temperature," *Int. J. Fatigue,* Vol. 19, No 8-9, 1997, p. 621.

34. A. J. Nachtigall, "Strain-Cycling Fatigue Behavior of Ten Structural Metals Tested in Liquid Helium (4 K), in Liquid Nitrogen (78 K) and in Ambient Air (300 K)," NASA TN D-7532, February 1974.

35. R. I. Stephens, J. H. Chung, S. G. Lee, H. W. Lee, A. Fatemi, and C. Vacas-Oleas, "Constant-Amplitude Fatigue Behavior of Five Carbon or Low-Alloy Cast Steels at Room Temperature and −45°C," *Fatigue at low Temperature,* R. I. Stephens, ed., ASTM STP 857, ASTM, West Conshohocken, PA, 1985, p. 140.

36. J. Polák and M. Klesnil, "The Dynamics of Cyclic Plastic Deformation and Fatigue Life of Low Carbon Steel at Low Temperatures," *Mater. Sci. Eng.,* Vol. 26, No. 2, 1976, p. 157.

37. M. Kikukawa, M. Jono, T. Kamato, and T. Nakano, "Low Cycle Fatigue Properties of Steels at Low Temperatures," *Proc. 13th Jap. Congr. Mater. Res.,* 1970, p. 69.

38. S. Ya Yarema, "Growth of Fatigue Cracks in Low Carbon Steel Under Room and Low Temperatures," *Probl. Prochn.,* No. 3, 1977, p. 21 (in Russian).

39. P. K. Liaw and W. A. Logsdon, "Fatigue Crack Growth Thresholds at Cryogenic Temperatures: A Review," *Eng. Fract. Mech.,* Vol. 22, No. 4, 1985, p. 585.

40. W. W. Gerberich and N. R. Moody, "A Review of Fatigue Fracture Topology Effects on Threshold and Kinetic Mechanism," *Symposium on Fatigue Mechanisms,* J. T. Fong, ed., ASTM STP 675, ASTM, West Conshohocken, PA, 1979, p. 292.

41. R. L. Tobler and R. P. Reed, "Fatigue Crack Growth Resistance of Structural Alloys at Cryogenic Temperature," paper presented at the Cryogenic Engineering Conference/International Cryogenic Materials Conference, University of Boulder, CO, August 1977.

42. T. Kawasaki, T. Yokobori, Y. Sawaki, S., Nakanishi, and H. Izumi, "Fatigue Fracture Toughness and Fatigue Crack Propagation in 5.5% Ni Steel at Low Temperature," *Fracture 1977,* ICF-4, Vol. 3, University of Waterloo Press, Waterloo, Canada, 1977, p. 857.

43. F. R. Stonesifer, "Effect of Grain Size and Temperature on Fatigue Crack Propagation in A533 B Steel," *Eng. Fract. Mech.,* Vol. 10, 1978, p. 305.

44. J. M. Cox, D. E. Pettit, and S. L. Langenbeck, "Effect of Temperature on the Fatigue and Fracture Properties of 7475-T761 Aluminum," *Fatigue at Low Temperature,* R. I. Stephens, ed., ASTM STP 857, ASTM, West Conshohocken, PA, 1985, p. 241.

45. P. R. Abelkis, M. B. Harmon, E. L. Hayman, T. L. Mackay, and J. Orlando, "Low Temperature and Loading Frequency Effects on Crack Growth and Fracture Toughness of 2024 and 7475 Aluminum," *Fatigue at Low Temperature,* R. I. Stephens, ed., ASTM STP 857, ASTM, West Conshohocken, PA, 1985, p. 257.

46. R. R. Stephens, R. I. Stephens, D. E. Lemm, S. G. Berg, H. O. Liknes, and C. J. Cousins, "Role of Crack Closure Mechanisms on Fatigue Crack Growth of Ti-62222 Under Constant-Amplitude and Transient Loading at −54, 25, and 174°C," *Advances in Fatigue Crack Closure Measurements and Analysis,* Vol. 2, R. C.

McClung and J. C. Newman, eds., ASTM STP 1343, ASTM, West Conshohocken, PA, 1999, p. 224.

47. R. I. Stephens, A. Fatemi, H. W. Lee, S. G. Lee, C. Vacas-Oleas, and C. M. Wang, "Variable-Amplitude Fatigue Crack Initiation and Growth of Five Carbon or low-alloy Cast Steels at Room and Low Temperatures," *Fatigue at Low Temperature,* R. I. Stephens, ed., ASTM STP 857, ASTM, West Conshohocken, PA, 1985, p. 293.

48. C. T. Liu and M. L. Duan, "Experimental Investigation on Low Temperature Fatigue Crack Propagation in Offshore Structural Steel A131 Under Random Sea Ice loading," *Eng. Fract. Mech.,* Vol. 53, No. 2, 1996, p. 231.

49. S. Suresh, *Fatigue of Materials,* 2nd ed., Cambridge University Press, Cambridge, 1998.

50. A. Saxena, *Nonlinear Fracture Mechanics for Engineers,* CRC Press, Boca Raton, FL, 1998.

51. R. H. Norris, P. S. Grover, B. C. Hamilton, and A. Saxena, "Elevated-Temperature Crack Growth," *Fatigue and Fracture, ASM Handbook,* Vol. 19, ASM International, Materials Park, OH, 1996, p. 507.

52. L. F. Coffin, "Fatigue at High Temperature," *Fracture 1977,* ICF-4, Vol. 1, University of Waterloo Press, Waterloo, Canada, 1977, p. 263.

53. F. H. Vitovec and B. J. Lazan, "Fatigue, Creep, and Rupture Properties of Heat Resistant Materials," WADC Technical Report No. 56-181, August 1956.

54. J. T. Berling and T. Slot, "Effect of Temperature and Strain Rate on Low Cycle Fatigue Resistance of AISI 304, 316, and 348 Stainless Steels," *Fatigue at High Temperatures,* ASTM STP 459, ASTM, West Conshohocken, PA, 1969, p. 3.

55. L. F. Coffin, "Fatigue at High Temperatures," *Fatigue at Elevated Temperatures,* ASTM STP 520, ASTM, West Conshohocken, PA, 1973, p. 5.

56. T. Berling and J. B. Conway, "Effect of Hold-Time on the Low-Cycle Fatigue Resistance of 304 Stainless Steel at 1200°F," *First Int. Conf. Pressure Vessel Tech., Part 2,* Delft, The Netherlands, 1969, p. 1233.

57. H. Sehitoglu, "Thermal and Thermomechanical Fatigue of Structural Alloys," *Fatigue and Fracture, ASM Handbook,* Vol. 19, ASM International, Materials Park, OH, 1996, p. 527.

58. G. R. Halford, M. A. McGraw, R. C. Bill, and P. Fanti, "Bithermal Fatigue: A Link between Isothermal and Thermomechanical Fatigue," *Low Cycle Fatigue,* ASTM STP 942, H. Solomon, G. Halford, L. Kaisand, and B. Leis, eds., ASTM, West Conshohocken, PA, 1988.

59. K. Kuwabara, A. Nitta, and T. Kitamura, "Thermal Mechanical Fatigue Life Prediction in High Temperature Component Materials for Power Plants," *Advances in Life Prediction,* D. A. Woodford and R. Whitehead, eds., ASME, New York, 1985, p. 131.

60. L. F. Coffin, "The Effect of Frequency on the Cyclic Strain and Low Cycle Fatigue Behavior of Cast Udimet 500 at Elevated Temperature," *Metall. Trans.,* Vol. 2, 1971, p. 3105.

61. L. F. Coffin, "The Concepts of Frequency Separation in Life Prediction for Time Dependent Fatigue," *ASME-MPC Symposium on Creep–Fatigue Interaction,* MPC-3, Metal Properties Council, New York, 1976, p. 349.

62. S. S. Manson, G. R. Halford, and M. H. Hirschberg, "Creep-Fatigue Analysis by Strain-Range Partitioning," *Symposium on Design for Elevated Temperature Environment,* ASME, New York, 1971, p. 12.

63. R. Viswanathan, *Damage Mechanisms and Life Assessment of High-Temperature Components,* ASM International, Metals Park, OH, 1989.

64. G. R. Halford, "Evolution of Creep-Fatigue Life Prediction Models," *Creep–Fatigue Interaction at High Temperature,* G. K. Haritos, ed., AD-Vol. 21, ASME, New York, 1991, p. 43.

65. L. A. James, "The Effect of Temperature Upon the Fatigue-Crack Propagation of Hastelloy X-280," Report HEDL-TME 76-40, 1976.

66. R. R. Stephens, L. Grabowski, and D. W. Hoeppner, "The Effect of Temperature on the Behavior of Short Fatigue Cracks in Waspaloy Using an In-Situ SEM Fatigue Apparatus," *Int. J. Fatigue,* Vol. 15, No. 4, 1993, p. 273.

67. M. A. Hicks and J. E. King, "Temperature Effects on Fatigue Thresholds and Structure Sensitive Crack Growth in a Nickel-Base Superalloy," *Int. J. Fatigue,* Vol. 5, No. 2, 1983, p. 67.

68. P. K. Liaw, A. Saxena, V. P. Swaminathan, and T. T. Shih, "Influence of Temperature and Load Ratio on Near-Threshold Fatigue Crack Growth Behavior of CrMoV Steel," *Fatigue Crack Growth Threshold Concepts,* D. L. Davidson and S. Suresh, eds., The Metallurgical Society of AIME, Warrendale, PA, 1984, p. 205.

69. R. R. Stephens, R. I. Stephens, D. E. Lemm, S. G. Berge, H. O. Liknes, and C. J. Cousins, "Role of Crack Closure Mechanisms on Fatigue Crack Growth of Ti-62222 Under Constant-Amplitude and Transient Loading at −54, 25, and 175°C," *Advances in Fatigue Crack Closure Measurement and Analysis,* Vol. 2, ASTM STP 1343, R. C. McClung and J. C. Newman, Jr., eds., ASTM, West Conshohocken, PA, 1999, p. 224.

70. H. D. Dudgeon and J. W. Martin, "Near Threshold Fatigue Crack Growth at Room Temperature and an Elevated Temperature in Al-Li Alloy 8090," *Mater. Sci. Eng.,* Vol. A150, 1992, p. 195.

71. L. A. James and R. L. Knecht, "Fatigue-Crack Propagation Behavior of Type 304 Stainless Steel in a Liquid Sodium Environment," *Metall. Trans. A,* Vol. 6A, 1975, p. 109.

72. L. A. James, "The Effect of Frequency upon the Fatigue-Crack Growth of Type 304 Stainless Steel at 1000°F," *Stress Analysis and Growth of Cracks,* ASTM STP 513, ASTM, West Conshohocken, PA, 1972, p. 218.

73. C. E. Jaske and J. A. Begley, "An Approach to Assessing Creep/Fatigue Crack Growth," *Ductility and Toughness Considerations in Elevated Temperature Service,* MPC-ASME-8, ASME, New York, 1978, p. 163.

74. A. Saxena and J. L. Bassani, "Time-Dependent Fatigue Crack Growth Behavior of Elevated Temperature," *Fracture: Interactions of Microstructure, Mechanisms and Mechanics,* TMS-AIME, Warrendale, PA, 1984, p. 357.

75. H. Hänninen, K. Törrönen, M. Kemppainen, and S. Salonen, "On the Mechanisms of Environment Sensitive Cyclic Crack Growth of Nuclear Reactor Pressure Vessel Steels," *Corrosion Sci.,* Vol. 23, No. 6, 1983, p. 663.

76. J. M. Beeston and C. R. Brinkman, "Axial Fatigue of Irradiated Stainless Steels Tested at Elevated Temperatures," *Irradiated Effects on Structural Alloys for*

Nuclear Reactor Applications, ASTM STP 484, ASTM, Conshohocken, PA, 1970, p. 419.

77. C. R. Brinkman, G. E. Korth, and J. M. Beeston, "Influence of Irradiation on the Creep/Fatigue Behavior of Several Austenitic Stainless Steels and Incoloy 800 to 700°C," *Effects of Radiation on Substructure and Mechanical Properties of Metals and Alloys,* ASTM STP 529, ASTM, West Conshohocken, PA, 1973, p. 473.

78. L. A. James, "Fatigue Crack Propagation in Neutron-Irradiated Ferritic Pressure-Vessel Steels," *Nuclear Safety,* Vol. 18, No. 6, 1977, p. 791.

79. L. A. James. "Effects of Irradiation and Thermal Aging Upon Fatigue-Crack Growth Behavior of Reactor Pressure Boundary Materials," *Time and Load Dependent Degradation of Pressure Boundary Materials,* IWG-RRPC-79/2, International Atomic Energy Agency, Vienna, 1979, p. 129.

80. W. H. Cullen, H. E. Watson, R. A. Taylor, and F. J. Loss, "Fatigue Crack Growth Rates of Irradiated Pressure Vessel Steels in Simulated Nuclear Coolant Environment," *J. Nuclear Mater.,* Vol. 96, No. 3, 1981, p. 261.

81. W. H. Cullen, "Fatigue Crack Growth Rates in Pressure Vessel and Piping Steels in LWR Environments," NUREG/CR-4724, U.S. Nuclear Regulatory Commission, 1987.

PROBLEMS

1. A mild steel circular stepped shaft is subjected to pure bending. The shaft diameters are 25 and 15 mm, and the notch root radius is 4 mm. The shaft is to operate in flowing seawater for 6 months. What fully reversed bending moment would you recommend for a life of 10^7 cycles? What moment would you recommend if the shaft were isolated from the water? How would you verify your decisions?

2. What materials or operations would you recommend for Problem 1 to increase the fully reversed bending moment 80 percent?

3. Solve Problem 19 in Chapter 7 if the component is to operate outdoors in a rain forest. Discuss your assumptions and indicate what types of errors you may have introduced.

4. Propose estimates of room temperature fully reversed fatigue strengths ($> 10^7$ cycles) for the mild steel shafts and supports shown in Fig. 11.9.

5. What methods can be used to improve the fretting fatigue resistance of the mild steel shafts shown in Fig. 11.9?

6. Solve Problem 19a in Chapter 7 assuming that the component is to be operated in liquid nitrogen at $-195°C$ (78 K), and the material is 18 Ni maraging steel, grade 300 as shown in Fig. 11.11. For this temperature $E = 194$ GPa and K' and n' are estimated to be about 2000 MPa and 0.1 respectively.

7. Solve Problem 2 of Chapter 9 if the component is to operate at −40°C (−40°F). Discuss the significance of your assumptions and the accuracy of your results.

8. If a wide plate has a through-thickness edge crack of 3 mm and the applied $R = 0$ stress range causes an initial $\Delta K = 18$ MPa$\sqrt{\text{m}}$, how many constant amplitude cycles of this stress range can be applied before fracture at room temperature and −160°C for the steel in Fig. 11.12a? Comment on your calculation procedures and results.

9. Discuss the ideas of Problem 13 in Chapter 9 for a temperature of −40°C (−40°F) and a liquid nitrogen temperature of −195°C (78 K).

10. Determine the effect of temperature on K_f in Fig. 11.17. Consider all four temperatures and $A = \infty$, 2.0, and 0.25.

11. Solve Problems 19a and 19b in Chapter 7, assuming that $\Delta\varepsilon/2 = 0.01$ and $P_a = 100$ kN and the component is annealed 304 stainless steel operating at 7 Hz and 500°C. Comment on the accuracy of your predictions.

12. A CrMoV steel tested at 550°C is subjected to fatigue cycles similar to that shown in Fig. 11.20b. Based on existing experimental data, the following relationships were determined: $\Delta\varepsilon_{cc} = 0.14(N_{cc})^{-0.45}$, $\Delta\varepsilon_{pp} = 0.48\,(N_{pp})^{-0.52}$, and $\Delta\varepsilon_{pc} = 0.27(N_{pc})^{-0.49}$. From the hysteresis loop, the following information was obtained: $\Delta\varepsilon_{cc} = 0.0085$, $\Delta\varepsilon_{pp} = 0.0123$, and $\Delta\varepsilon_{pc} = 0.0012$.
   (a) Draw the strain-range partitioning relationships for this alloy,
   (b) Determine the total number of cycles for the above strains that will cause failure.

13. Solve Problem 12a–c of Chapter 6, assuming that the plate is to operate at 93°C (200°F) where $K_{Ic} = 44$ MPa$\sqrt{\text{m}}$. Assume that the Paris relationship for long crack behavior at $R = 0$ and 93°C is $da/dN = 8.5 \times 10^{-11}(\Delta K)^{3.85}$ where da/dN is in m/cycle and ΔK is in MPa$\sqrt{\text{m}}$. Also, do you expect creep to be of significance in this problem? Explain your reasoning.

CHAPTER 12

FATIGUE OF WELDMENTS

Parts and structures are often welded together in some fashion, usually due to cost and weight effectiveness. Steels, followed by aluminum alloys, are the most frequently welded metals, while some metals cannot be effectively welded. Weldments present difficulties because of macro and micro discontinuities, residual stresses, and posssible misalignment, all of which may vary between nominally equal parts. Weldments are frequently the prime location for fatigue failures. Welding itself is a complex procedure that can result in a wide range of fatigue resistance. The quality of workmanship and the design determine the fatigue resistance of weldments. A carefully designed and processed weldment can develop the same fatigue strength as a part forged and machined from one piece, and at far less cost. An aircraft part that incorporates a nose wheel spindle and a landing gear piston in one piece of high-strength steel may serve as an example of a carefully designed weldment having good fatigue resistance. The parts are rough machined from a material like 4340 steel, welded together, finish machined, heated in a controlled atmosphere, quenched in oil, tempered to achieve the desired hardness, and shot-peened. By contrast, an attachment to a machine using an untreated fillet weld can substantially reduce the fatigue resistance of the machine.

We consider different weldments from a fatigue design viewpoint by indicating the different macro and micro discontinuities that can exist, typical fatigue behavior, and methods for improving weldment fatigue resistance. We then briefly discuss the four fatigue design procedures (S–N, ε–N, da/dN–ΔK, and the two-stage method) outlined in previous chapters for application to weldments, along with current weldment design codes.

12.1 WELDMENT NOMENCLATURE AND DISCONTINUITIES

Several typical weldments are shown schematically in Fig. 12.1 with accompanying fatigue strengths for structural steel at 2×10^6 cycles with $R = 0$ [1]. These are rather simplified welded joints, but they do represent many real

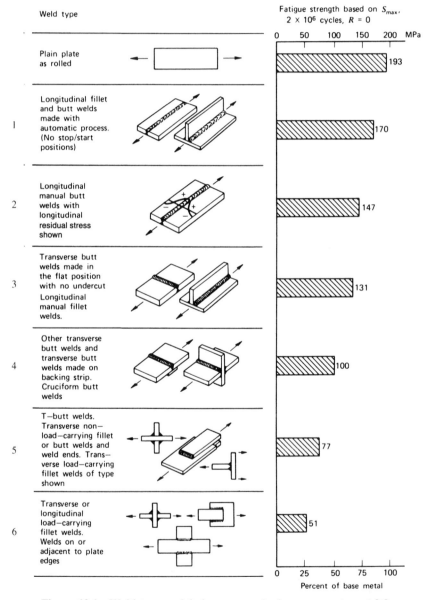

Figure 12.1 Weld type and fatigue strengths for structural steel [1].

parts and structures. In general there are butt, fillet, and spot welds with many different weldment shapes and configurations. Both transverse butt and longitudinal butt welds are common. Fillet welds, however, are more common and, as shown in Fig. 12.1, may be load carrying, as in the left of row 6, or non-load carrying, as in the left of row 5. The limited fatigue strengths given in Fig. 12.1 range from about 25 to 90 percent of the unnotched as-rolled base plate fatigue strength and provide a reasonable guide for actual variation in weldment fatigue strengths, depending on the weldment. The range from 25 to 90 percent indicates a significant difference in weldment fatigue strengths.

Because of nonuniform temperature gradients and thermal expansion and contraction that cause local elastic/plastic deformations during the welding and cooling process, biaxial or triaxial residual stresses are formed in all welds. The residual stress profiles and their magnitudes are difficult to quantify. Figure 8.14a and Fig. 12.1, row 2, show a typical longitudinal residual stress distribution at a transverse section. The residual stresses are tensile in the weld region, which must be balanced by compressive residual stresses away from the weld. The tensile stresses may reach values equal to the yield strength and can contribute to the lower fatigue resistance of some weldments. Residual stresses in weldments are considered in more detail in Section 12.3.

A photograph of a polished and etched longitudinal section of a cruciform fillet weldment obtained by Albrecht [2] is shown in Fig. 12.2. The toe and root of the weldment are indicated, along with three basic weldment regions:

1. Parent or base metal (BM)
2. Deposited weld metal (WM)
3. Heat affected zone (HAZ)

In addition, a fourth region, a fusion zone, exists between the deposited weld metal and the heat affected zone. These four regions can have different microstructures, residual stresses, discontinuities, and monotonic strength, ductility, and fracture toughness properties. The heat-affected zone is the base metal, which is subjected to high-temperature gradients during welding. The higher temperature adjacent to the fusion line causes recrystallization, while other regions of the heat-affected zone may not completely recrystallize. Thus, mechanical properties and metallurgical structures may vary across the width of the heat-affected zone and may be similar to, or different from, the base metal. For welded aluminum alloys, the heat-affected zone thermally softens to a condition similar to an annealed base metal condition.

Figure 12.3 shows transverse sections of full and partial penetration butt and fillet weldments. Common locations for cracks to nucleate and/or grow are shown for each weldment. Stress concentrations, which occur at the toe of butt and fillet welds, are common locations for fatigue cracks. The toe is also at the surface, where bending stresses are the largest. K_t at the toe depends on the geometry of the weld, as defined in Fig. 12.3b, and is usually larger in

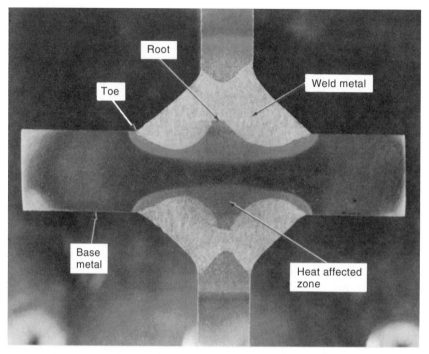

Figure 12.2 Polished and etched longitudinal section through a non-load-carrying cruciform weldment [2] (reprinted by permission of the American Society for Testing and Materials).

fillet welds than in butt welds. Values of K_t have ranged from essentially 1 for butt welds with the reinforcement removed to 3 to 5 for sharp geometrical changes in fillet welds. The word "reinforcement" is a misnomer since it implies a positive effect. Actually, the "reinforcement" acts as a stress concentration, is detrimental in fatigue, and should be called "overfill" or perhaps "excess" weld metal. Fatigue cracks also grow at the weld root, as shown in Fig. 12.3 for fillet welds and partial penetration butt welds.

There are always macro and/or micro discontinuities in weldments that provide sites for cracks to nucleate. Some of these discontinuities may actually be planar, as in the case of cracks, and hence crack nucleation fatigue life may be zero or small in this case. Fatigue life in weldments is then considered to involve only fatigue crack growth life [3,4]. When crack-like discontinuities do not exist, weldment fatigue life may be considered to consist of fatigue crack nucleation and fatigue crack growth of small cracks leading to the growth of large cracks [5,6]. Whichever philosophy or situation is most relevant, significant agreement exists that fatigue crack growth plays a dominant role in fatigue of weldments. An important key aspect of weldments then is the reduction of macro and micro discontinuities which will enhance fatigue crack nucleation life and thus enhance total fatigue life.

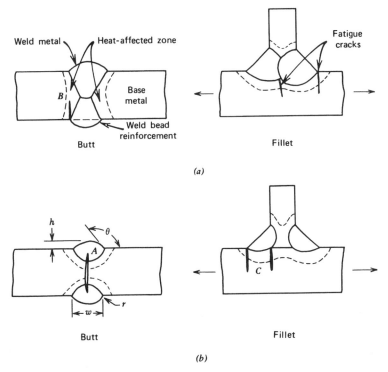

Figure 12.3 Weldment nomenclature and fatigue crack nucleation and/or growth sites. (*a*) Full penetration welds. (*b*) Partial penetration welds.

Weldment discontinuities can be classified as planar, volumetric, or geometric as follows:

Planar discontinuities: solidification cracks, shrinkage cracks, hydrogen-induced cracks, lamellar tears, lack of fusion, partial penetration, or sharp oxide inclusions.

Volumetric discontinuities: porosity or slag inclusions.

Geometric discontinuities: stop/start/terminations/intermittent locations, surface ripples, reinforcement, toe radius, toe undercutting, overlaps, section changes, misalignment, stray arc strikes, or spatter.

Solidification cracks may occur in the deposited weld metal and are caused by excessive restraint on adjacent material as the metal cools from liquid to solid. Shrinkage cracks are due to cooling temperature gradients and can occur in metal that has not been melted. Hydrogen can be introduced during welding from surface water, oil, grease, or paint and welding flux and can cause cracking in high cooling rate regions containing higher carbon contents.

Lamellar tears can develop during welding at prior elongated inclusions formed during rolling operations. Porosity is caused primarily by trapped gases during solidification, and slag inclusions form from the electrode coating. Many of these discontinuities occur at the surface or intersect the surface, and many are also subsurface. The planar discontinuities are crack-like and are thus considered to be cracks. The shapes of porosity and slag inclusions are often somewhat rounded, and hence may not represent crack-like disconti-nuities and are thus considered less harmful. Geometric discontinuities more closely resemble macro stress concentrations.

Figure 12.3 indicates that fatigue cracks may nucleate and/or grow in the weld metal (A) or in the heat-affected zone (B) or through the heat-affected zone and the base metal (C). Thus, we should have information about fatigue behavior in all three zones.

12.2   CONSTANT AMPLITUDE FATIGUE BEHAVIOR OF WELDMENTS

12.2.1   Stress–Life (*S–N*) Behavior

Most steels used in weldments have yield strengths below 700 MPa (100 ksi). Even with strengths above this value, much information indicates that for a given transverse butt or fillet weldment, as-welded constant amplitude fatigue strengths at 10^6 cycles or more are rather independent of material ultimate tensile strength. This was shown by Reemsnyder [7] for transverse butt weld-ments using a number of steels in different conditions with tensile strengths varying from 400 to 1030 MPa (58 to 148 ksi). The geometric notch severity of the weldment, residual stresses which may or may not relax, other disconti-nuities, loss of heat treatment, and cyclic softening or hardening are the causes of this behavior. As we see later, however, methods do exist to improve weldment fatigue resistance.

Geometrical stress concentrations have a significant effect on weldment fatigue resistance. The influence of the reinforcement angle for transverse butt welds is shown in Fig. 12.4 [1]. Here we see that a factor of almost 2 exists for the fatigue strength at 2×10^6 cycles as the reinforcement angle varies from 100° to 150°. Reemsnyder [7] showed that decreasing the height h of the transverse butt weld reinforcement continually improved the fatigue strengths of 785 MPa (114 ksi) ultimate tensile strength steel, as shown in Fig. 12.5. Fatigue notch factors, K_f, at 2×10^6 cycles were 4.5, 2.5, 1.5, and essen-tially 1.0 as the height was decreased from 3.8 mm to complete reinforcement removal. Thus, S_f for the transverse butt weldment with reinforcement re-moved is very similar to that for the base metal. However, care must be taken during grinding of the reinforcement so as not to introduce excess grinding notches that could be more detrimental than the reinforcement. Sanders and Lawrence [8] also showed significant improvement in fatigue strength in alumi-

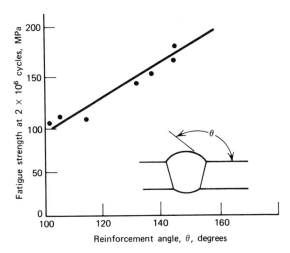

Figure 12.4 Influence of reinforcement shape on fatigue strength of transverse butt welds [1].

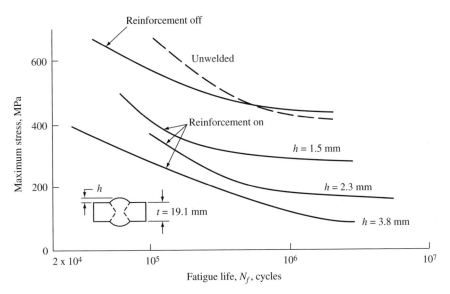

Figure 12.5 *S–N* curves for transverse butt welds, Q&T carbon steel, S_u = 785 MPa (114 ksi), $R = 0$ [7] (reprinted by permission of the American Society for Testing and Materials).

num alloy transverse butt welds when the reinforcement was removed. Thus, it should be clear that reducing stress concentrations in weldments is a major factor in improving fatigue resistance.

Mean stress influence on fatigue resistance of weldments is similar to that of other severely notched components. A constant-life diagram with S_a versus S_m for carbon steel butt welds is shown in Fig. 12.6 [7]. The resemblance of Fig. 12.6 to the Haigh diagram for notched parts (e.g., Fig. 7.10) is quite strong. Figure 7.10 for notches indicates appreciable influence from compressive mean stress and significantly smaller influence from tensile mean stress. Tensile mean stresses in weldments do not have an appreciable effect on the allowable alternating stress, and therefore it has become common practice to disregard tensile S_m in weldment fatigue design codes. This small tensile S_m effect is due to local plasticity at the weld toe or root caused by superposition of applied stresses with high tensile residual stresses remaining after the welding process. Compressive mean stresses are found to enhance fatigue resistance, as shown in Fig. 12.6. However, they also are commonly neglected in most weldment fatigue design codes.

12.2.2 Strain–Life (ε–N) Behavior

Cyclic stress–strain data and strain–life (ε–N) fatigue data for the three weldment regions have been determined by Higashida et al. [9] for A36 steel, A514 steel, and 5083 aluminum alloy. Cyclic softening was predominant in the A36 steel WM and HAZ metal. Even greater cyclic softening occurred in the A514 WM, BM, and HAZ metal. For the aluminum alloy, cyclic hardening occurred in the WM and BM. Strain–life (ε–N) curves for the A36 steel weldment materials are shown in Fig. 12.7 and monotonic, cyclic stress–strain, and ε–N properties for the different materials are given in Table 12.1. The general trend found for the ε–N curves for the three weldment regions is that

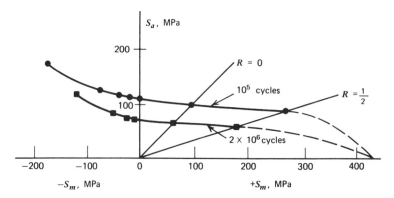

Figure 12.6 Constant-life diagram for carbon steel butt welds [7] (reprinted by permission of the American Society for Testing and Materials).

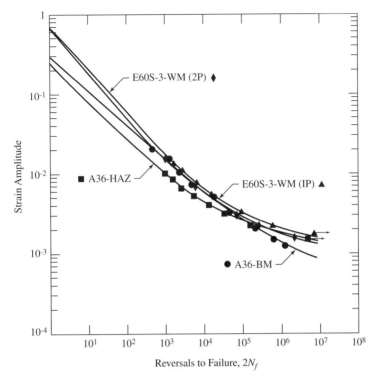

Figure 12.7 Strain-controlled fatigue behavior of A36 steel weldment materials [9] (reprinted with permission of the American Welding Society).

the softer materials have better fatigue resistance at short lives, while the harder materials have better fatigue resistance at long lives. This finding agrees with the ε–N data described in Fig. 5.13.

12.2.3 Crack Growth (*da/dN–*ΔK) Behavior

Maddox [10] obtained region II fatigue crack growth data for weld metals with S_y ranging from 386 MPa (56 ksi) to 636 MPa (92 ksi), a simulated heat-affected zone, and C-Mn base metal, as shown in Fig. 12.8. Eleven different conditions are superimposed in this figure. The scatter band for all stress intensity factor ranges varied from 2 to 1 and from 3 to 1 for crack growth rates between 10^{-8} and 3×10^{-6} m/cycle (4×10^{-7} and 1.2×10^{-4} in./cycle), respectively. This small scatter band implies that region II fatigue crack growth behavior is similar in sound weldments. James [11] also indicated that this same behavior occurs in pressure vessel steel weldments. Fatigue crack growth in weldments is also somewhat independent of tensile mean stress in region II. The small scatter in *da/dN–*ΔK behavior and the small influence from S_m for

TABLE 12.1 Monotonic and Cyclic Strain Properties of Some Weld Materials: SI Units [9]

	S_u (MPa)	S_y/S_y' (MPa/MPa)	K/K', (MPa/MPa)	n/n'	$\varepsilon_f/\varepsilon_f'$	σ_f/σ_f' (MPa/MPa)	b	c
A36 base metal	414	225/230	780/1100	/0.25	1.19/0.27	950/1015	−0.132	−0.45
A36 HAZ	667	530/400	980/1490	0.102/0.215	0.74/0.22	920/720	−0.070	−0.49
E60S weld metal	710	580/385	990/1010	0.098/0.155	0.59/0.61	990/900	−0.075	−0.55
E60 weld metal	580	410/365	850/1235	0.130/0.197	0.93/0.60	1015/1030	−0.090	−0.57
A514 base metal	938	890/600	1190/1090	0.060/0.091	0.99/0.97	1490/1305	−0.080	−0.70
A514 HAZ	1408	1180/940	2110/1765	0.092/0.103	0.75/0.78	2250/2000	−0.087	−0.71
E110S weld metal	1035	835/650	1560/2020	0.092/0.177	0.86/0.85	2210/1890	−0.115	−0.73
E110 weld metal	910	760/600	1290/1670	0.085/0.166	0.90/0.59	1660/1410	−0.079	−0.59
5083-0 aluminum base metal	294	130/290	300/580	0.129/0.114	0.36/0.40	415/710	−0.122	−0.69
5183 aluminum weld metal	299	140/270	310/510	0.133/0.072	0.40/0.58	420/640	−0.107	−0.89

Reprinted with permission of the American Society for Testing and Materials.

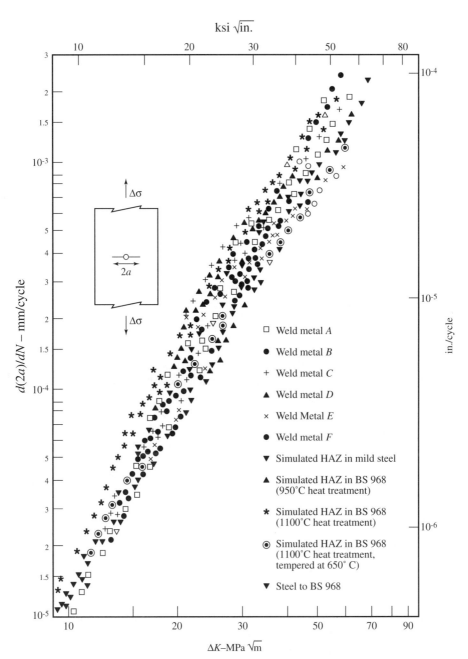

Figure 12.8 Fatigue crack growth data for structural C–Mn steel weld metals, HAZ, and base metals [10] (reprinted with permission of the American Welding Society).

a variety of steel weld metals and base metals have been important influences in formulating steel weldment design codes.

12.2.4 Spot Welds

Spot welds usually fail in fatigue through the sheet at long lives and through the nugget at short lives or in monotonic loading. Failure through the nugget occurs by nugget shear or nugget pullout from the heat-affected zone. Fatigue failure through the sheet involves initial cracking at the nugget edge in the heat-affected zone, with cracks then growing through the base metal. Fatigue strengths can be substantially less than those for the base metal alone, and this can be significantly attributed to high multiaxial stress concentrations. Typical spot weld lap joint fatigue strengths can range from 15 to 30 percent of the base metal fatigue strength with single or multiple rows of spots. The number of spot weld rows and their alignment can alter fatigue resistance. The ultimate tensile strength of the base metal, as with other weldments, does not have a major influence on fatigue of spot welds [6]. Fatigue strength of spot welds decreases with greater sheet thickness due to the usual thickness effects and additional secondary bending stresses from out-of-plane loading of lap joints. S–N, ε–N, da/dN–ΔK, and the two-stage methods have been used for spot weld fatigue design. Modifications of stress intensity factors for spot welds have been made to incorporate the stress ratio, weldment geometry, and the presence of mixed-mode fatigue crack growth. See [12] for additional information on spot weld design.

12.3 IMPROVING WELDMENT FATIGUE RESISTANCE

There are essentially four basic ways of improving weldment fatigue resistance:

1. Improve the actual welding procedure.
2. Alter the material microstructure.
3. Reduce geometrical discontinuities.
4. Induce surface compressive residual stresses.

These four methods actually overlap, since improving the welding procedure can improve the microstructure, reduce the stress concentrations, and alter residual stresses. We mentioned earlier that substantial differences in weld metal ultimate tensile strengths have not provided accompanying large differences in the fatigue resistance of as-welded joints. Thus, in improving weldment fatigue resistance, the major emphasis is on reducing geometrical discontinuities and the use of compressive residual stresses. Many of these improvements involve post weld treatments that add cost to the product and, therefore, perhaps have not received enough use during manufacture. In many instances, careful consideration of applying improvement methods only to critical areas may significantly justify the expense. Post weld improvement methods have also been used for modifications, repairs, and when higher applied loads are planned.

Reducing Geometrical Discontinuities   Removing the reinforcement from butt welds is the most obvious improvement. Careful grinding can be the quickest way to accomplish this. Fillet weld toe profiles can be improved by careful grinding, but since fatigue cracks often grow from either the toe or the root, this may sometimes shift the failure location without increasing fatigue resistance, particularly with only partial penetration welds. Tungsten arc inert gas (TIG) dressing of welds causes local remelt and can improve the weld profile by rounding the fillet toe and removing trapped inclusions, porosity, surface cracks, and undercuts at the weld toe. Large changes in stiffness at welds should be avoided. Choose butt welds rather than lap joints if possible. Start/stop positions, weld ends, and even arc strikes are common locations for fatigue cracks to nucleate and grow. Careful local grinding can improve the fatigue resistance of these regions. All of the above grinding and dressing operations must be done with care. Otherwise, stress concentrations greater than that resulting from the original welding can occur, lowering fatigue resistance. Intersecting welds and excess surface ripples should be avoided. Partial penetration essentially provides a crack-like planar discontinuity that should be avoided. Thus, good detail design can appreciably increase weldment fatigue resistance. Additional beneficial design details can be found in references and design codes discussed in Section 12.4.

Altering Residual Stresses   Common methods used to induce desirable surface compressive residual stresses in weldments include shot- and hammer-peening, surface rolling, spot heating, and tensile or proof overloading. Thermal stress relieving has been used to reduce tensile residual stresses. Each of these operations can provide a substantial increase in fatigue resistance or can be ineffective. This inconsistency is due to relaxation of the residual stresses in the lower-strength weldment materials caused by repeated loading. For example, a single high compressive load can remove much of the desirable compressive residual stress. Since residual stresses can be as high as the yield strength, better improvement from compressive residual stresses can be expected with the higher-strength weldments.

It appears that mild usage of the above methods for inducing compressive residual stresses will not substantially improve fatigue resistance. Reemsnyder [7] indicated that shot-peening of non-load-carrying fillet welded carbon steel and butt welded Q&T alloy steel produced 20 to 40 percent increases in S_f. However, peening to half of the arc height for the above welds resulted in only 7 percent improvement. For significant and repeatable improvements, shot-peening must be closely controlled. Hammer-peening requires even greater quality assurance control in order to ensure repeatable and adequate

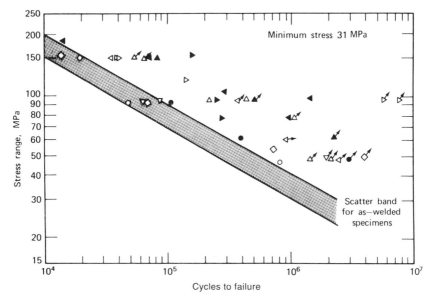

Figure 12.9 The effect of potential life improvement methods of non-load-carrying Al–Zn–Mg (7005) fillet welds [13] (reprinted by permission of the American Society for Testing and Materials) (○) shot-peened, (▽) ground, (●) ground and shot-peened, (△) hammer-peened, (▲) ground and hammer-peened, (◁) one preload, (▷) ten preloads, (◀) one overload every 1000 cycles, (▶) 10 overloads every 1000 cycles, (◇) grit blast and paint, (↗) unfailed or failure remote from the transverse weld toe.

compressive residual stress fields and depths. The fatigue resistance of spot welds can be increased by static overloading or by applying compression (coining) directly to the spot.

Weber [13] determined the effect of various methods of inducing compressive residual stresses in 7005 aluminum alloy weldments. These results are shown in Fig. 12.9. Light shot-peening or grit blasting did not improve fatigue resistance, but a more severe hammer-peening was effective. Multiple tensile overloads were better than single overloads.

12.4   WELDMENT FATIGUE LIFE ESTIMATION

12.4.1   General Weldment Fatigue Life Models

General methods for estimating fatigue life are given in previous chapters. One method involves using S–N behavior and the Haigh or modified Goodman diagrams for mean stress. Another method uses ε–N behavior with a notch strain analysis model and Morrow or SWT models for mean stress. A third method involves da/dN–ΔK integration with Forman, Walker, or other models for mean stress. The two-stage method using both ε–N and da/dN–ΔK to include both fatigue crack nucleation and fatigue crack growth can also be used. The S–N approach developed from fatigue tests of weldments is historically, and currently, the most common weldment fatigue design method. The aforementioned four methods are applicable to both constant and variable amplitude loading of weldments. The later also involves cycle counting and cumulative damage in which rainflow counting and the Palmgren-Miner linear damage rule are most commonly used. These same procedures can be, and have been, used in weldment fatigue life estimates. However, additional complications exist, involving the following:

1. Determining realistic values of the fatigue notch factor, K_f.
2. Incorporating the local multiaxial stress state caused by the weldment notch geometry even under uniaxial loading.
3. Determining fatigue strength, S_f, strain–life fatigue properties, σ_f', ε_f', c, and b (Eq. 5.14), da/dN–ΔK properties, A and n (Eq. 6.19), or equivalent expressions, and fracture toughness for weldment materials.
4. Including welding and deliberate multiaxial residual stresses when relaxation of residual stresses may be present.
5. Making basic assumptions about what size discontinuities, including cracks, may exist in weldments following welding.
6. Determining realistic stress intensity factors for small cracks in weldments.

Item 1 could be handled by assuming a value of K_f. However, K_f can vary from about 1 up to about 5 for $R = -1$ and 10^6 or 10^7 cycles. Thus, some information on welding quality for each product would be needed.

Item 2 could be handled by using an equivalent nominal stress or an equivalent stress intensity factor. The multiaxial stress state resulting from weld geometry is accounted for in the S–N curve generated for a specific weldment geometry and loading.

The fatigue properties specified in item 3 are often similar to base metal values, which can be used as a starting point in the analysis. If actual weldment fatigue properties are known, they of course should be used.

Residual stress relaxation (item 4) in low-strength weldments is quite common. For higher-strength weldment materials, however, stress relaxation is much less. Hence, surface compressive residual stresses can be one of the best ways to improve the fatigue resistance of high-strength weldment materials.

Item 5 is quite controversial, but since crack-like discontinuities may exist in weldments, calculations have been made based only on fatigue crack growth from an initially small crack-like discontinuity. Successful estimates of fatigue life have also been made using S–N or ε–N data. The two-stage method is also relevant to weldments.

A primary problem of item 6 is deciding what stress to use in the vicinity of a weld notch that contains a steep stress gradient. A simple approach is to use the product of nominal stress and K_t while the crack tip is near the notch discontinuity. However, K_t values in weldments are difficult to obtain and involve significant variability. Stress intensity factor solutions for cracks emanating from notches, although not extensive for either applied or residual stresses, are given in stress intensity factor handbooks [14,15] and in welding design codes and handbooks. The surface elliptical crack is of greatest importance in weldments.

12.4.2   Weldment Fatigue Design Codes and Standards

The above complications have been somewhat overcome with many years of national and international cumulative experience involving many industries and professional societies devoted to formulating and updating weldment fatigue design codes and standards. Examples of these fatigue design codes or standards are given in [16–24]. These codes or standards provide significant quantitative information on weldment fatigue design and inspection and are based primarily on S–N curves developed from tests of component-type welded specimens. BS 7608 [22], published by the British Standards Institution, is a comprehensive weldment design code that has incorporated fatigue provisions of standards for highway and railway bridges, cranes, and offshore structures. It is applicable to many fields of engineering and provides eight classes of S–N design curves for many different steel weldments. The classes are B, C, D, E, F, F2, G, and W, and typical descriptions and class examples are given in Table 12.2. Quantitative S–N design curves for variable amplitude loading are given in Fig. 12.10. These eight S–N curves are recommended for fatigue design of most steel weldments. They are based upon many fatigue tests involving many different weldment configurations and a wide variety of steels, ultimate tensile strengths, and mean stresses. Based upon the small influence of ultimate strength and mean stress, as discussed in Section 12.2, the fatigue design S–N curves shown in Fig. 12.10 are independent of S_u and S_m. The S–N curve for a given class, B to W, is applicable to many different weldment configurations and is based on mean S–N data or mean minus 1 (mean − 1SD) or two standard deviations (mean − 2SD). The classes do not relate to weldment quality, but more to fatigue cracking locations. The curves of Fig. 12.10 are based upon recommended usage of mean fatigue behavior (50 percent survival) minus two standard deviations (mean − 2SD). Assuming a log-normal distribution on life, this results in a calculated survival of over 97 percent with a confidence level of 50 percent.

The S–N curves for each of the eight classes shown in Fig. 12.10 are modeled as log-log straight lines, analogous to Basquin's equation. However, the inverse slope, m, as defined in Fig. 12.10 is used along with the nominal stress range near the weld detail, ΔS, not S_a. For fatigue life $N_f \leq 10^7$ cycles the S–N equation for a given class is

$$N_f\,(\Delta S)^m = C \quad \text{for } N_f \le 10^7 \text{ cycles} \tag{12.1}$$

where $m = 3$, 3.5, or 4 according to the class. C is obtained from the fatigue strength at 10^7 cycles. Values of m and C for $N_f \le 10^7$ cycles are given in Table 12.2 in SI units for each class for both mean and (mean − 2SD) curves. Note that BS 7608 uses the coefficient C as C_0, C_1, and C_2 for mean (mean − 1SD), and (mean − 2SD) S–N curves, respectively. We shall use C as a generic coefficient but recognizing that the value of C depends upon the chosen probability of survival. The equations can be extrapolated to higher fatigue strengths until S_{max} reaches a static design criterion such as yield strength or ultimate strength. For constant amplitude loading, a fatigue limit is assumed at 10^7 cycles and the curves become horizontal (not shown in Fig. 12.10). However, for variable amplitude loading, a continued decrease in fatigue resistance is assumed for $N_f \ge 10^7$ cycles. The inverse slope for this region is taken as $m + 2$, as shown in Fig. 12.10, i.e., 5, 5.5, and 6, resulting in the equation

$$N_f\,(\Delta S)^{m+2} = C \quad \text{for } N_f \ge 10^7 \text{ cycles} \tag{12.2}$$

The coefficients C for $N_f \ge 10^7$ cycles will be different than C for $N_f \le 10^7$ cycles. A thickness correction is also recommended, since weldment fatigue resistance has been found to decrease with an increase in plate thickness. The right side of Eqs. 12.1 and 12.2 are multiplied by the thickness correction, which for nontubular welded joints is

$$(22/t)^{0.25} \tag{12.3}$$

where the plate thickness, t, is in mm. If $t < 22$ mm, no correction is made.

Values of ΔS at 10^7 cycles are given in Table 12.2 as $2S_f$ for the eight different classes. Since the S–N curves are independent of mean stress, the values at 10^7 cycles can represent fully reversed, $R = -1$, constant amplitude fatigue limit ranges. Values of $2S_f$ or S_f can then be compared within the classes or with other notched components. The extremes of ΔS at 10^7 cycles, i.e., $2S_f$, for the recommended (mean − 2SD) curves, are 100 MPa (14.5 ksi) for class B and 25 MPa (3.6 ksi) for class W. This represents a 4 to 1 difference in weldment fatigue strength. At $\Delta S = 50$ MPa (7.25 ksi), a difference in fatigue life of two orders of magnitude occurs, and at $\Delta S = 100$ MPa (14.5 ksi) a 70 to 1 ratio in fatigue life exists for the recommended (mean − 2SD) curves. Thus, based on fatigue strength or fatigue life, a significant difference in fatigue resistance exists in these eight steel weldment classes. The recommended (mean − 2SD) fatigue strength extremes, $2S_f$, at 10^7 cycles yield values of S_f between 50 MPa (7.2 ksi) and 12.5 MPa (1.8 ksi). These values are only a fraction of the ultimate tensile strengths and have values similar to S_{CAT} from Section 7.2.3 and to threshold stress ranges obtained from region I ΔK_{th} values, similar to Fig. 6.16. This indicates that S_f for

TABLE 12.2 BS 7608 Standard Weldment Classes, Selected Examples of Each Class, and Fatigue Design Curve Values: Units for MPa and Cycles [4,22]

Class	Description and Class Example (Arrows Indicate Direction of Loading)	Curve	$N_f \leq 10^7$		$N_f = 10^7$ $2S_f$ MPa
			m	C	
B	Butt weld with penetration and no backing strip. Weld reinforcement dressed flush.	Mean	4.0	2.34×10^{15}	124
		Mean $-$ 2SD		1.01×10^{15}	100
C	Fillet and butt welded joints made from one or both sides. Automatic weld with no stop/starts.	Mean	3.5	1.08×10^{14}	102
		Mean $-$ 2SD		4.22×10^{13}	78
D	Transverse butt weld joining two single plates together. Overfill profile $\mu > 150°$.	Mean	3.0	3.99×10^{12}	74
		Mean $-$ 2SD		1.52×10^{12}	53
E	Intermittent fillet welds with g/h < 2.5.	Mean	3.0	3.29×10^{12}	69
		Mean $-$ 2SD		1.04×10^{12}	47

	Description				
F	Fillet weld at cope hole, for weld continuing around plate ends or not.		Mean	1.73×10^{12}	56
		3.0	Mean − 2SD	6.33×10^{11}	40
F_2	Partial penetration butt or fillet weld.		Mean	1.23×10^{12}	50
		3.0	Mean − 2SD	4.3×10^{11}	35
G	Fillet welded lap joints between plates or sections.		Mean	5.66×10^{11}	38
		3.0	Mean − 2SD	2.50×10^{11}	29
W	Partial penetration weld. Refers to fatigue failure across weld throat based on stress range on weld throat area.		Mean	3.68×10^{11}	33
		3.0	Mean − 2SD	1.6×10^{11}	25

Reprinted with permission of the British Standards Institute.

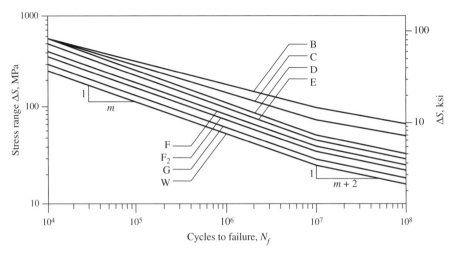

Figure 12.10 BS 7608 standard S–N weldment fatigue design curves (mean - 2SD) for variable amplitude loading of steels [22] (reprinted by permission of the British Standards Institute).

weldments is closely associated with threshold stresses for nonpropagating cracks. Fatigue design codes for aluminum alloys are also very similar to these described for steels, except that different values of C and m exist with the S–N curves. The models in Fig. 12.10 and Table 12.2 can be used in variable amplitude loading using rainflow counting and the Palmgren-Miner linear damage rule, as discussed in Chapter 9. These models can be incorporated into commercial or in-house computer fatigue software. For unprotected seawater environments, Eq. 12.1 for each weldment class is reduced by a factor of 2 on life in BS 7608 and then is linearly extrapolated to all values of N_f with a slope of m.

Example Problem Using BS 7608 A full penetration butt weld joins two 10×50 mm cross section steel plates. A constant amplitude axial load with an R ratio of -0.5 is applied to the joint. It is desired to apply 5×10^5 cycles of this load to the joint. What values of P_{max} and P_{min} can be applied to achieve this life?

From Table 12.2, the full penetration butt weld for the reinforcement (overfill) profile angle θ greater than $150°$ is given as class D. The weldment must be inspected to determine if this is reasonable. If it is not reasonable, we must refer to BS 7608 for the proper class. We will assume that $\theta > 150°$ and use the class D recommended (mean $-$ 2SD) S–N curve with $m = 3.0$ and $C = 1.52 \times 10^{12}$. From Eq. 12.1, which is independent of mean stress,

$$\Delta S = (C/N_f)^{1/m} = (1.52 \times 10^{12}/5 \times 10^5)^{1/3} = 145 \text{ MPa}$$

This value can also be obtained from Fig. 12.10 for the class D curve. For $R = -0.5$

$$P_{min} = -0.5P_{max}$$
$$\Delta S = \Delta P/A = (P_{max} - P_{min})/A = (P_{max} + 0.5P_{max})/A$$
$$= 1.5P_{max}/(50)(10) = 145 \text{ MPa}$$

resulting in

$$P_{max} = \underline{48 \text{ kN}} \quad \text{and then} \quad P_{min} = \underline{-24 \text{ kN}}$$

Thus, according to the BS 7608 recommended procedure, a force varying from -24 to 48 kN could be applied to this axial loaded butt welded joint without a fatigue failure in 5×10^5 cycles.

Hot-Spot Stress/Strain Range The nominal value of ΔS near the weld detail in Eqs. 12.1 and 12.2 may be difficult to obtain in complex weldments, and/or a particular class of weldments may be difficult to ascertain. To circumvent these complexities, a "hot-spot stress range" has been defined to use with a special class T S–N curve similar to class D for all weldments [22]. The hot-spot stress range is computed or measured adjacent to and perpendicular to the weld toe. It is a linear extrapolation of the maximum nominal principal stress in the weld vicinity to the weld toe. It does not incorporate the weld geometrical stress concentration at the toe, since this is already incorporated into the recommended empirical S–N design curve. A hot-spot strain range can be measured adjacent to the weld toe. This measurement should be taken after stable hysteresis loops are formed. These hot-spot stress or strain ranges can be used under constant amplitude loading or with variable amplitude loading by using rainflow cycle counting and the Palmgren-Miner linear damage model. Additional information on use of the hot-spot stress or hot-spot strain range can be found in [16,20,22].

Fitness-for-Purpose Initial planar or volumetric discontinuities in weldments discussed in Section 12.1 play a key role in weldment fatigue design. Nondestructive inspection (NDI) can provide quantitative information on their size and location. These discontinuities may or may not be acceptable for a given service. This acceptance determination for planar (crack-like) discontinuities involves fracture mechanics concepts presented in Chapter 6 that are incorporated within a fitness-for-purpose design philosophy established by the welding profession. This could also be called "fitness-for-service." The British Standards Institution guide PD 6493 [24] provides significant guidelines for evaluating discontinuity acceptability in weldments including fracture, plastic collapse, and fatigue crack growth. Acceptable or unacceptable fatigue crack growth life is determined by integrating the sigmoidal *da/dN*–Δ*K* curve from the initial crack size to the final crack size using realistic service spectra. The Paris fatigue crack growth equation can also be used. For most ferritic steels, PD 6493 uses a Paris equation based on results by Maddox [10], as given in Fig. 12.8. This equation with units of mm/cycle for *da/dN* and MPa√mm for Δ*K* is

$$da/dN = 3 \times 10^{-13} \, (\Delta K)^3 \tag{12.4}$$

and for threshold behavior, the equations are

$$\Delta K_{th} = 170 - 214R \text{ MPa}\sqrt{\text{mm}} \quad \text{for } 0 \leq R \leq 0.5 \tag{12.5a}$$

or

$$\Delta K_{th} = 63 \text{ MPa}\sqrt{\text{mm}} \quad \text{for } R > 0.5 \tag{12.5b}$$

or

$$\Delta K_{th} = 170 \text{ MPa}\sqrt{\text{mm}} \quad \text{for } R < 0 \tag{12.5c}$$

where *R* is the stress ratio based on the combined applied and residual stresses. The integration results, along with a safety factor, can be used for acceptance, rejection, or repair of a specific weldment planar defect. This can also be used to establish NDI inspection periods. For unprotected seawater conditions, Eq. 12.4 is taken as

$$da/dN = 2.3 \times 10^{-12} \, (\Delta K)^3 \tag{12.6}$$

where again, *da/dN* is in mm/cycle and Δ*K* is in MPa√mm. Equations 12.5 a–c remain unchanged. For nonferrous metals with Young's modulus, *E*, PD 6493 uses

$$da/dN = 3 \times 10^{-13} \, (\Delta K)^3 (E_{steel}/E)^3 \tag{12.7}$$

with units of mm/cycle for *da/dN* and MPa√mm for Δ*K* and

$$\Delta K_{th} = \Delta K_{th,steel}(E/E_{steel}) \tag{12.8}$$

For volumetric discontinuities (porosity or slag inclusions) PD 6493 uses *S–N* curve comparisons relating to different levels of porosity or inclusion severity. This provides an acceptance, rejection, or repair criterion.

Additional sources of information on the many aspects of weldment fatigue design are [25–31].

12.5 SUMMARY

Lower-strength steels and aluminum alloys are the most common metals welded. The fatigue resistance of these weldments is rather independent of the base metal's ultimate tensile strength and applied mean stress. However, weldment fatigue life is strongly dependent upon applied stress range, ΔS, weld geometry, and the size and distribution of microscopic and macroscopic discontinuities. The geometry and other discontinuities are influenced by welding procedures. The discontinuities can be planar, such as shrinkage cracks and partial penetration; volumetric, such as porosity and slag inclusions; or geometric, such as weld toe radius, undercutting, and misalignment. Fatigue cracks nucleate and/or grow from these discontinuities, usually at the weld toe or weld root, and can grow in the base metal, weld metal, and/or heat-affected zone. The material fatigue properties of these three regions are often similar. Residual stresses can also be important in weldment fatigue behavior and are usually tensile in the weld region after welding. They can be as high as the yield strength. However, since many weldments are of lower strength, relaxation of residual stresses under service loading is common. This is beneficial for relaxing undesirable welding tensile residual stresses, but it is undesirable for relaxing deliberate post weld compressive residual stresses induced by shot-peening or other methods. Careful control of shot-peening has provided significant improvements in weldment fatigue resistance. Overload stressing for one-direction loading has also been very beneficial. Thermal stress relief has provided some improvements in long-life fatigue resistance, along with little effect if tensile loads dominate. Reducing the geometric stress concentration of weldments by methods such as carefully grinding off the weld reinforcement in butt welds and carefully grinding the toes of fillet welds or using TIG dressing at weld toes may significantly improve fatigue resistance.

Weldment fatigue design can, and has, involved $S-N$, $\varepsilon-N$, $da/dN-\Delta K$, and the two-stage methods. However, weldments present additional fatigue complexities in comparison to other components since each specific weldment contains different macro and micro discontinuities and residual stresses. Discontinuities may be crack-like; hence, fatigue life is entirely fatigue crack growth. Or they may be negligible or more rounded, providing both fatigue crack nucleation and fatigue crack growth lives. To circumvent these weldment complexities, national and international design codes have been established based upon enormous weldment test programs conducted over many years. These weldment design codes are primarily based on $S-N$ methods and to a lesser extent on $da/dN-\Delta K$ methods. Fatigue strength design amplitudes at 10^7 cycles from BS 7608 for steels vary from 12.5 to 50 MPa ($\sim$2 to 7 ksi), which is a very small percentage of the ultimate tensile strength of the base metal. These values are consistent with threshold stress amplitudes for non-propagating cracks.

To overcome the difficulty of determining nominal applied stress ranges and weldment classes, a hot-spot stress or hot-spot strain range has been incorporated into some design codes for use with a single specific $S-N$ curve. Fitness-for-purpose is a key methodology for determining if a specific defect can be accepted, rejected, or repaired. Service inspection periods using NDI can also be established using fitness-for-purpose procedures.

12.6   DOS AND DON'TS IN DESIGN

1. Do recognize that $S-N$, $\varepsilon-N$, $da/dN-\Delta K$, and the two-stage methods outlined in previous chapters for constant and variable amplitude loading can be applied to weldments.
2. Do refer to the many weldment fatigue design codes available.
3. Do recognize that weldment fatigue resistance is less dependent on ultimate tensile strength of the base metal and mean stress and more dependent on applied stress range and class of weld.
4. Do reduce stress concentrations by grinding (significant care is needed) butt welds flush and smooth, dressing fillet welds, and avoiding undercuts and stray arc strikes.
5. Don't locate joints in regions of high tensile stress, and do avoid large variations in stiffness.
6. Don't use intermittent welds, and use butt welds rather than fillet welds.
7. Do consider methods such as shot- or hammer-peening, surface rolling, tensile overloading, and local heating to induce desirable surface compressive residual stresses. Careful control is needed.

REFERENCES

1. K. G. Richards, "Fatigue Strength of Welded Structures," The Welding Institute, Cambridge, May 1969.
2. P. Albrecht, "A Study of Fatigue Striations in Weld Toe Cracks," *Fatigue Testing of Weldments*, D. W. Hoeppner, ed., ASTM STP 648, ASTM, West Conshohocken, PA, 1978, p. 197.
3. T. Jutla, "Fatigue and Fracture Control of Weldments," *Fatigue and Fracture, ASM Handbook*, Vol. 19, ASM International, Materials Park, OH, 1996, p. 434.
4. S. J. Maddox, *Fatigue Strength of Welded Structures*, 2nd ed., Abington Publishing, Cambridge, England, 1991.
5. R. C. Rice, ed., *SAE Fatigue Design Handbook*, AE-22, 3rd ed., SAE, Warrendale, PA, 1997.
6. F. V. Lawrence, S. D. Dimitrakis, and W. H. Munse, "Factors Influencing Weldment Fatigue," *Fatigue and Fracture, ASM Handbook*, Vol. 19, ASM International, Materials Park, OH, 1996, p. 274.

7. H. S. Reemsnyder, "Development and Application of Fatigue Data for Structural Steel Weldments," *Fatigue Testing of Weldments,* D. W. Hoeppner, ed., ASTM STP 648, ASTM, West Conshohocken, PA, 1978, p. 3.

8. W. W. Sanders, Jr., and F. V. Lawrence, Jr., "Fatigue Behavior of Aluminum Alloy Weldments," *Fatigue Testing of Weldments,* D. W. Hoeppner, ed., ASTM STP 648, ASTM, West Conshohocken, PA, 1978, p. 22.

9. Y. Higashida, J. D. Burk, and F. V. Lawrence, Jr., "Strain-Controlled Fatigue Behavior of ASTM A36 and A514 Grade F Steels and 5083-0 Aluminum Weld Materials," *Welding J.,* Vol. 57, 1978, p. 334s.

10. S. J. Maddox, "Assessing the Significance of Flaws in Welds Subject to Fatigue," *Welding J.,* Vol. 53, 1974, p. 401s.

11. L. A. James, "Fatigue-Crack Propagation Behavior of Several Pressure Vessel Steels and Weldments," *Welding J.,* Vol. 56, 1977, p. 386s.

12. H. S. Reemsnyder, "Modeling of the Fatigue Resistance of Single-Lap Spot-Welded Steel Sheet," Document XIII-1469-92, International Institute of Welding, 1992. Available through the American Welding Society, Miami, FL.

13. D. Weber, "Evaluation of Possible Life Improvement Methods for Aluminum-Zinc-Magnesium Fillet-Welded Details," *Fatigue Testing of Weldments,* D. W. Hoeppner, ed., ASTM STP 648, ASTM, West Conshohocken, PA, 1978, p. 73.

14. H. Tada, P. C. Paris, and G. R. Irwin, *The Stress Analysis of Cracks Handbook,* 2nd ed., Paris Productions, St. Louis, MO, 1985.

15. Y. Murakami, ed. in chief, *Stress Intensity Factors Handbook,* Vol. 2, Pergamon Press, Oxford, 1987.

16. *Structural Welding Code—Steel,* ANSI/AWS D1.1-98, American Welding Society, Miami, FL, 1998.

17. "Specifications for Structural Steel Buildings—Allowable Stress Design, Plastic Design," AISC S335, American Institute of Steel Construction, Chicago, 1989.

18. "Guide, Specifications for Fracture Critical Non-redundant Steel Bridge Members," American Association of State Highway and Transportation Officials, Washington, DC, 1991.

19. "Fatigue Design of New Freight Cars," *Manual of Standards and Recommended Practice,* Section C, Vol. 2, Chapter 7, Association of American Railroads, Washington, DC, 1994.

20. "API Recommended Practice for Planning, Designing and Constructing Fixed Offshore Platforms—Working Stress Design," API 2A-WSD, 20th ed., American Petroleum Institution, Washington, DC, 1996.

21. *ASME Boiler and Pressure Vessel Code,* Section IX, American Society of Mechanical Engineers, New York, 1998.

22. "Code of Practice for Fatigue Design and Assessment of Steel Structures," BS 7608:1993, British Standards Institution, London, 1993.

23. "Offshore Installations: Guidance on Design Construction and Certification," 4th ed., U.K. Department of Energy, Her Majesty's Stationery Office, London, 1993.

24. "Guidance on Methods for Assessing the Acceptability of Flaws in Fusion Welded Structures," PD 6493:1991, 2nd ed., British Standards Institution, London, 1991.

25. W. H. Munse and L. Grover, *Fatigue of Welded Structures,* Welding Research Council, New York, 1964.

26. T. R. Gurney, *Fatigue of Welded Structures,* 2nd ed., Cambridge University Press, Cambridge, 1979.

27. S. J. Maddox, ed., *Proceedings of the International Conference on Fatigue of Welded Construction,* Brighton, England, The Welding Institute, Cambridge, 1987.

28. D. Radaj, *Design and Analysis of Fatigue Resistant Welded Structures,* John Wiley and Sons, New York, 1990.

29. A. Hobbacher, *Fatigue Design of Welded Joints and Components,* The International Institute of Welding, Abington Publishing, Cambridge, 1996.

30. M. L. Sharp, G. E. Nordmark, and C. C. Menzemer, *Fatigue Design of Aluminum Components and Structures,* McGraw-Hill Book Co., New York, 1996.

31. D. Radaj and C. M. Sonsino, *Fatigue Assessment of Welded Joints by Local Approaches,* Abington Publishing, Cambridge, 1998.

PROBLEMS

1. A longitudinal butt welded structural steel plate with width $w = 50$ mm and thickness $t = 8$ mm is subjected to an axial load $P_{min} = 3$ kN. What value of P_{max} do you recommend such that 10 million cycles can be applied without fracture for full penetration with weld bead excess material ground off?

2. Repeat Problem 1 if only 50 000 cycles are required.

3. A plate with width $w = 50$ mm and thickness $t = 8$ mm is welded to a wider plate as a lap joint. If P_{min} is 3 kN, what P_{max} can be applied for (a) 10 million cycles, and (b) 50 000 cycles?

4. Determine the fatigue coefficient C in Eq. 12.2 for $N_f \geq 10^7$ cycles using class F_2 weldment.

5. A partial penetration T joint is subjected to the following variable amplitude nominal stress spectrum (one block). How many blocks of this stress spectrum can be applied without fatigue failure?

n (cycles applied)	10^3	10^4	10^3	10^5
S_{max} (MPa)	50	25	100	50
S_{min} (MPa)	10	-25	50	25

6. A butt welded A36 steel plate is subjected to fully reversed bending. A small strain gage is bonded to the plate directly at the weld toe and reads a stable strain range $\Delta\varepsilon = 0.02$. Assuming strain–life fatigue concepts, determine the expected number of cycles to the appearance of a small crack. Repeat the problem if the strain gage reads $\Delta\varepsilon = 0.005$.

7. A full penetration transverse butt welded structure of steel plate with $w = 50$ mm and $t = 10$ mm was subjected to axial repeated loads. A nondestructive inspection indicated that a crack existed at the weld bead toe. It was less than 1 mm deep across the entire section. If $P_{min} = 0$, estimate the value of P_{max} that can be applied without the crack growing to fracture for 50 000 additional cycles. Use a safety factor of 2 on life. List all of your assumptions and comment on the validity of your solution.

8. What can you do to increase the life of the component in Problem 7 assuming that the operating load P_{max} found in Problem 7 cannot be reduced?

CHAPTER 13

STATISTICAL ASPECTS OF FATIGUE

Scatter in fatigue testing or in component fatigue life is a very important consideration in using test data. A variety of factors contribute to scatter. These include inherent material variability (i.e., variations in chemical composition, impurity levels, and discontinuities), variations in heat treatment and manufacturing (i.e., surface finish and hardness variations), variations in specimen or component geometry (i.e., differences in notch radii and weld geometry), and variability from differences in the testing conditions (i.e., environment and test machine alignment variations). In addition, there are sources of uncertainty arising from measured or applied load history variations, as well as from the analytical methods used, such as using the Palmgren-Miner linear damage rule to assess damage for a variable amplitude load history. These variations and uncertainties can result in significant fatigue life variations in the specimen, component, or machine.

Statistical analysis can be used to describe and analyze fatigue properties, as well as to estimate the probability associated with fatigue failure or product life. Such analysis allows us to evaluate component or product reliability quantitatively and to predict service performance for a given margin of safety. Statistical analysis can also be used for the design of experiments such that confounding of sources of variability is avoided and the number of specimen or component tests required for a given reliability and confidence level can be determined.

This chapter first presents basic definitions and concepts in statistical analysis and then discusses commonly used probability distributions for fatigue analysis. Brief discussions of tolerance limits, regression analysis, and reliability analysis are also included. When fatigue life variations are described, the analysis and discussion presented apply equally to fatigue crack nucleation life, fatigue crack growth life, and total life. For more expanded discussion of these and other related topics, books on probability and statistics such as [1], and books on applications of probability and statistics to fatigue analysis such as [2–5], are recommended.

13.1 DEFINITIONS AND QUANTIFICATION OF DATA SCATTER

A quantity such as fatigue strength or fatigue life that has a statistical variation is called a "stochastic variable," x. Characteristics of such a variable for a population are usually obtained from a small part of the population, called a "sample." The mean or average for a sample size n is defined by

$$\bar{x} = \frac{1}{n} \sum_{i=1}^{n} x_i \tag{13.1}$$

The mean value gives a measure of the central value of the sample. Another measure of the central value is the median, which is the middle value in an ordered array of the variable x in the sample. If the sample contains an even number of values, the median is the mean value of the two middle values.

The spread or dispersion of the values in the sample is measured by the sample standard deviation, given by

$$S = \sqrt{\frac{1}{n-1} \sum_{i=1}^{n} (x_i - \bar{x})^2} \tag{13.2}$$

Standard deviation gives a measure of the magnitude of the variation. Other measures of dispersion in the sample are the variance, given by the square of the standard deviation, S^2, and the coefficient of variation, given by

$$C = \frac{S}{\bar{x}} \tag{13.3}$$

The coefficient of variation normalizes the standard deviation with the mean and is therefore a dimensionless quantity, often given as a percentage. The smaller the standard deviation or the smaller the coefficient of variation, the smaller the variability of the sample values (i.e., x values) and the closer the values in the sample are to the sample mean.

13.2 PROBABILITY DISTRIBUTIONS

Variation of the variable x in its range can be described quantitatively by a probability function, $f(x)$. The probability that the variable x assumes any particular value in the range can be specified by this function. This function is also called the "probability density function" or the "frequency function." The probability that the variable x is less than or equal to a particular value in its range of values is given by the "cumulative probability distribution function." This function usually has a sigmoidal shape. The important statistical distributions often used in fatigue and durability analysis are the normal, log-normal, and Weibull distributions. These will now be discussed.

13.2.1 Normal and Log-Normal Distributions

If variations of the variable x are symmetric with respect to the mean, the probability density function, $f(x)$, is represented by a normal or Gaussian distribution expressed as

$$f(x) = \frac{1}{S\sqrt{2\pi}}\, e^{\,-\frac{1}{2}\left(\frac{x-\bar{x}}{S}\right)^2} \tag{13.4}$$

The normal distribution is described in terms of the mean, $\bar{x}$, and standard deviation, S, and has a bell-shaped curve, as shown in Fig. 13.1a. The area under the curve in Fig. 13.1a is unity. If a set of data conforms to a normal distribution, then 68.3 percent of the data fall within $\pm S$ from the mean, 95.5 percent of the data fall within $\pm 2S$ from the mean, and 99.7 percent of the data fall within $\pm 3S$ from the mean. The cumulative frequency function for this distribution is shown in Fig. 13.1b and is given by

$$F(x) = \int f(x)\, dx \tag{13.5}$$

Sinclair and Dolan [6] performed an extensive statistical fatigue study involving 174 so-called identical highly polished, unnotched 7075-T6 aluminum

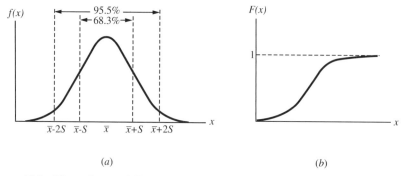

(a)   (b)

Figure 13.1   Plots of normal distribution. (a) Probability density function. (b) Cumulative distribution function.

alloy specimens. They used six different alternating stress levels under fully reversed ($S_m = 0$) conditions. Figure 13.2 shows histograms of the fatigue life distribution of 57 specimens tested at a 207 MPa (30 ksi) stress level. A stepped histogram can be replaced by a frequency distribution curve (dashed lines in Fig. 13.2). When normalized to the unit area, this becomes the probability density function. The curve on the left, Fig. 13.2a, which is based on cycles to failure, is skewed and hence is not a normal or Gaussian distribution. The curve on the right, Fig. 13.2b, is based on the logarithm of cycles to failure and reasonably approximates a normal or Gaussian distribution. This is called a "log-normal distribution," and its probability density function is the same as that in Eq. 13.4, with $x = \log N_f$. Figure 13.3 includes all 174 tests plotted on log-normal probability paper for each value of S_a. If the data are truly log-normal, the data points for each value of S_a will be on a straight line. As seen, a log-normal distribution for the actual data region at each stress level appears reasonable. Based on these and many other statistical test results, a log-normal distribution of fatigue life is often assumed in fatigue design. From the probability distribution functions for each stress level, a family of S–N curves at different probabilities of failure can then be constructed. Such a plot for the fatigue data in Fig. 13.3 is shown in Fig. 13.4.

An analysis of Fig. 13.3 reveals that less scatter occurred at the higher stress levels, as indicated by the steeper slopes. At the highest stress level the life varied from about 1.5×10^4 to 2×10^4 cycles, or a factor of less than 2. At the lowest stress level the life varied from about 2×10^6 to 7×10^7 cycles, or a factor of about 35. Factors of 100 in life are not uncommon for very low stress level fatigue tests. Scatter is usually greater in unnotched, polished, specimens than in notched or cracked specimens. As discussed previously, scatter can be attributed to testing techniques, specimen preparation, and variations in the material. The greater scatter at low stress levels in these

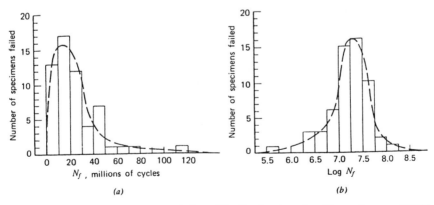

Figure 13.2 Histograms showing fatigue life distributions for 57 specimens of 7075-T6 aluminum alloy tested at 207 MPa (30 ksi) [6] (reprinted with permission of the American Society of Mechanical Engineers).

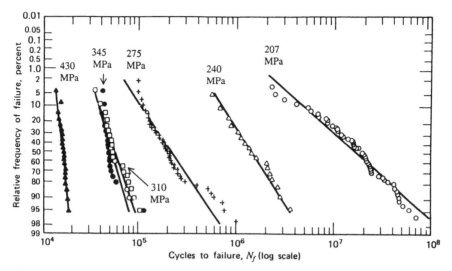

Figure 13.3   Log-normal probability plot at different stress levels for 7075-T6 aluminum alloy [6] (reprinted with permission of the American Society of Mechanical Engineers).

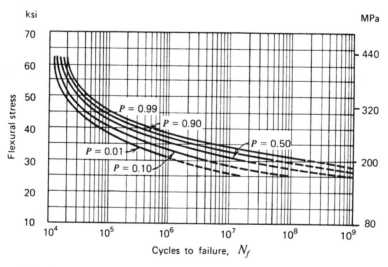

Figure 13.4 S–N curves for different probabilities of failure for specimens of 7075-T6 aluminum alloy [6] (reprinted with permission of the American Society of Mechanical Engineers).

smooth, unnotched specimens can be attributed to the greater percentage of life needed to nucleate small microcracks and then macrocracks. At higher stress levels a greater percentage of the fatigue life involves growth of macrocracks. Tests involving only fatigue crack growth under constant amplitude conditions usually show scatter factors of 2 or 3 or less for identical tests. Thus, the greatest scatter in fatigue involves the nucleation of microcracks and small macrocracks. In notched specimens and components, cracks form more quickly, and subsequently a greater proportion of the total fatigue life involves crack growth that has less scatter.

13.2.2 Weibull Distributions

Weibull distributions are often used in preference to the log-normal distribution to analyze probability aspects of fatigue results. Weibull developed this engineering approach [7] and applied it to the analysis of fatigue test results [8]. Both two- and three-parameter Weibull distribution functions exist, but the two-parameter function is most frequently used in fatigue design and testing. It assumes that the minimum life, N_{fo}, of a population is zero, while the three-parameter function defines a finite minimum life other than zero. The three-parameter Weibull model is

$$F(N_f) = 1 - e^{-\left(\frac{N_f - N_{fo}}{\theta - N_{fo}}\right)^b} \qquad (13.6)$$

where

$$F(N_f) = \text{fraction failed in time or cycles, } N_f$$
$$N_{fo} = \text{minimum time or cycles to failure}$$
$$\theta = \text{characteristic life (time or cycles when 63.2 percent have failed)}$$
$$b = \text{Weibull slope or shape parameter}$$

The terms N_{fo}, θ, and b are three Weibull parameters. The two-parameter Weibull model has $N_{fo} = 0$ and hence

$$F(N_f) = 1 - e^{-\left(\frac{N_f}{\theta}\right)^b} \qquad (13.7)$$

The slope, b, gives a measure of the shape or skewness of the distribution. Two-parameter Weibull distributions for several values of b are shown in Fig. 13.5 [9]. For b between 3.3 and 3.5, the Weibull distribution function is approximately normal or Gaussian, and for $b = 1$ it is exponential. The coefficient of variation (standard deviation/mean) is approximately $C \approx 1/b$ for the two-parameter Weibull distribution. For values of b typical of fatigue, i.e., between 3 and 6, the error from this approximation is about 10 to 15 percent. It should be noted that here we have used $x = N_f$, since Weibull

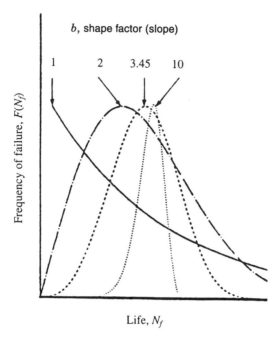

Figure 13.5   Two-parameter Weibull distribution for different values of shape parameter b [9].

distributions are often used for statistical treatment of fatigue life. If a quantity other than fatigue life is being considered, x can be used as the variable in Eqs. 13.6 and 13.7, similar to Eqs. 13.4 and 13.5 for normal or log-normal distributions.

On Weibull probability paper, percent failed is plotted against time or cycles to failure. In general, to plot the data on Weibull probability paper, the array of n data points must first be ordered or ranked from smallest to largest (i.e., $i = 1, 2, 3, \ldots , n$). Next, a plotting position in terms of percent failure is determined for each data point. This can be done by dividing each data point rank by $n + 1$, $i/(n + 1)$. An alternative plotting position is recommended in [10], which is independent of the shape of the distribution and is given by $(i - 0.3)/(n + 0.4)$. This plotting position reduces the bias in estimation of the standard deviation from small samples. Once a plotting position is determined, each data point is plotted at its proper position on the probability paper. This procedure is illustrated by the example problem in Section 13.6 for a two-parameter Weibull probability distribution.

Figure 13.6 is a three-parameter Weibull distribution plot of 1814 fatigue tests of mild steel thin sheet specimens tested at one constant amplitude condition [11]. Data shown represent only partial data from the extensive test program. Several points have been labeled for better clarity, indicating the number of specimens that failed at that life. The minimum life, N_{fo}, of 80 987

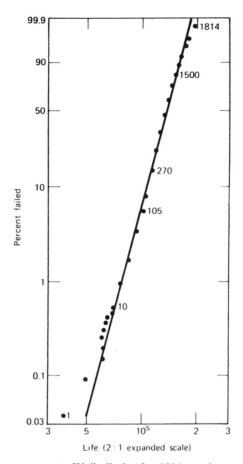

Figure 13.6 Three-parameter Weibull plot for 1814 specimens of mild steel [11].

cycles must be added to the lives in Fig. 13.6 to obtain the total life. On Weibull paper, most data can be plotted as straight lines by judicious choice of N_{fo}. However, to use a three-parameter Weibull distribution, we must be certain that a non-zero minimum life exists below which fatigue failure is unlikely and that all of the data are from a homogeneous sample (i.e., with the same failure mode, the same testing system, the same environment, etc.). Also, as with all distribution function models, the tails at either end may deviate from the model. This is seen in Fig. 13.6. Additional numerical data on scatter in fatigue and its treatment by statistical methods are given in [12].

13.2.3 Estimating Low Probabilities of Failure

The data in Fig. 13.3 for 7075-T6 aluminum are quite comprehensive, yet they do not provide adequate engineering information for fatigue life decisions based on real-life desired probability of failure. Data do not extend past 2 percent probability of failure. Engineers are most interested in 0.01 percent probability of failure or less. This means that extrapolation is needed, for which no accurate method or mathematical justification exists. Only engineering judgment justifies extrapolation. For instance, in Fig. 13.3, if the six stress level curves were extrapolated to desirable low probabilities of failure, they would intersect, which implies that a higher stress level would give longer fatigue life than a lower stress level, which is unreasonable. Great caution must be used in extrapolating fatigue data to low probabilities of failure.

To be able to estimate low probabilities of failure, Abelkis [13] pooled data from more than 6600 specimens of aluminum alloys tested in 1180 groups or samples. He derived a three-term exponential equation:

$$F(z) = A_1 e^{s_1 z} + A_2 e^{s_2 z} + A_3 e^{s_3 z} \qquad \text{for } z < 0 \qquad (13.8)$$

where $z = [\log N_f - \log (\text{median } N_f)]/S$
   N_f = number of cycles to failure
   S = standard deviation of log N_f
   F = probability of failure (F = 50 percent for $z = 0$)

The coefficients and exponents are

$A_1 = 1.687\sqrt{S}$ $s_1 = 1.3 + 0.86\sqrt{S}$
$A_2 = 0.015$ $s_2 = 0.28 + 0.44\sqrt{S}$
$A_3 = 0.485 - 1.687\sqrt{S}$ $s_3 = 1.09 + 2.16\sqrt{S}$

Near the mean, this distribution is almost the same as the log-normal distribution, but for low probabilities of failure it predicts a much lower life than the log-normal distribution. For example, Abelkis shows test data from 2103 specimens in 375 groups. The standard deviation of log N_f is $S = 0.175$. Both the log-normal distribution and his distribution show that 16 percent of specimens failed at $z = -1$ or at a life $1/10^{0.175} = 67$ percent of the median life. Table 13.1 shows values that apply for lower fractions of failure. Use of the log-normal distribution would have predicted five times longer life than Abelkis found for the first five failures of 10 000 specimens. The formula and data from Abelkis [13] are quoted not to recommend them as the best distribution function, but as an example of a function that is more realistic than the log-normal function for estimating low probabilities of failure.

13.3   TOLERANCE LIMITS

As discussed in Section 13.1, statistical variations are usually obtained from a random sample, which is often a small part of the population. As a result,

TABLE 13.1   Expected Fractions of Median Life for Various Probabilities of Failure

	Expected z and Percent of Median Life			
	From Log-Normal		From Abelkis	
Fraction Failed, Percent	z	Fraction of Med. Life, Percent	z	Fraction of Med. Life, Percent
2	-2	45	-2.2	41
0.2	-2.9	31	-4.7	15
0.05	-3.3	26	-7.3	5

the sample parameters such as its mean and variance are not the same as the population mean and variance. Obviously the larger the sample size, the better the quality of the estimate for the population. Tolerance limits are used to better estimate population parameters from sample parameters.

Specifying a confidence level provides a quantitative measure of uncertainty or confidence. It is possible to determine a factor, k, by which a probability of survival, p, with a confidence level, γ, for a sample size, n, can be predicted. The expected value of the variable, x, is then given by

$$x = \bar{x} - k S \tag{13.9}$$

where $\bar{x}$ and S are the sample mean and standard deviation, respectively. The factor k is called the "one-sided tolerance limit factor," and its values, based on normal distributions for various sample sizes, confidence levels, and probabilities of survival, are listed in Table 13.2.

TABLE 13.2   Values of Factor k for One-Sided Tolerance Limits Assuming Normal Distribution [12,14]

	$\gamma = 50\%$			$\gamma = 90\%$			$\gamma = 95\%$		
n	$p = 90\%$	99%	99.9%	90%	99%	99.9%	90%	99%	99.9%
4	1.42	2.60	3.46	3.19	5.44	7.13	4.16	7.04	9.21
6	1.36	2.48	3.30	2.49	4.24	5.56	3.01	5.06	6.61
8	1.34	2.44	3.24	2.22	3.78	4.95	2.58	4.35	5.69
10	1.32	2.41	3.21	2.07	3.53	4.63	2.35	3.98	5.20
20	1.30	2.37	3.14	1.77	3.05	4.01	1.93	3.30	4.32
50				1.56	2.73	3.60	1.65	2.86	3.77
100				1.47	2.60	3.44	1.53	2.68	3.54
500				1.36	2.44	3.24	1.39	2.48	3.28
∞				1.28	2.33	3.09	1.28	2.33	3.09

Example   Ultimate tensile strength values obtained from eight specimens of a material were measured to be 616, 669, 649, 600, 658, 629, 684, and 639 MPa. Determine the tensile strength for 99 percent reliability with 95 percent confidence level.

The mean tensile strength is found to be $\bar{x}$ = 643 MPa, and the standard deviation is calculated from Eq. 13.2 to be S = 27.8 MPa. Assuming normal distribution, with p = 99 percent, γ = 95 percent, and n = 8, Table 13.2 gives k = 4.35. The predicted tensile strength is then calculated from Eq. 13.9 to be

$$x = \bar{x} - k\,S = 643 - 4.35\,(27.8) = 522 \text{ MPa}$$

We can also predict the value of variable x to be within the interval $(\bar{x} - k\,S)$ and $(\bar{x} + k\,S)$ with a specific probability and confidence level. In this case, the factor k used is for two-sided tolerance limits, and its values are different from the one-sided tolerance limits listed in Table 13.2. Values of k for two-sided tolerance limits can be found in [14]. Based on these values, tolerance limits for the variable x can be established.

Tolerance limits can also be established for a Weibull distribution. Lower and upper tolerance limits for the two-parameter Weibull distribution are given by

$$\text{Lower limit} = F(N_f) - k \qquad\qquad (13.10)$$

$$\text{Upper limit} = F(N_f) + k \qquad\qquad (13.11)$$

where k is a function of the sample size, n, given in Fig. 13.7 for a 90 percent confidence level. Establishing the lower and upper tolerance intervals for the two-parameter Weibull distribution is illustrated by the example problem in Section 13.6.

13.4   REGRESSION ANALYSIS OF FATIGUE DATA

Regression analysis is used to obtain a curve that best fits a set of data points. In fatigue data analysis, a linear or linearized regression is often employed using a least squares fit, where the square of the deviations of the data points from the straight line is minimized. The equation of the straight line can be expressed by

$$y = a + b\,x \qquad\qquad (13.12)$$

where x is the independent variable and y is the dependent variable. Fitting constants a and b are the regression coefficients, where b is the slope of the line and a is the y intercept. Such linear regression is often used for analysis of S–N, ε–N, and da/dN–ΔK fatigue data and curves. For example, the coeffi-

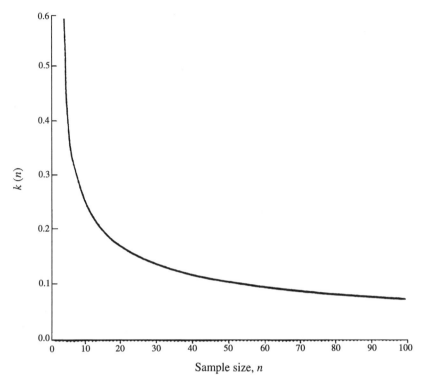

Figure 13.7 Variation of factor k with sample size n for a 90 percent confidence level for the Weibull distribution [9].

cients A and B in Basquin's log-log straight line S–N relationship (Eq. 4.7) are found by least squares fit of (log S_a) versus (log N_f) data. It should be recognized that since the stress amplitude, S_a, is the input or controlled parameter in S–N tests, it is the independent variable. Fatigue life, N_f, is considered to be the response or observed parameter in the test and is therefore the dependent variable. For S–N data, Eq. 13.12 is then written as

$$\log N_f = a + b \ (\log S_a) \tag{13.13}$$

The fitting constants a and b in Eq. 13.13 are related to the coefficient A and exponent B in Basquin's Eq. 4.7 by $a = -(1/B)\log A$ and $b = 1/B$. Similarly, linear regressions are used to obtain the strain–life fatigue properties σ_f' and b, and ε_f' and c in Eqs. 5.15 and 5.16, respectively. Examples of such a regression for $\Delta\sigma/2$ versus $2N_f$ and $\Delta\varepsilon_p/2$ versus $2N_f$ for a 4340 steel are shown in Figs. 5.12b and 5.12c, respectively. Similar linear regression analysis is used for da/dN versus ΔK data in the Paris equation regime to find the coefficient A and slope n in Eq. 6.19 or other similar equations for fatigue crack growth rates.

The degree of correlation between two variables such as x and y in Eq. 13.12 can be quantified by the correlation coefficient, r. The correlation coefficient has a range of values between -1 and $+1$. A value of $r = 0$ indicates no correlation, whereas $r = \pm 1$ indicates perfect correlation. A negative value of r indicates that the regression line has a negative slope. ASTM Standard E739 provides additional details on statistical analysis of linear or linearized stress–life (S–N) and strain–life (ε–N) fatigue data [15].

13.5 RELIABILITY ANALYSIS

In most cases, the service load spectra for a component or structure are probabilistic rather than deterministic. Therefore, statistical analysis is often required to obtain the stress distribution. In addition, as discussed in previous sections, the fatigue strength of the component or structure has a statistical distribution due to the many sources of variability. Schematic representations of probability density functions for both service loading and fatigue strength are shown in Fig. 13.8. If there were no overlap between the two distributions, failure would not occur. However, there is often an overlap between the two distributions, as shown in Fig. 13.8, indicating the possibility of fatigue failure. The overlap can increase as the strength decreases due to accumulation of damage with continued use of the component or structure. Reliability analysis provides an analytical tool to quantify the probability of failure due to the overlapping of the two distributions.

The reliability or probability of survival, R, is obtained from

$$R = \int_0^\infty F_1(x)\, F_2(x)\, dx \tag{13.14}$$

where $F_1(x)$ and $F_2(x)$ are the cumulative density functions of the service stress and fatigue strength, respectively. A detailed presentation and discussion

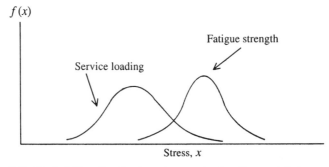

Figure 13.8   Probability distributions for service loading and fatigue strength.

of reliability analysis is beyond the scope of this book, and the interested reader is referred to books such as [16–18].

13.6 EXAMPLE PROBLEM USING THE WEIBULL DISTRIBUTION*

A durability test on 10 units resulted in failures at 140, 90, 190, 220, 270, 200, 115, 170, 260, and 330 hours. Assume the life distribution to be a two-parameter Weibull. Determine (a) the slope, b, and characteristic life, θ, for the distribution, (b) the median and B_{10} lives, (c) the percentage of the population that would be expected to fail in 300 hours with 50 percent confidence, and (d) the 90 percent tolerance interval.

The data are first rank ordered in ascending order, and the plotting position for percent failure is determined from $[(i - 0.3)/(n + 0.4)]$ 100 percent, as listed below

Rank, i	Life to Failure (hours)	Plotting Position (%)
1	90	6.7
2	115	16.2
3	140	25.9
4	170	35.5
5	190	45.2
6	200	54.8
7	220	64.5
8	260	74.1
9	270	83.8
10	330	93.3

Life to failure versus percent failure data are then plotted on the Weibull probability plot shown in Fig. 13.9. A straight line is passed through the data.

(a) A line parallel to the line through the data and passing through the index point shown in the plot by $\oplus$ intersects the slope scale on top of the figure and gives the Weibull slope, $b = 2.77$. The characteristic life, θ, is found from the line passing through the data at 63.2 percent failure, $\theta = 225$ hours. The Weibull distribution function is then given as

$$F(N_f) = 1 - e^{-\left(\frac{N_f}{225}\right)^{2.77}}$$

* This example problem is courtesy of Dr. H. S. Reemsnyder and is used in the SAEFDE/ University of Iowa short course on Fatigue Concepts in Design. Data for this example were supplied by H. R. Jaeckel.

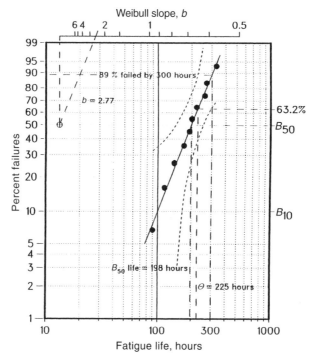

Figure 13.9 Example of two-parameter Weibull plot showing median line and 90 percent tolerance interval [9].

(b) The median life is the B_{50} life, which is found from the plot to be $B_{50} = 198$ hours. The B_{10} life is also found from the plot to be $B_{10} = 100$ hours. This is the life corresponding to a 10 percent failure rate.

(c) The line fitted through the data corresponds to a 50 percent confidence level. Therefore, the percentage of the population that would be expected to fail in 300 hours with 50 percent confidence is found from this line for a fatigue life of 300 hours to be 89 percent.

(d) The 90 percent tolerance interval for the data is obtained by computing and plotting the lower and upper tolerance limits from Eqs. 13.10 and 13.11, respectively. The value of k for $n = 10$ is found from Fig. 13.7 to be 0.25. Therefore,

$$\text{Lower limit} = F(N_f) - k = 0.75 - e^{-\left(\frac{N_f}{225}\right)^{2.77}}$$

$$\text{Upper limit} = F(N_f) + k = 1.25 - e^{-\left(\frac{N_f}{225}\right)^{2.77}}$$

The lower and upper tolerance limits for various lives are calculated from the above equations, multiplied by 100 to convert them to a percentage, and then plotted, as shown by the dashed curves in Fig. 13.9.

13.7 SUMMARY

Many factors contribute to scatter in fatigue life and properties. Statistical analysis is used to describe scatter quantitatively and to estimate the probability associated with fatigue failure or predicted life. Basic characteristics of data scatter for a sample of data are measured from the sample's mean and standard deviation, and data variation is described by a probability distribution function. Common probability distribution functions in fatigue analysis are the normal, log-normal, and Weibull distributions. The normal distribution assumes symmetric variation about the mean. When $\log(x)$ of a variable x has a normal distribution, the variable x is said to have a log-normal distribution. Distributions of fatigue life data are often skewed, and two- and three-parameter Weibull distributions are most commonly used to analyze probability aspects of fatigue life. At low probabilities of failure the distribution function for fatigue life may differ from the distribution function near the mean. Extrapolations to low probabilities of failure may therefore not be appropriate. Tolerance limits are used to quantify the uncertainty in estimating the population's statistical characteristics from smaller sample statistical parameters. Regression analysis is used to obtain best fit curves to fatigue data, and reliability analysis is used to estimate the probability of failure associated with overlapping of the applied stress and the material or component strength probability distributions.

13.8 DOS AND DON'TS IN DESIGN

1. Don't ignore data scatter in fatigue testing and analysis, as it can result in significant variations in predicted fatigue life.
2. Do use great caution in extrapolating fatigue data to low probabilities of failure, as the probability distribution function for low failure probabilities may be different than for higher failure probabilities.
3. Do use tolerance limit analysis when extrapolating statistical parameters and characteristics from a small sample of data to a much larger population.
4. Do recognize that in regression analysis of fatigue data the fatigue life should be treated as the dependent variable (x), while load, stress, or strain is the independent variable (y).

REFERENCES

1. R. A. Johnson, *Miller and Freund's Probability and Statistics for Engineers,* 5th ed., Prentice-Hall, Upper Saddle River, NJ, 1994.

2. R. R. Little, *Manual on Statistical Planning and Analysis for Fatigue Experiments,* ASTM STP 588, ASTM, West Conshohocken, PA, 1975.

3. L. G. Johnson, *The Statistical Treatment of Fatigue Experiments,* Elsevier Science, New York, 1964.

4. J. L. Bogandoff and F. Kozin, *Probabilistic Models of Cumulative Damage,* John Wiley and Sons, New York, 1985.

5. W. Nelson, *Accelerated Testing: Statistical Models, Test Plans, and Data Analysis,* John Wiley and Sons, New York, 1990.

6. G. M. Sinclair and T. J. Dolan, "Effect of Stress Amplitude on Statistical Variability in Fatigue Life of 75S-T6 Aluminum Alloy," *Trans. ASME,* Vol. 75, 1953, p. 867.

7. W. Weibull, "A Statistical Distribution Function of Wide Applicability," *J. Appl. Mech.,* Vol. 73, 1951, p. 293.

8. W. Weibull, *Fatigue Testing and Analysis of Results,* Pergamon Press, London, 1961.

9. H. S. Reemsnyder, SAEFDE/The University of Iowa Short Course Notes on Fatigue Concepts in Design, 1999.

10. A. Bernard and E. D. Bos-Levenbach, "The Plotting of Observations on Probability Paper," *Statistica,* Vol. 7, 1953, p. 163.

11. W. E. Hering and C. W. Gadd, "Experimental Study of Fatigue Life Variability," Research Publication, GMR-555, General Motors Corp., 1966.

12. *A Guide for Fatigue Testing and the Statistical Analysis of Fatigue Data,* 2nd ed., ASTM STP 91-A, ASTM, West Conshohocken, PA, 1963.

13. P. R. Abelkis, "Fatigue Strength Design and Analysis of Aircraft Structures, Part I: Scatter Factors and Design Charts," AFFDL-TR 66-197, June 1967.

14. R. E. Odeh, D. B. Owen, Z. W. Birnbaum, and L. Fisher, *Pocket Book of Statistical Tables,* Marcel Dekker, New York, 1977.

15. *Standard Practice for Statistical Analysis of Linear or Linearized Stress–Life (S–N) and Strain–Life (ε–N) Fatigue Data,* ASTM Standard E 739, Vol. 3.01, ASTM, West Conshohocken, PA, 2000, p. 631.

16. N. R. Mann, N. D. Singpurwalla, and R. E. Schafer, *Methods for Statistical Analysis of Reliability and Life Data,* John Wiley and Sons, New York, 1974.

17. W. Nelson, *Applied Life Data Analysis,* John Wiley and Sons, New York, 1982.

18. B. S. Dhillon, *Design Reliability: Fundamentals and Applications,* CRC Press, Boca Raton, FL, 1999.

PROBLEMS

1. The following data from 43 specimens were obtained from rotating bending fatigue tests at one constant stress level and were grouped as follows:

Samples failed, n	3	8	6	10	8	5	3
Cycles to failure, N_f	1600	1900	2200	2500	2800	3100	3400

(a) Calculate the mean, median, standard deviation, and coefficient of variation for the fatigue lives. (b) Show the data in histograms of fatigue life distribution using both N_f and log N_f. (c) Fit the stepped histograms in part (b) by frequency distribution curves, assuming normal and lognormal distributions. (d) Plot the cumulative histograms and distribution functions. (e) Determine B_{10} and B_{50} lives for these data, assuming normal distribution.

2. Fatigue testing of eight similar components at a given load level has produced the following fatigue lives: 14 700, 94 700, 31 700, 27 400, 17 300, 70 100, 21 800, and 39 800. (a) Compute the mean life and standard deviation for this load level. (b) Plot the data on normal probability paper, lognormal probability paper, and Weibull probability paper. Which probability distribution do the data fit best? (c) Obtain the B_{50} life from each distribution in part (b).

3. A fatigue life test program on a new product consisted of 10 units subjected to the same load spectrum. The units failed in the following number of hours in ascending order: 75, 100, 130, 150, 185, 200, 210, 240, 265, and 300. Obtain the percent median rank values for the 10 failures. Plot the percent median rank values versus hours to failure on Weibull paper. Draw the best fit (usually a straight line) curve through the data and find (a) 10 percent (B_{10}) expected failure life and (b) 50 percent (B_{50}) expected failure life. (c) What percentage of the population will have failed in 230 hours? (d) Consider what the 1.0 or 0.1 percent expected life would be.

4. Plane strain fracture toughness tests of six identical samples from a metal produced the following values of K_{Ic}: 57.2, 54.0, 59.7, 55.5, 62.4, and 58.6 MPa$\sqrt{m}$. Assuming a normal distribution, estimate the plane strain fracture toughness of the metal with 90 percent reliability and a 90 percent confidence level.

5. Fatigue strength, S_f, at 10^6 cycles versus Brinell hardness (HB) for seven grades of SAE 1141 steel are given as follows:

HB	223	277	199	241	217	252	229
S_f (MPa)	286	433	276	342	287	332	296

(a) Using linear regression, find the slope, intercept, and correlation coefficient for these data. (b) Is there a good correlation between the hardness and fatigue strength of these steel grades? (c) What fatigue strength corresponds to a Brinell hardness of 225?

6. Constant amplitude crack growth rate data in the threshold region for a cast aluminum alloy are listed as a function of stress intensity factor range, ΔK:

da/dN ($\times 10^{-10}$ m/cycle)	8.6	5.9	6.8	3.9	4.5	1.6	2.2
ΔK (MPa$\sqrt{m}$)	1.91	1.84	1.89	1.79	1.84	1.67	1.73

Fit these data with a linear regression and determine the threshold stress intensity factor range, ΔK_{th}, defined at a 10^{-10} m/cycle fatigue crack growth rate.

APPENDIX: MATERIAL PROPERTIES

TABLE A.1 Monotonic Tensile Properties and Fully Reversed, Bending Unnotched Fatigue Limits, S_f, of Selected Engineering Alloys

Material	Process Description	Hardness	S_u MPa (ksi)	S_y MPa (ksi)	%El	%RA	S_f^a MPa (ksi)
Steels[b]			S_f based on 10^6 to 10^8 cycles to failure				
1020	Annealed	111*HB*	393 (57)	296 (43)	36	66	138 (20)
1020	Hot Rolled	143*HB*	448 (65)	331 (65)	36	59	241 (35)
1040	Annealed	149*HB*	517 (75)	351 (51)	30	57	269 (39)
1050	Annealed	187*HB*	634 (92)	365 (53)	24	40	365 (53)
4130	Normalized	197*HB*	668 (97)	434 (63)	26	60	324 (47)
4130	WQ&T 650°C	245*HB*	809 (118)	703 (102)	22	64	489 (71)
4140	OQ&T 650°C	285*HB*	758 (110)	655 (95)	18	53	420 (61)
4140	OQ&T 540°C	358*HB*	1137 (165)	985 (143)	15	50	455 (66)
4340	OQ&T 540°C	380*HB*	1261 (183)	1171 (170)	14	52	668 (97)
4340	OQ&T 425°C	430*HB*	1530 (222)	1378 (200)	12	47	468 (68)
5140	OQ&T 540°C	311*HB*	1068 (155)	923 (134)	17	53	620 (90)
5140	OQ&T 425°C	375*HB*	1309 (190)	1164 (169)	12	42	565 (82)
8640	OQ&T 540°C	331*HB*	1068 (155)	944 (137)	17	56	537 (78)
HY-140	Q&T 540°C	34*Rc*	1027 (149)	978 (142)	20	65	482 (70)
H-11	Q&T	52*Rc*	1791 (260)	1447 (210)	8	52	634 (92)
300M	Q&T 260°C	52*Rc*	1791 (260)	1585 (230)	12	37	620 (90)
D6AC	Q&T 260°C	54*Rc*	1998 (290)	1722 (250)	9	36	689 (100)
9Ni–4Co–25	T 540°C	36*Rc*	1378 (200)	1309 (190)	17	70	758 (110)
9Ni–4Co–45	T 315°C	48*Rc*	1929 (280)	1757 (255)	5	35	620 (90)
18Ni 200 marage	Aged 480°C	43*Rc*	1550 (225)	1481 (215)	11	55	689 (100)
18Ni 250 marage	VM Aged 480°C	50*Rc*	1764 (256)	1633 (237)	11	62	689 (100)
18Ni 300 marage	VM Aged 480°C	55*Rc*	1984 (288)	1922 (279)	7	50	758 (110)
18Ni 350 marage	VM Aged 480°C	59*Rc*	2425 (352)	2377 (345)	8	44	758 (110)

Material	Condition						S_f based on 5×10^8 cycles to failure
302	Annealed	80Rb	640 (93)	276 (40)	68	65	234 (34)
302	CR 40%	35Rc	1040 (151)	909 (132)	13	8	517 (75)
304	Annealed	80Rb	599 (87)	234 (34)	57	67	241 (35)
304	CW 10%	10Rc	675 (98)	482 (70)	35		413 (60)
304	CW 40%	35Rc	1006 (146)	930 (135)	12		634 (92)
316	Annealed	77Rb	586 (85)	262 (38)	61	67	269 (39)
403	Annealed	155HB	517 (75)	310 (45)	30	70	276 (40)
403	T 650°C	97Rb	758 (110)	585 (85)	23	65	379 (55)

Aluminum Alloys[c]

Material	Condition						
1100-0	Annealed	23HB	90 (13)	35 (5)	45		35 (5)
2014-T6	Sol. Treat Aged	135HB	482 (70)	413 (60)	13		124 (18)
2024-T3	Sol. Treat CW Aged	120HB	482 (70)	345 (50)	18		138 (20)
2024-T4	Sol. Treat Aged	120HB	468 (68)	324 (47)	19		138 (20)
2219-T851	Sol. Treat CW Aged		455 (66)	351 (51)	10		103 (15)
3003-H16	Strain Hardened	47HB	179 (26)	172 (25)	14		69 (10)
3004-H36	Strain Hardened	70HB	262 (38)	227 (33)	9		110 (16)
6061-T4	Sol. Treat Aged	65HB	242 (35)	145 (21)	25		96 (14)
7075-T6	Sol. Treat Aged	150HB	572 (83)	503 (73)	11		158 (23)

Others

Material	Condition						
Ti[d]	Annealed		520 (75)				330 (52)
Ti-6A1-4V[e]			1190 (172)	1090 (158)			365 (53)
Copper[d]	Annealed		235 (34)	75 (11)			75 (11)
60/40 brass[d]	Annealed		315 (46)	95 (14)			85 (12)
Phospher bronze[d]	Annealed		340 (49)	150 (22)			170 (25)

[a] Incomplete information on surface finish. These values *do not* represent design fatigue limits.
[b] Structural Alloys Handbook, CINDAS/Purdue University, West LaFayette, IN, 1996.
[c] Aluminum Standards and Data 1997, The Aluminum Association, Washington, DC, 1997.
[d] Metals Handbook, Vol. 1, ASM, Materials Park, OH, 1961.
[e] Mil-Handbook 5, Department of Defense, Washington, DC.

TABLE A.2 Monotonic, Cyclic, and Strain–Life Properties of Selected Engineering Alloys[a-c]

Material	Process Description	S_u MPa (ksi)	HB	E GPa (ksi·10^3)	%RA	S_y/S_y' MPa (ksi)	K/K' MPa (ksi)	n/n'	$\varepsilon_f/\varepsilon_f'$	σ_f/σ_f' MPa (ksi)	b	c
								Steel				
1010	HR sheet	331 (48)	—	203 (29.5)	80	200/— (29)/—	534/867 (78)/(126)	0.185/0.244	1.63/0.104	—/499 —/(72)	−0.100	−0.408
1020	HR sheet	441 (64)	109	203 (29.5)	62	262/— (38)/—	738/1962 (107)/(284)	0.190/0.321	0.96/0.377	—/1384 —/(201)	−0.156	−0.485
1038[c]	Normalized	582 (84)	163	201 (29.5)	54	331/342 (48)/(50)	1106/1340 (160)/(195)	0.259/0.220	0.77/0.309	898/1043 (130)/(151)	−0.107	−0.481
1038[c]	Q&T	649 (94)	195	219 (31.5)	67	410/364 (60)/(53)	1183/1330 (172)/(193)	0.221/0.208	1.10/0.255	1197/1009 (174)/(146)	−0.097	−0.460
Man-Ten	HR sheet	510 (74)	—	207 (30)	64	393/372 (57)/(54)	—/786 —/(114)	0.20/0.11	1.02/0.86	814/807 (118)/(117)	−0.071	−0.65
RQC-100	HR sheet	931 (135)	290	207 (30)	64	883/600 (128)/(87)	1172/1434 (170)/(208)	0.06/0.14	1.02/0.66	1330/1240 (193)/(180)	−0.07	−0.69
1045	Annealed	752 (109)	225	—	44	517/— (75)/—	—/1022 —/(148)	—/0.152	0.58/0.486	—/916 —/(133)	−0.079	−0.520
1045	Q&T	1827 (265)	500	207 (30)	51	1689/— (245)/—	—/3371 —/(489)	0.047/0.145	0.71/0.196	—/2661 —/(386)	−0.093	−0.643
1090[c]	Normalized	1090 (158)	259	203 (29.5)	14	735/545 (107)/(79)	1765/1611 (256)/(234)	0.158/0.174	0.15/0.250	—/1310 —/(190)	−0.091	−0.496
1090[c]	Q&T	1147 (166)	309	217 (31.5)	22	650/627 (94)/(91)	1895/1873 (275)/(272)	0.165/0.176	0.24/0.700	—/1878 —/(273)	−0.120	−0.600
1141[c]	Normalized	789 (115)	229	220 (32)	47	493/481 (72)/(70)	1379/1441 (200)/(209)	0.187/0.177	0.64/0.602	1117/1326 (162)/(192)	−0.103	−0.581
1141[c]	Q&T	925 (134)	277	227 (33)	59	814/591 (118)/(86)	1205/1277 (125)/(185)	0.074/0.124	0.88/0.309	1405/1127 (204)/(164)	−0.066	−0.514
4142	Q&T	1413 (205)	380	207 (30)	48	1378/— (200)/—	—/2266 —/(387)	0.051/0.124	0.65/0.637	—/2143 —/(311)	−0.094	−0.761
4142	Q&T	1929 (280)	475	207 (30)	35	1722/— (250)/—	—/2399 —/(348)	0.048/0.094	0.43/0.331	—/2161 —/(314)	−0.081	−0.854
4340	HR	827 (120)	243	193 (28)	43	634/— (92)/—	—/1337 —/(194)	—/0.168	0.57/0.522	—/1198 —/(174)	−0.095	−0.563

Material	Condition											
4340	Q&T	1240 (180)	350	193 (28)	57	1178/— (171)/—	1580/1887 (229)/(274)	0.066/0.137	0.84/1.122	—/1917 —/(278)	-0.099	-0.720
4340	Q&T	1468 (213)	409	200 (29)	38	1371/— (199)/—	—/1996 —/(290)	—/0.135	0.48/0.640	—/1879 —/(273)	-0.086	-0.636
0030	Cast	496 (72)	137	207 (30)	46	303/320 (44)/(46)	—/738 —/(107)	—/0.136	0.62/0.280	750/655 (109)/(95)	-0.083	-0.552
8630	Cast	1144 (166)	305	207 (30)	29	985/682 (143)/(99)	—/1502 —/(218)	—/0.122	0.35/0.420	1268/1936 (184)/(281)	-0.121	-0.693
304	Annealed	572 (83)	—	190 (27.5)	—	276/— (40)/—	—/2275 —/(330)	—/0.334	—/0.174	—/1267 —/(184)	-0.139	-0.415
304	CD	951 (138)	327	172 (25)	69	744/— (108)/—	—/2270 —/(329)	—/0.176	1.16/0.554	—/2047 —/(297)	-0.112	-0.635
Aluminum												
2024-T3	—	469 (68)	—	70 (10)	24	379/427 (55)/(62)	455/655 (66)/(95)	0.032/0.065	0.28/0.22	558/1100 (81)/(160)	-0.124	-0.59
5456-H311	—	400 (58)	95	69 (10)	35	234/— (34)/—	—/817 —/(118)	—/0.145	0.42/1.076	—/826 —/(120)	-0.115	-0.797
7075-T6	—	579 (84)	—	70 (10)	34	469/524 (68)/(76)	827/— (120)/—	0.11/0.146	0.41/0.19	745/1315 (108)/(191)	-0.126	-0.52
A356	Cast	283 (41)	93	70 (10)	5.7	229/295 (33)/(43)	388/379 (56)/(55)	0.083/0.043	0.06/0.027	274/594 (40)/(86)	-0.124	-0.530
Others												
AZ91E-T6	Cast Mg.	318 (46)	—	45 (6.5)	13	142/180 (21)/(26)	639/552 (92)/(80)	0.137/0.184	0.14/0.089	356/831 (52)/(121)	-0.148	-0.451
Incon 718	Aged	1304 (189)	—	204 (29.5)	—	1110/— (161)/—	—/1986 —/(288)	—/0.112	—/3.637	—/2295 —/(333)	-0.100	-0.894

[a] These values do not represent final fatigue design properties. J1099 states, "Information presented here can be used in preliminary design estimates of fatigue life, the selection of materials and the analysis of service load and/or strain data."

[b] "Technical Report on Low Cycle Fatigue Properties, Ferrous and Non-Ferrous Materials," SAE J1099, 1998 and 1975. With permission of the Society of Automotive Engineers.

[c] M. L. Roessle and A. Fatemi, "Strain-Controlled Fatigue Properties of Steels and Some Simple Approximations," *Int. J. Fatigue*, Vol. 22, No. 6, 2000, p. 495.

TABLE A.3 Plane Strain Fracture Toughness, K_{Ic}, for Selected Engineering Alloys (Plate Stock, L-T Direction Unless Otherwise Specified)[a,b,c,d]

Material	Process Description	S_y		K_{Ic}	
		MPa	(ksi)	MPa$\sqrt{\text{m}}$	(ksi$\sqrt{\text{in}}$)
		Steel			
11V41[c]	666°C temper	670	(97)	113	(104)
11V41[c]	As-forged	524	(76)	67	(62)
4340	425°C temper	1360–1455	(197–211)	79–89	(72–81)
4340	350°C temper	1380	(200)	66–68	(60–62)
4340	260°C temper	1495–1640	(217–238)	50–63	(45–57)
4330V	275°C temper	1400	(203)	85–92	(77–84)
300M	315°C temper	1625	(236)	56–57	(51–52)
300M	245°C temper	1655	(240)	37–38	(34–35)
D6AC	540°C temper	1495	(217)	102	(93)
HP 9-4-20	550°C temper	1280–1310	(186–190)	132–154	(120–140)
HP 9-4-30	540°C temper	1320–1420	(192–206)	90–115	(82–105)
10Ni (vim)	510°C temper	1770	(257)	54–56	(49–51)
18Ni (200)	Marage	1450	(210)	110	(100)
18Ni (250)	Marage	1785	(259)	88–97	(80–88)
18Ni (300)	Marage	1905	(277)	50–64	(45–58)
Ph13–8Mo	Annealed	1380–1420	(200–206)	113–141	(103–128)

Aluminum

2014-T651		435–470	(63–68)	23–27	(21–25)
2020-T651		525–540	(76–78)	22–27	(20–25)
2024-T351		370–385	(54–56)	31–44	(28–40)
2024-T851		455	(66)	23–28	(21–25)
2124-T851		440–460	(64–67)	27–36	(25–33)
2219-T851		345–360	(50–52)	36–41	(33–37)
7050-T73651		460–510	(67–74)	33–41	(30–37)
7075-T651		515–560	(75–81)	27–31	(25–28)
7075-T7351		400–455	(58–66)	31–35	(28–32)
7079 T651		525–540	(76–78)	29–33	(26–30)
7178-T651		560	(81)	26–30	(24–27)
A356–T6[d]	Cast	217–229	(31–33)	17–18	(15.5–16.5)

Titanium

Ti-6Al–4V	Mill annealed	875	(127)	123	(112)
Ti-6Al–4V	Recrystallized Annealed	815–835	(118–121)	85–107	(77–97)
Ti-6Al–6V–2Sn	Mill annealed	1165	(169)	41–51	(37–47)
Ti-6Al–6V–2Sn	Mill annealed	990–1070	(144–155)	48–67	(44–61)

[a] *Damage Tolerant Design Handbook*, CINDAS/Purdue University, Lafeyette, IN, 1994.
[b] R. W. Hertzberg, *Deformation and Fracture Mechanics of Engineering Materials*, 4th ed., John Wiley and Sons, New York, 1996. Reprinted by permission of John Wiley & Sons, Inc.
[c] Courtesy of A. Fatemi.
[d] Courtesy of R. I. Stephens.

TABLE A.4 Fatigue Crack Growth Threshold, ΔK_{th}, for Selected Engineering Alloys

Material	S_u, MPa (ksi)	$R = K_{min}/K_{max}$	ΔK_{th} MPa$\sqrt{m}$ (ksi$\sqrt{in}$)
Mild steel[a]	430 (62)	0.13	6.6 (6.0)
		0.35	5.2 (4.7)
		0.49	4.3 (3.9)
		0.64	3.2 (2.9)
		0.75	3.8 (3.5)
A533B[b]	—	0.1	8.0 (7.3)
		0.3	5.7 (5.2)
		0.5	4.8 (4.4)
		0.7	3.1 (2.8)
		0.8	3.1 (2.8)
A508[b]	606 (88)	0.1	6.7 (6.1)
		0.5	5.6 (5.1)
		0.7	3.1 (2.8)
18/8 stainless[a]	665 (97)	0	6.0 (5.5)
		0.33	5.9 (5.4)
		0.62	4.6 (4.2)
		0.74	4.1 (3.7)
D6AC[c]	1970 (286)	0.03	3.4 (3.1)
7050-T7[c]	497 (72)	0.04	2.5 (2.3)
2219-T8[d]	—	0.1	2.7 (2.5)
		0.5	1.4 (1.3)
		0.8	1.3 (1.2)
Titanium[a]	540 (78)	0.6	2.2 (2.0)
Ti–6A1–4V[e]	1035 (150)	0.15	6.6 (6.0)
		0.33	4.4 (4.0)
Copper[a]	215 (31)	0	2.5 (2.3)
		0.33	1.8 (1.6)
		0.56	1.5 (1.4)
		0.69	1.4 (1.3)
		0.80	1.3 (1.2)
60/40 brass[b]	325 (47)	0	3.5 (3.2)
		0.33	3.1 (2.8)
		0.51	2.6 (2.4)
		0.72	2.6 (2.4)
Nickel[c]	430 (62)	0	7.9 (7.2)
		0.33	6.5 (5.9)
		0.57	5.2 (4.7)
		0.71	3.6 (3.3)

[a] N. E. Frost, K. J. Marsh, and L. P. Pook, *Metal Fatigue,* Oxford University Press, London, 1974.
[b] P. C. Paris, R. J. Bucci, E. T. Wessel, W. G. Clark, and T. R. Mager, in *Stress Analysis and Growth of Cracks,* Part 1, ASTM STP 513, ASTM, West Conshohocken, PA, 1972.
[c] J. Mautz and V. Weiss, in *Cracks and Fracture,* ASTM STP 601, ASTM, West Conshohocken, PA, 1976.
[d] S. J. Hudak, A. Saxena, R. J. Bucci, and R. C. Malcolm, "Development of Standard Methods of Testing and Analyzing Fatigue Crack Growth Data," Westinghouse Research Labs, Pittsburgh, PA, March 1977.
[e] R. J. Bucci, P. C. Paris, R. W. Hertzberg, R. A. Schmidt, and A. F. Anderson, in *Stress Analysis and Growth of Cracks,* Part 1, ASTM STP 513, ASTM, West Conshohocken, PA, 1972.

TABLE A.5 Corrosion Fatigue Strength In Water or Salt Water for Life $\geq 10^7$ Cycles for Selected Engineering Alloys[a]

Material	Process Description	S_u, MPa (ksi)	Frequency, Hz	Corrosive Media	Fatigue Limit in Air, MPa (ksi)	Corrosion Fatigue Strength, MPa (ksi)	Corrosion Fatigue Strength ÷ Air Fatigue Limit
Mild steel	Normalized			River water	260 (38)	32 (4.6)	0.12
0.21% C steel	Annealed	490 (71)	22	Sea H$_2$O	220 (32)	30 (4.3)	0.13
1035		600 (87)	29	6.8 percent salt water; complete immersion	280 (41)	170 (25)	0.61
1050		650 (94)			220 (32)	140 (20)	0.63
1050	Q&T	900 (130)			415 (60)	170 (25)	0.42
4130	Q&T	880 (128)			480 (70)	190 (27)	0.38
9260	Normalized	985 (143)			500 (72)	170 (25)	0.35
						125 (18)	0.49
12.5Cr steel	Annealed	1000 (146)	6	Fresh water in torsion	225 (37)	125 (18)	0.49
18/8 stainless	Annealed	1300 (188)			195 (28)	85 (12)	0.44
18.5Cr steel	Annealed	770 (112)			240 (35)	195 (28)	0.79
Aluminum	Annealed	90 (13)	24	River water with saline solution about one-half of sea water	19 (2.7)	7.6 (1.1)	0.41
Copper	Annealed	215 (31)			69 (10)	70 (10.5)	1.04
Monel	Annealed	565 (82)			250 (36)	200 (29)	0.81
60/40 brass	Annealed	365 (53)			150 (22)	130 (19)	0.85
Nickel	Annealed	530 (77)			235 (34)	160 (23)	0.68
Phosphor bronze	Normalized	430 (62)	37	3 percent salt spray	150 (22)	180 (26)	1.20
Aluminum–3 Mg	Heat treated	—	83	3 percent salt spray	125 (18)	48 (7)	0.38
Aluminum–7 Mg	Heat treated	—	83	3 percent salt spray	110 (16)	48 (7)	0.45
Aluminum–Cu Mg	Heat treated	—	83	3 percent salt spray	180 (26)	85 (12)	0.47
Mg–Al–Zn AZG		—	—	Tap H$_2$O	75 (11)	40 (6)	0.49
Mg–Al–Mn AZ537		—	—	Tap H$_2$O	75 (11)	55 (8)	0.68

[a] Reprinted from P. G. Forrest, *Fatigue of Metals*, Pergamon Press, Oxford, 1962 (with permission of Pergamon Press.) All test specimens rotating bending unless specified. These results have been collected from various sources. They illustrate only corrosion fatigue effects and are not to be used as design values.

AUTHOR INDEX

SUBJECT INDEX